Y0-ARO-733

Computers in Chemical Engineering Education

Editor

Brice Carnahan

University of Michigan

CACHE

Austin

This monograph may be ordered directly from:

CACHE
P. O. Box 7939
Austin, TX 78713-7939

Telephone:	(512) 471-4933
FAX:	(512) 295-4498
Email:	cache@uts.cc.utexas.edu
WWW:	http//www.che.utexas.edu:80/cache

ISBN: 0-9655891-0-2
LCCN: 0-96-072249

PREFACE

The first general-purpose electronic digital computer, the ENIAC, executed its first instruction in a laboratory at the University of Pennsylvania more than fifty years ago. From that seminal moment, the performance (choose almost any metric) of the digital computer and its impact on the life of the nation have grown year by year, virtually unchecked.

Most technological developments of consequence pass through the familiar S curve of slow initial growth, then a period of rapid acceleration, and finally another slow-growth phase of important, but marginal, improvement. For the digital computer, the slow growth period lasted about 15 years. The acceleration phase began with the introduction of the transistor in the late 1950s, received additional thrust with each new transforming technology (time-sharing, integrated circuits, real-time minicomputers, networking, the microprocessor, interactive graphical operating system interfaces and programming environments, supercomputers and parallel machines, vastly improved communication, the Internet, the World Wide Web, etc.), and continues to this day.

Clearly there is no end in sight. *Business Week* (October 1996) predicts that possibly in 1997, and certainly no later than 1998, the computing industry and its ancillary products and services (software, communications, etc.) will supplant the automobile industry and its ancillary products and services as the largest contributor to the US Gross Domestic Product. The only certainty is that the future, driven by the core technologies of computing and communication, will be a digital one, and that the centrality of "computing" in society, business, government, and yes, education, is assured.

In many respects, the impact of computing on education and academic life has paralleled that in the world outside the academy. Substantial basic research that feeds the computer revolution is performed by academics and their students. The University of the present looks quite different from the University of even a decade ago. Virtually every desk supports a networked desktop computer, the University library is "on-line," and every dorm room is (or soon will be) connected to the rest of the electronic world. The computer has brought with it systemic changes in the ways the University conducts its business and research and interacts with its students, graduates, faculty, and staff.

The impact of the computer *in* the classroom has, to date, been less dramatic than in other areas of the academy. An 1896 still photo of an engineering classroom, professor lecturing with chalk in hand, would look remarkably similar to most engineering classrooms in 1996. Will that paradigm last for yet another century? Not likely.

Starting about 1960, computing in chemical engineering education began its period of slow but steady growth. By the late 1960s, it was clear to many chemical engineering faculty that the computer could no longer be ignored because of (among others) the important role "computing" would play in the professional lives of chemical engineering graduates. In 1969, a few of the most computer-active faculty from several chemical engineering departments in the US and Canada formed CACHE (Computer Aids for Chemical Engineering Education) at a meeting in Ann Arbor. The mission of the new group was to promote both computer use in the chemical engineering curricula and cooperation (involving chemical engineering computing) among industry, academia, and government. CACHE has continued its original mission to the present, serving as a catalyst for introducing software and other instructional aids to chemical engineering faculty and students in both the US and abroad.

This monograph is an outgrowth of the 25th anniversary celebration for CACHE that included a Faculty Reception and an afternoon session at the November 1994 Meeting of the AIChE in San Francisco. Professors J. D. Seader (University of Utah) and Warren Seider (University of Pennsylvania), two founding members of CACHE, subsequently prepared an archival paper on CACHE activities and parallel developments in computing in chemical engineering education during the twenty-five year period 1969-94. I was asked by the CACHE Board to solicit other papers on computer-related topics from prominent chemical engineering faculty and to serve as editor of this monograph, *Computers in Chemical Engineering Education*. The twenty papers included cover a wide range of subjects, from the impact of computing in specific chemical engineering courses to more general topics, such as accreditation, multimedia instruction, numerical software, and laboratory automation. Most of the papers were written in 1995 and 1996 from the perspectives of past developments, present activities, and future directions.

I thank the thirty five authors of these papers for their outstanding contributions, and also for their patience with me in editing, compiling, and publishing this monograph. I also thank CACHE for distributing the monograph to individual chemical engineering departments and the AIChE Publications Office for including it as a bonus offering to academic and industrial libraries that subscribe to the AIChE Package Plan.

Brice Carnahan

CONTENTS

ROLE AND IMPACT OF COMPUTERS IN ENGINEERING EDUCATION

Richard S. H. Mah
Northwestern University
Evanston, IL 60208

David M. Himmelblau
University of Texas
Austin, TX 78712

Abstract

After three and one-half decades of uninterrupted development, the computing environment is now highly interconnected. Networks proliferate between computers, laboratories, buildings, campuses and across continents and oceans. Use of computers is integrated with many chemical engineering courses in teaching, learning, and communication. Many pioneers' dreams are already a reality. With computers, one can cover more course material using more realistic illustrations.

The continuing decline in the computer price/performance ratio makes it affordable to create software which is not just functional, but also fault-tolerant and user-friendly, making it useful and accessible to a wide community of users who have limited or no formal training in computers. Enhanced capabilities of general purpose software, like Matlab, diminish the need for chemical engineers to program in Fortran and other procedural languages. With ever-improving software, the bottleneck on process analysis rests once again on the quality and fidelity of the model, and we are back to the basics.

One important impact of computers on engineering education is to broaden the access to teaching and learning styles. New pedagogy creates opportunities for curriculum revamping which will surely be needed at some point, since we cannot go on adding new material to the existing courses without deleting other topics. By broadening our choices in pedagogy, we may also make our profession more accessible to a wider range of candidates.

Historically, the path to progress is strewn with expensive wreckage. Megabuck investment does not ensure that a project will succeed, and today's success is no guarantee for tomorrow. But there is no sign that the pace of development in computing and information technology is slowing down. How engineering education can continue to make use of these rapid changes remains a challenge. An education built on sound fundamentals and in-depth understanding is the best strategy to allow one's knowledge base to evolve and grow with changing times. While hands-on practical experience is indispensable to engineers, one must avoid over-specialization. Kilobit education is dangerous in a world of gigabyte technology.

On the other hand, history also shows that the momentum generated by a real winner can go a long way. Fortran, LP, word processors, and E-mail are some examples. We now have a global market for buyers and sellers of information technology, vast capital and financial institutions, vast trained manpower, and many potential winners. A list of promising developments include networks, optical and parallel computers, CD ROMs, satellite broadcasting and reception, personalized portable phones and pagers, notebook computers, and high definition television. There is a great likelihood for information technology mergers, and high potential that such mergers will create new products and technology which will further enhance the use of computers in engineering education.

We are almost at the dawn of the 21st Century. Looking back along the pathway leading to the present, we realize how far we have traveled in a journey propelled by the success of a few key inventions, and how many more wonders lie ahead of us to be discovered, invented and applied to engineering education in the decades ahead. The prospect is truly exhilarating and exciting.

Current Status

The Computing Environment

When Northwestern University is in session, the chances are that the lights are on, the computers are running in the Computer Teaching Lab, and students are using the computers in various ways, some of which their older brothers or sisters, just a few years ago, could not have done. The Computer Teaching Lab is now easily the most used facility in the chemical engineering department. System crashes are now rare events. The opening hours are only dictated by security and maintenance considerations. There is no full time staff associated with this facility. It is user-serviced with a half-time teaching assistant acting as the Lab Manager. Only policy guidance and planning are provided by a faculty director. Fourteen hours a day during the week and eight hours on Saturday and on Sunday the micros slave tirelessly at the friendly commands of users. The micros are connected in a local area network (LAN), served by a file server, printers and other peripherals. The LAN is linked to the campus fiber optic backbone, and through it, to the Internet worldwide.[1]

Access to the Internet is the most significant step forward in the empowerment of faculty and students, which has already taken place on many campuses, while the process continues in others. Give and take a few details such as types of hardware and software, the physical dimensions of the lab, and the size of the student population, the environment at Northwestern described above is the computing and information processing environment currently existing in many universities, and the computing facilities available for chemical engineering education.

Impact of Computers on Chemical Engineering Education

We are concerned here with engineering education, with specifics taken primarily from chemical engineering. How have computers affected the learning and teaching of engineering? To continue with our example, the use of computers is now closely integrated in most of our current undergraduate courses, beginning with material balances and stoichiometry (analysis of chemical process systems), thermodynamics, and equilibrium separations, continuing with

---

1. In this simplified description we omitted a few hardware and software details, which are transparent to the user.

process dynamics and control, process design, process optimization, and the chemical engineering lab, and ending with electives such as statistics in process modeling. Significant changes have already taken place in content, and in the teaching and learning of these subjects. For instance, linearization and Laplace transformation play a ubiquitous role in classical process control. In the days before computers were available, much time was spent on inverse transformations, and the preparation of Bode and Nyquist diagrams in stability analysis. Now, with Program CC, we simply input the appropriate polynomials in the numerator and denominator of the transfer function in the Laplace domain, and let the computer, the program, and the graphics do the tedious work. Parametric studies are easy to carry out. Understanding and insight, which used to take a long time to develop, are now acquired rapidly and enthusiastically. Similarly, TK-solver and Lotus 1-2-3 take a lot of drudgery and mystery out of balances and stoichiometry. With flowsheet simulators and property libraries, the dual role of thermodynamics in process analysis and in property estimation becomes very much easier to teach and explain. In statistics, by using Monte Carlo simulation, the instructor can readily demonstrate and verify, for instance, the Central Limit Theorem, and display plots in vivid color graphics in dimensions which "will cross a rabbi's eyes" (*Fiddler on the Roof*).

The upshot is that by using computers one can cover more territory and tackle more realistic problems with less time and more fun. With the availability of these new tools and techniques, it is possible to begin experimenting with new pedagogy (Felder and Silverman 1988, Schank 1994, and Stice 1987), which, in time, may profoundly change the ways students learn and instructors teach these subjects. This is particularly true with subjects involving many elements, complex structures and closely knit relationship, such as systems engineering, which would be difficult to demonstrate experimentally. With computer simulation we can reproduce precisely controlled "misbehavior" to study its impact on every aspect of the system.

Communication and Productivity Tools

Equally remarkable are advances which have taken place in communication and personal productivity tools. Students are expected to acquire serviceable skills in word processing, graphics, desktop publishing, database and E-mail with only a modicum of formal instruction. With spell checkers there is no excuse not to get the spelling right.

By making it fun to prepare texts and illustrations, not only do the reports and illustrations begin to look more professional, but the substance and style also improve in due course. With universal access to computer networks, everyone can send a message or be reached via E-mail without having to play phone tag. Through remote access the instructor could just as easily review class records and assign homework problems as he or she could conduct an electronic dialog with a colleague at another location - all without leaving the physical environment of home or office. Last but not least, by greatly simplifying the protocol, distribution and delivery, E-mail lowers the threshold of communication and shrinks the physical and psychological distances of an organization, be it a corporation, a government or a university.

To appreciate that profound and pervasive changes are rapidly taking place in information technology in general, and computers in particular, we need only look back to the path of progress which has led us to the present state of development.

Highlights of the Past

By most reckonings we are in the fourth decade of computer applications, even though there may not be an exact point of origin. The first two decades were dominated by mainframes and minis. In chemical engineering much of the initial programming efforts were directed at replacing repetitive calculations. Taking 1958 as our reference point, the establishment of FORTRAN as the universal high level programming language for quantitative computation must rank among the foremost achievement of the first decade. By the second decade, LP and more specifically, codes based on the Simplex Method, had become the single largest user of computer time in the process industries. Time-sharing, on-line terminals, and flowsheeting programs were some of the other notable developments of that decade. The year 1978 heralded the introduction of the first commercial scale microcomputers, the Apple II, followed three years later by the IBM PC and the mass marketing of software (e.g., word processor, spreadsheet, and database), which fueled the microcomputer revolution. By 1990, personal computers (PCs, IBM clones, and Macintoshes) became widely owned, exceeding, for the first time, the number of television sets sold during the 1995 calendar year. It is notable that E-mail and networking did not gain popularity until well into the third decade.

One of the most remarkable characteristics of the computer industry is the continual improvement in performance in relation to price, which has been sustained for over three decades already. Figure 1 and Figure 2 show that the price of computing has dropped by one-half every 2 to 3 years ever since computers were marketed commercially. A $3,000 PC now is comparable in computing speed to a million dollar mainframe a decade ago. If progress in the rest of the economy had matched progress in the computer sector, a Cadillac would cost $4.98, while 10 minutes' worth of labor would buy a year's worth of groceries (Brynjolfsson, 1993).

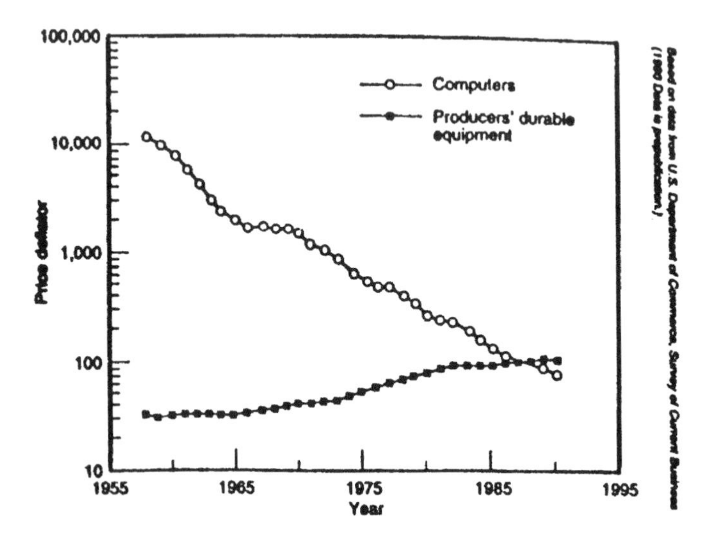

Figure 1. The cost of computing has declined substantially relative to other capital purchases.

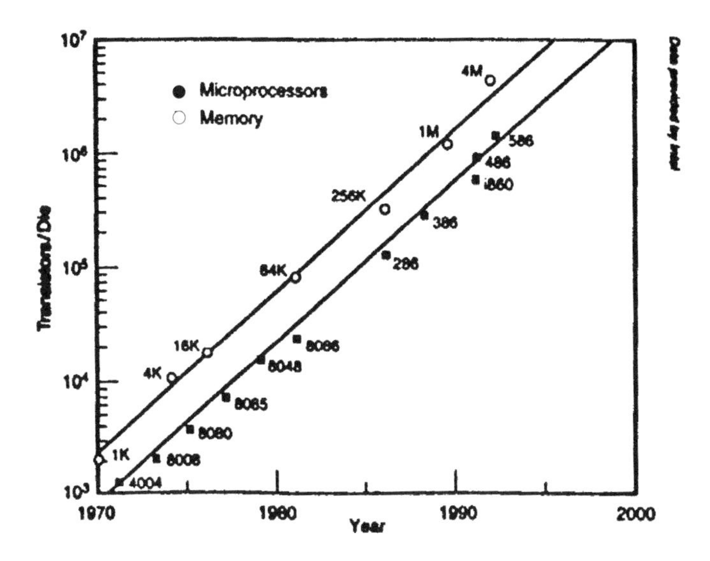

Figure 2. Microchip performance has shown uninterrupted exponential growth.

What Has Already Happened or is Happening

One impact of the changing price/performance ratio is improved user-friendliness. In early 1980s a word processor ran on a 64 KB memory microcomputer. By the mid 1980s it requires 640 KB of memory. In 1994 no respectable application software requires less than several megabytes of memory. However, with this extravagance in memory requirements came a much more fault-tolerant and user-friendly interface, and the same "look and feel" (under Windows, for instance) which make the task of assimilation much less formidable for lay users. In fact, previous emphasis on learning to program in FORTRAN and similar high level languages has diminished, as MATLAB, MathCad, Polymath, Mathematica and their class of applications enhance their capabilities, and relieve users of the requirement to program, raising anew the question whether there is a need to teach programming, except to computer engineering and computer science majors.

With ever-improving computing capabilities and solution methods, the bottleneck in process analysis is once again the quality and fidelity of models suitable for different applications. Attention is already being directed to applications typically ignored by educators, such as modeling less-structured, "fuzzy" problems, and applications involving noisy and correlated data.

Cheap information storage and improved means of transmission and distribution have already changed the modus operandi of traditional institutions such as libraries and publishing houses. Journal abstracts and even articles are sold on compact disks which can be searched at will at nominal cost, and with an E-mail address and access to databases students can download information just as easily as they can send electronic files to friends. Textbooks have changed substantially. Readers are expected to have access to a computer to solve exercises and problems. Disks containing pertinent software are commonly found in inserts in the back of books.

The technology exists today to customize, assemble, and electronically deliver textbooks for each student. It will take time to resolve all the copyright issues and to provide suitable marketing mechanisms. How to provide teaching material for engineering courses will almost certainly be a major issue in the next decade, and the opportunities for innovation will be limited only by our imagination.

One important impact of computers on engineering education is to broaden the access to teaching and learning styles. In a few instances, computer-aided learning has completely replaced the lecture-recitation format for learning. But in most universities, changes have occurred in a more limited way over a period of many years, as the role of computers in education became better appreciated by the faculty. Such changes as have taken place are often caused more by the influx of young faculty members who have hands-on knowledge in using computers than by the action of accreditation committees or university guidelines. So retooling of tenured faculty may well be one limiting factor in introducing new information technology in our pedagogy. Nonetheless, the rate of technological innovation will continue to be rapid, and equipment will be technologically obsolete when it is still in good mechanical condition. Short life cycles in computing technology will continue to be a fact of life. To stay in the competition, schools must have plans and funding to rejuvenate programs and facilities. Those with a foresight to anticipate will have a competitive advantage.

Curriculum revamping will surely be needed at some point, since we continue to add new material to existing courses without deleting some other topics. This will create opportunities for experimenting with new pedagogy, which may in turn make our profession more accessible to a wider range of candidates, thereby contributing to the retooling of the national workforce.

Historically, the path to progress is strewn with expensive wreckage. Megabuck investment does not ensure that a project will succeed, and today's success is no guarantee for tomorrow. An example of innovative educational software is the PLATO system, which reportedly cost CDC hundreds of millions of dollars in the 1970s, but which has left no lasting imprint on engineering education today. However, we did learn some valuable lessons. Most potential users cannot visualize how to use unfamiliar technology in large mental steps. If the context of the new technology is sufficiently dissimilar to the current context, rejection is likely. Thus, quantum leaps often fail where incremental changes may succeed. Another lesson for developers of new computing technology is to focus on the relevance to the educational needs and not be carried away by clever, exciting or imaginative technology. Changing a curriculum solely to take advantage of computing technology is usually a waste of resources.

The Future

There is no sign that the pace of development in computing and information technology is slowing down. An education built on sound fundamentals and in-depth understanding is the best strategy to allow one's knowledge base to evolve and grow with changing times. While hands-on practical experience is indispensable to engineers, one must avoid over-specialization. Paraphrased differently, kilobit education is dangerous in a world of gigabyte technology.

On the other hand, history also shows that the momentum generated by a real winner can carry development a long way. Fortran, LP, word processor, and E-mail are some examples.

Word processing was probably the single largest application which spearheaded the commercialization of personal computers. Figure 3 (Alspach, 1993) shows that it continues to be the dominant application of microcomputer users even today.

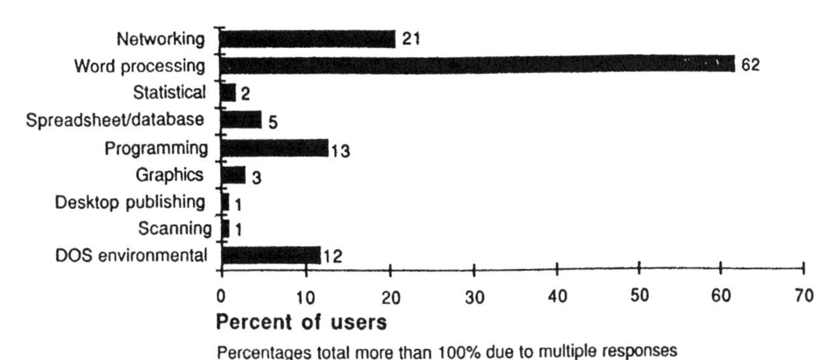

Graphic by Steve Alspach

Figure 3. Types of software and usage in computer laboratories.

Compared with earlier decades, when IBM accounted for 3 of every 4 computers sold, we now have a global market for buyers and sellers of information technology, vast capital and financial institutions, vast trained manpower, and many potential winners. A list of promising developments include networks, optical and parallel computers, CD-ROMs, satellite broadcasting and reception, personalized portable phones and pagers, notebook computers, and high definition television. The potential for information technology mergers which will further enhance the use of computers in engineering education is very large and very likely.

We are almost at the dawn of the 21st Century. Looking back along the pathway leading to the present, we realize how far we have traveled in a journey propelled by just a few key inventions, and how many more wonders lie ahead of us to be discovered, invented and applied to engineering education in the decades ahead. The prospect is truly exhilarating and exciting.

References

Alspach, S. (1993). *News and Views*, NU Information Systems and Technology, Northwestern University, Evanston, Illinois, **2**, 3, Spring 1993.

Augustine, N.R. (1994). Socioengineering Age. *ASEE Prism*, February, 24-26.

Brynjolfsson, E. (1993). The productivity paradox of information technology. *Comm. ACM*, December 1993, **36**, 12, 67-77.

Carnahan, B., and J.O. Wilkes (1995). *The Macintosh, the IBM PC, and Unix Workstations: Operating Systems and Applications*. College of Engineering, The University of Michigan, Ann Arbor, Michigan.

Felder, R.M., and L.K. Silverman (1988). Learning and teaching styles in engineering education. *Engineering Education*, April, 674-681.

Schank, Roger (1994). How Students Learn - Educational Software and the Future of Education. Sponsored by Searle Center for Teaching Excellence, April 21.

Stice, J.E. (1987). Using Kolb's Learning cycles to improve student learning. *Engineering Education*, February, 291-296.

COMPUTING SKILLS IN THE CHEMICAL ENGINEERING CURRICULUM

Jeffrey C. Kantor
University of Notre Dame
Notre Dame, IN 46556

Thomas F. Edgar
University of Texas
Austin, TX 78712

Abstract

The practice of chemical engineering has become strongly dependent on computing, but with the exception of courses in design and control, the chemical engineering curriculum generally does not utilize computer-based methods in order to enhance students' problem-solving skills. One reason is that popular textbooks tend not to be computer-oriented. Previous efforts by CACHE in 1985 to influence accreditation guidelines for computing skills in chemical engineering have been valuable, as have various projects aimed at making computer tools available throughout the curriculum. It is time to re-evaluate the list of preferred computer skills for chemical engineers, and to integrate computer-enhanced problem-solving skills throughout the curriculum. We propose de-emphasizing computer programming language skills (such as FORTRAN) in favor of using professional software packages based on higher level languages such as Matlab.

Introduction

There is no doubt that the revolution in computing and information technology during the past 40 years has changed the industrial world. In contrast, the typical engineering educator, rather than being on the cutting edge of those developments, has been slow to incorporate new computer-based ideas in curriculum, teaching methodologies, and educational materials. With a few notable exceptions, mainly in the area of process systems engineering, computing and information technologies have not had a major impact on the chemical engineering canon. Word processors and spreadsheets are used to prepare lab reports, but there has not been a substantial shift in the core curriculum. At most institutions, thermodynamics is not taught any differently than it was 30 years ago. If computers are so important, why is Bird, Stewart and Lightfoot, now in its fourth decade, still the model for teaching transport phenomena? It is apparent that in the eyes of a large number of faculty the investment made by students in learning computer languages and programming does not yield any discernible advantage in the training of chemical engineers. The questions of how extensively computers should be used and which computing skills should be taught in the undergraduate chemical engineering curriculum are difficult to answer for several reasons:

- There is no generally agreed upon 'core' set of computing skills necessary for being a productive engineer, either in academia or in industry.
- The availability of professional software now makes many engineering tasks, including computer programming, simple to carry out.
- Providing and maintaining a state-of-the-art facility for engineering computation is expensive, in terms of both capital and human resources.
- Development of high quality computer-based lessons is quite costly in terms of faculty and staff time.

This chapter discusses these points in the context of the historical evolution of computer usage in chemical engineering. Key influences include chemical engineering textbooks, computing aids disseminated by the CACHE Corporation, industrial practices, several prior reports produced by the CACHE Corporation, and the ABET-AIChE accreditation criteria. Unfortunately, the current ABET-AIChE criteria provide only general guidance, and there is a significant mismatch between the traditional academic view of computing and the needs articulated by practicing engineers.

After analyzing the state of affairs in computer-based education, we suggest some strategies for outfitting graduating engineers with practical, computer-enhanced problem-solving skills. These strategies de-emphasize the mechanics of computer use and the traditional reliance on scientific programming languages such as FORTRAN in favor of integrating higher-level, mathematically-oriented software systems into the curriculum.

Evolution of Computing Skills Since 1960

The capability of computing hardware has improved by orders of magnitude over the past forty years (Seader, 1989), evolving from mainframes to today's multifunctional personal computer/workstation. These striking developments have occurred while cycle costs were actually reduced, and the computer has become an omnipresent tool for increased productivity in engineering practice. Prior to the mid-1980s, the lack of professional software and inexpensive computing equipment limited computing experiences for undergraduate engineers, but no such constraints exist today. However, the ubiquitous nature of PCs on university campuses has not caused a quantum change in the way computing is taught or applied in the typical chemical engineering department. In fact, instead of computing practices influencing the way textbooks are written, the opposite effect seems to have occurred, namely popular textbooks have dictated the computing activities (or the lack thereof) in a given course.

The solution of chemical engineering problems in the 1950s and 1960s was limited to graphical techniques or manual manipulation of a few algebraic equations, often solved by trial and error. Very little matrix analysis was employed, except perhaps the crude solution of linear equations by determinants. Ordinary differential equations (almost always initial value problems) were solved by analytical or graphical integration. The textbook by Hougen, Watson, and Ragatz (1943) was exemplary of this approach. It is important to realize that the inertia brought on by the use of popular textbooks over as long as twenty years without revision has the tendency to stultify teaching practices and limit innovations in areas such as computing. For example, the books by Hougen et al. (1943); Bird, Stewart and Lightfoot (1960); McCabe, Smith and Harriott (1993); Levenspiel (1962); and Coughanowr and Koppel (1965) are texts that dominated various core courses in sophomore through senior years for long periods of time. The publication of a new undergraduate textbook in chemical engineering during the 1990s has become a relatively rare occasion, and factors that could encourage more textbook writing, such as direct encouragement from university administration or less emphasis on research, are unlikely to change during the rest of this century.

The first chemical engineering textbooks with strong ties to computer applications were the numerical analysis books by Lapidus (1962) and Carnahan et al. (1969). These books served as important bridges between the developments in other engineering disciplines, mathematics, computer science, and chemical engineering. However, they probably were perceived in the 1960s and 1970s as research-oriented texts rather than serving a standard undergraduate course in chemical engineering. Today about half of the U.S. chemical engineering departments teach their own numerical analysis course with specific chemical engineering applications, while the other half utilize a course outside the department. The choice is usually determined by the size of the department, with larger departments customizing their own course.

While a few innovative educators introduced some applications of numerical analysis into core courses, a fragmented state of affairs still existed in the early 1970s. In order to provide some assistance to faculty who wanted to share computer software for specific courses, the CACHE Committee, with financial support by the National Science Foundation, published a seven-volume set of books entitled *Computer Programs for Chemical Engineering Education.* These volumes were distributed to all departments in the U.S. and Canada and covered the curriculum areas of stoichiometry, kinetics, control, transport phenomena, thermodynamics, design, and stagewise computations. Included were 97 debugged and tested FORTRAN programs written by a large group of chemical engineering educators. It is interesting to note that even in the late 1980s, these books were still being distributed, although usually outside the United States.

The next large push in the area of computer applications was in the area of design, led again by the CACHE group. In the early 1970s, process simulation programs for computer-aided design of continuous steady-state vapor-liquid and liquid-liquid processes were beginning to be used at a number of companies. With built-in libraries of subroutines for equipment and thermodynamic property models and techniques for converging iterative computations when nonlinear equations and/or recycle loops were involved, these so-called flowsheeting programs performed material and energy balance calculations and, in some cases, sized and determined the cost of equipment. Before using the program, the user was required to develop a flowsheet and a set of design specifications. In 1974, the FLOWTRAN simulator of Monsanto, consisting of 160 subroutines and 60,000 lines of FORTRAN code and data, was made available through CACHE for educational use by departments of chemical engineering via a national computer network. Subsequently FLOWTRAN modules for different mainframe computers and workstations became available, thus reducing the cost of utilization for most universities. During the past 20 years FLOWTRAN has been employed mainly in senior design courses, with 190 tapes of the code disseminated to departments (141 in the United States). Sales of the FLOWTRAN textbook (Seader, Seider and Pauls, 1974) totaled 15,000 during that same period. In the late

1970s it became generally accepted in academic circles that the senior design experience should be computer-intensive. More recently, commercial simulators such as ASPEN, DESIGN II, PROCESS, CHEMCAD, and HYSIM also became readily available to universities for a modest fee. User-friendly interfaces (front-ends) and PC-based simulators have eliminated many of the barriers to using such packages. The use of computer-aided simulation in the capstone senior design course can certainly be characterized as a major success story in the education of chemical engineers.

However, integration of computing skills throughout the curriculum (from the sophomore year to the senior capstone design course) was rarely done in 1970s and 1980s. In fact, attempts to introduce a high level of computing in the sophomore material and energy balance course did not receive general acceptance at all. In 1969, the first published textbook to treat stoichiometry using digital computation was authored by Henley and Rosen (1969). Although manual methods were also included, the emphasis was on machine methods. Particular attention was paid to techniques for solving systems of nonlinear equations and converging recycle loops. The book appeared before most educators felt comfortable with numerical methods for digital computers, hence it never received widespread adoption. More recently, a similar fate befell the textbook by Reklaitis (1983). While it is now feasible to introduce some form of flowsheet simulation in this course, apparently very few educators use such software.

Similarly, in courses such as thermodynamics, transport phenomena, unit operations, separations, and reactor design, there is still only a modest level of computation in each of these courses at most universities. Certainly in the thermodynamics and separations area, there is a lot to be gained by introducing simulation packages and VLE subroutines. Sandler's second edition (1989) text on thermodynamics does have a set of computer disks including equations of state and connections to TK Solver. Reactor design is a particularly interesting case, in that powerful numerical solution methods for reactor design, ordinary and partial differential equations, and parameter estimation for these systems have not been utilized in textbook presentations. The current leading textbook in reactor design (Fogler, 1986) has introduced interactive computer exercises for demonstration of important concepts, but reactor simulation is not emphasized.

CACHE recognized the need to stimulate more activity outside design and control; in 1990 CACHE published a set of six chemical engineering lessons, which exploited the powerful features of the IBM PC. Written by six well-known educators and their students, these lessons covered the areas of fluid mechanics, thermodynamics, reactor engineering, and separations. The lessons were distributed to 100 departments of chemical engineering with unlimited copying privileges. This effort has been continued up to the present time by NSF grants at the University of Michigan, Purdue University, and the University of Washington, in cooperation with CACHE.

Some courses in chemical engineering, such as process control and optimization, are computer-intensive by their nature, and there are quite a few professional (and semi-professional) software packages that are available for student use. It has only been since the mid-1980s that user-friendly efficient computer codes (often PC-based) have become available for optimization, as reviewed by Edgar and Himmelblau (1988). Today packages such as LINDO, GINO, and GAMS offer easy-to-use interfaces to solve almost any linear or nonlinear programming problem of reasonable size. Similarly in the process control field, packages such as CC, Matrix-X, and Matlab make dynamic open and closed-loop simulation quite easy to perform. Unfortunately, none of the commonly used textbooks (Seborg, Edgar and Mellichamp, 1989; Smith and Corripio, 1985; and Stephanopoulos, 1984) present homework problems using such general purpose software.

Position Papers on Computing Skills

In 1985, the CACHE corporation recognized the tremendous lack of uniformity in computing experiences across the U.S. and authored a position paper specifying a particular set of computing competencies that all chemical engineering graduates should possess. This was the first effort in chemical engineering to influence training of undergraduates *vis-a-vis* computing skills. The main recommendations from the body of the report are as follows:

1. The chemical engineering graduate must be familiar with at least one operating system for personal and mainframe computers. This implies ability to manipulate files, perform text editing, and develop graphic displays.
2. The chemical engineering graduate must be competent in the use of at least one scientific programming language, i.e. have a sufficient understanding of programming logic to test and adapt programs written by others. In addition, he or she should be able to evaluate programs written by others in order to perform software selection.
3. The chemical engineering graduate must have experience in the computer-aided acquisition and processing of information.
4. It is desirable for each student to conduct at least one search using information retrieval from electronic data bases such as Chemical Abstracts Service and Scientific Information Systems.
5. It is desirable for the chemical engineering graduate to have experience in the use of word processors and graphics programs as well as spreadsheets for the generation of reports.
6. Although it is too early to require that all graduates have experience with electronic mail and external data bases, such a requirement should be feasible in the next 2-5 years.
7. It would be desirable for the student to have an appreciation of principles of numerical analysis (including convergence and stability) and non-numeric programming (such as used in artificial intelligence).

The CACHE position paper had a direct impact on accreditation requirements in chemical engineering, as discussed in the next section.

About the time the CACHE position paper was issued, a study on changes in the curriculum of chemical engineering was published by the Septenary Committee at the University of Texas. This committee, composed of industrial leaders from a wide variety of companies which employ chemical engineers, made extensive recommendations on how the teaching of chemical engineering should be changed so that the B.S. chemical engineer is better prepared for the competitive environment in today's (1985) workplace. It is interesting that they reached many of the same conclusions as did CACHE. Additional recommendations beyond those proposed by CACHE included:

8. All courses involving calculations should make extensive use of the computer and the latest software. Application should be frequent as students progress in the curriculum. Clearly, adequate computer hardware and software must be freely available to the student either through superior centralized facilities, individual PC's, or both. Extensive development of professionally written software for chemical engineering should be pursued.

9. A great deal of time can be saved in addressing designed equipment such as reactors, fractionators, and absorbers by emphasizing rigorous computer calculations and the simplest shortcut procedures. Valid shortcut methods require a solid conceptual base that assists in developing an intuitive feel for a problem. However, most intermediate (manual) calculation procedures should be eliminated unless they have real conceptual value. Existing software for algebraic and differential equation solving make simulations and design calculations quite straightforward.

10. Laboratory reports should use word processing when possible, and graphical presentation of data should be emphasized, using computers with modern graphics capabilities.

11. Some hands-on experience using current practices of computer data acquisition and control with industrial-type consoles should also be encouraged.

12. In the design course in engineering, students learn the techniques of complex problem-solving and decision-making within a framework of economic analysis. The very nature of processes requires a systems approach, and the ability to analyze a total system is one of the desirable attributes of chemical engineers. Rigorous economic analysis and predictive efforts should be required in all decision processes, given the availability of modern simulation tools.

Current Accreditation Requirements

Other engineering fields were addressing the need for computer skills in roughly the same time frame as chemical engineering, described in the previous section. The Accreditation Board for Engineering and Technology (ABET) is responsible for the accreditation of engineering programs in the United States and publishes general criteria for accreditation (1993). The current ABET position on computing skills in engineering curricula was proposed in the mid-1980s (new criteria will be in place in the near future):

> "Appropriate computer-based experience must be included in the program of each student. Students must demonstrate knowledge of the application and use of digital computational techniques for specific engineering problems. The program should include, for example, the use of computers for technical calculations, problem solving, data acquisition and processing, process control, computer-assisted design, computer graphics, and other functions and applications appropriate to the engineering discipline. Access to computational facilities must be sufficient to permit students and faculty to integrate computer work

into coursework whenever appropriate throughout the academic program."

A later section of the same document offers additional specific criteria developed by AIChE for chemical engineering programs, which was largely influenced by the 1985 CACHE statement presented earlier. The part discussing computing skills for chemical engineers is as follows:

> "Computer Use. Appropriate use of computers must be integrated throughout the program. Acceptable computer use will include most of the following: (1) programming in a high-level language; (2) use of software packages for analysis and design; (3) use of appropriate utilities such as editors; (4) simulation of engineering problems."

A more recent addition is that students must be exposed to the use of statistics in data analysis and problem solving.

ABET has encouraged a broader view of design to include more open-ended problems throughout the curriculum, rather than concentrating these kinds of problems in the capstone design course only. This view suggests that computer-based interactive problem solving can be employed to explore various options and alternative solutions for a given problem. For example, in the calculation of pressure drop in a pipeline, one can explore the effect of pipe diameter on velocity and pressure drop by use of the computer and also evaluate the operating costs of pumping vs. pipe capital costs. The personal computer or workstation has become an effective device for exposing students to open-ended problems.

Discussion

The general engineering accreditation statement on computing seems reasonable and consistent with what chemical engineers appear to be doing on the job. The criteria focus on what is done with computers rather than on the tools used to accomplish the task. However, because the criteria are so general, they do not offer a lot of guidance other than to give some examples of routine engineering activities.

The second passage relevant to chemical engineering has a somewhat different character. Items 1 and 2 presumably address the use of a language like FORTRAN for routine computation; however, recent CACHE surveys (discussed later in this paper) have shown that FORTRAN programming is not a critical skill for engineering practice. Items 3 and 4 are rather vague, but could be interpreted to include the spreadsheet analysis of process flowsheets, or the use of process flowsheet simulators.

Some of the items discussed in the 1985 CACHE position paper have lost their significance in the face of wide-spread adoption of computing technology on campuses. For example, E-mail has gone from the exotic to mundane in less than ten years. Specifying that students should know how to use word processors and email today is akin to asking students to learn to use the telephone or copy machine. Students accomplish these things on their own for very practical and immediate needs. Learning to use these tools well is another matter, and, to this end, there is no substitute for critical evaluation by faculty. Nevertheless, items 5 and 6 of the 1985 CACHE recommendations are probably no longer necessary.

On many campuses, libraries are providing an increasing number of electronic services for all of their patrons. For example, electronic bibliographic services are now incorporated as part of the electronic card catalog. Similar services are now available through the Internet.

Clearly, engineers must learn how to access information, and that now implies electronic access and database searching. If we insist that students learn how to use the library and reference materials, there may no longer be a need to specify a specific computing competency in electronic data access. Rather, we can assume that student will learns (or figure out) the relevant computing techniques as they learn to use a modern technical library.

Of the seven computing competencies proposed by CACHE, the first three probably take care of themselves in the normal course of a rigorous undergraduate education. While they are important professional engineering skills, there is not immediate need today to list them separately as desired computing competencies.

The remaining items concern more specific skills traditionally associated with scientific computing, and retain much of their significance. However, these items must be interpreted in terms of current computing technology. For example, on most campuses and at an increasing number of companies, the traditional mainframe has given way to a distributed model for computation. For routine work, most users have access to a graphical user interface (GUI) that simplifies routine tasks. Nevertheless, engineering users must have a sound understanding of how to set up and structure their files and directories, and how to access resources distributed over a network.

In a similar vein, there has been much change in what is meant by a scientific programming language. The authors primarily had FORTRAN in mind when item 2 was formulated. A 1993 CACHE survey discussed in the next section has shown that FORTRAN and other comparable general-purpose programming languages such as C and Pascal are not often used by most engineers once they are on the job. While it is still true that FORTRAN and mathematical subroutine libraries are very important for numerically intensive work, the use of FORTRAN for more routine uses has been supplanted by general purpose applications software.

Spreadsheets are now pervasive in engineering use. The newer commercial spreadsheets are easy to use, allow the user to easily document the results, and offer sophisticated equation solving and optimization functions. Spreadsheets should be viewed as a very useful item in the engineer's toolbox.

Current Use of Computers by Practicing Engineers

In 1993, CACHE sponsored a survey on the use of computing by engineers in the process industries (Davis, Blau and Reklaitis, 1993). Engineering management and recent B. S. chemical engineering graduates were surveyed at four major companies. The recent graduates were primarily in technical positions (83%) and in their jobs for less than five years (75%). Their managers typically had over 15 years experience (75%) with job descriptions that included technical management (76%). An additional survey was conducted of instructors at a number of universities. A total of 367 responses were obtained from these three groups. One of the goals of the survey was to compare the perceptions of the three groups regarding computer usage.

Among the results, the surveys of engineering management and recent graduates found that most of the engineers average 20-40% of their time at the computer, and another third spend 40-60% of their time at the computer. Academics significantly underestimated this level of usage, estimating that 70% of practicing engineers would spend 20% or less of their time at the computer. These statistics are a remarkable indication of how the computer has become a central feature of the technical workplace.

What tools do chemical engineers use when working at their computers? The surveys found that large fractions of the recent graduates make frequent use of spreadsheets (74%), presentation graphics (80%), database systems (70%), and electronic communications (89%). Managers and academic instructors heavily underestimated this pattern of computer use among recent graduates.

How do engineers use these tools? The recent graduates said their primary use of spreadsheets was for data analysis (75%), material balances (40%), economic studies (25%), and numerical analysis (24%). Databases were used for access to project information and process data. The use of computers to prepare reports and technical documentation was acknowledged by virtually everyone.

The surveys also revealed some tools that recent graduates do not use. Most (92%) say they never or seldom program in FORTRAN or another computer language. The overwhelming majority (85%) say they never use numerical method libraries such as IMSL or mathematical packages (86%).When asked if their company expected them to be literate in different computer language paradigms, 86% of the recent graduates said no. In fact, many companies explicitly tell their engineers *not* to write software because of the difficulty of maintaining such programs written by individuals.

Some rather obvious conclusions can be drawn from these survey results. Foremost is the simple observation that the set of tools chemical engineers use to do their jobs has changed significantly over last few decades. FORTRAN, compilers, and subroutine libraries have given way to spreadsheets, databases, and professionally prepared applications software. These tools are used to perform traditional engineering tasks, such as flowsheet analysis and data manipulation.

Beyond the traditional engineering tasks of quantitative analysis, computers have become a key tool for report writing, data access, and electronic communications. Because the computer is used for so many functions, many engineers are spending a significant amount of their time working at their machines.

Computer Skills for the Next Decade

When CACHE was formed twenty-five years ago, computers were expensive resources that were accessible on a relatively limited basis. Users had to be efficient, and they only had a limited set of tools to accomplish their tasks. That picture has changed dramatically. Computers are now ubiquitous and cheap, and there is a rich set of tools for doing routine tasks. Since computers are used for so many things, we can assume that students will learn on their own the routine skills of word processing, electronic mail, and file management.

What can we say about the needed computing skills in the undergraduate curriculum, both now and in the next century? We believe the ten-year old statement on computing skills is now off the mark. The focus should now be on what kinds of experiences and computer-enhanced problem-solving abilities chemical engineers must have when they graduate. The list below is not radically different from the previous position papers, and perhaps illustrates that changes in academia come about with great difficulty and on a time scale approaching glacial phenomena. Graduating engineers should:

1. know how to use a modern technical library to search for information located in electronic databases, and how to access electronic information services through the World Wide Web.
2. understand the implementation of elementary algorithms for the numerical solution of engineering problems. These algorithms should include algebraic and differential equation solving, linear algebra, and optimization.
3. be able to solve more sophisticated engineering problems using appropriate applications software. The types of problems include material and energy balances, optimization problems with constraints, and statistical data analysis.
4. be familiar with software for computer-aided process design and analysis.
5. have experience with computer-based instrumentation, process control, data collection, and analysis.

The difficulty comes when trying to reduce these general principles to practice. Some of these items can be incorporated into the standard chemical engineering canon. For example, a modern unit operations laboratory would likely have at least several pieces of equipment with computer-based instrumentation, and perhaps a process control experiment. The last item above could be addressed in this framework. Similarly, the items regarding information access and computer-aided design could be integrated as part of traditional process design course, or even introduced into the sophomore level course on material and energy balances.

The items regarding problem-solving are more difficult to integrate into the curriculum without students first having some sort of systematic introduction. Clearly there is mathematical component that requires students to have had an exposure to algebraic and differential equations, calculus, and linear algebra.

The key is for the freshman and sophomore years to integrate rudimentary computer skills into courses such as mathematics and physics. Unfortunately, at many universities science departments lag behind engineering in introducing such technology. If problem-solving skills are to be developed and used in the other courses, a sophomore level course may be the most practical option in which students would learn several of these techniques.

How should this material be taught? One purpose is to teach how to implement elementary algorithms for problem-solving. While practicing engineers might use spreadsheets for this purpose, teaching requires a vehicle well-suited to efficient development of numerical algorithms. A second purpose is to teach students how to formulate and solve larger scale problems that might involve differential equations or constrained optimization. For teaching purposes, the most useful tools are numerically oriented and allow students to explore the use of different algorithms, problem formulation, and means to visualize the results.

Note that we do not include programming language expertise in the above list. FORTRAN coupled with mathematical libraries has been a traditional tool. However, there are several excellent higher-level language alternatives, including Matlab, Mathematica, and Maple. These tools allow one to script solution algorithms very efficiently, and come with excellent visualization and problem-solving toolboxes. Our opinion is that the case for using these tools is now so overwhelming, that it would be legitimate to entirely omit FORTRAN from the undergraduate curriculum in favor of these alternatives.

The resistance to adding another course to the curriculum is well-known. Many departments currently require their students to take a computer programming course, and then follow that with a numerical analysis course, for a total of five to seven semester credit hours. If one de-emphasizes teaching of a traditional programming language in favor of higher level software combined with numerical analysis, it may be possible to recover as much as three credit hours for other needs. Such a combined course is already being taught at a few departments such as at the University of Texas (UT).

Algorithm development and programming structures are both covered in the UT course (four credit hours), and a native programming language contained in the software package Octave (authored by Eaton, 1994) is used to introduce programming structures to the students[1]. Other programming concepts taught include flowcharts, if-then-else structure, the looping (iteration) structure, and comparison operators. It is still necessary to spend about a month teaching students the operating system (Unix workstations are employed) and covering utilities such as file and directory manipulation, text editing, and job control. Given the choice between writing FORTRAN programs and using Octave to solve an ODE, the students invariably select the latter approach. The second half of the course surveys numerical analysis topics (with applications), which include finite difference methods, matrix algebra, and linear and nonlinear equation solving.

Ideally, students would enhance their computer-based problem-solving skills continually as they pass through the standard curriculum. Thermodynamics, fluid mechanics, and heat and mass transfer allow many opportunities for students to solve problems involving algebraic equations, integration, data regression, and challenges in visualizing solutions. Reaction engineering and process control courses offer opportunities for dynamic simulation.

Conclusions

In order to develop the new paradigm of problem-solving skills, it will be necessary to develop more computer-based lessons in subjects such as transport, thermodynamics and separations. Improvements in standard textbooks to include a stronger computer orientation will also be valuable. However, strong leadership in many departments is needed to increase communication among faculty who teach core courses and encourage more integration of computer skills. One or two champions for computing in a department can facilitate the necessary chang-

1. Available via anonymous ftp from ftp.che.utexas.edu in the directory /pub/octacve

es. Multimedia computing may offer new ways to enhance the traditional lecture format. For example, the concept of a "studio" style of teaching using computers in order to cluster students into small groups has been implemented in large classes such as freshman physics at some universities. Such teaching techniques can make new computer-based problem-solving tools a major focus in each course and offer an enhanced learning experience for engineering students.

References

Bird, R.B., W.E. Stewart, and E.N. Lightfoot (1960). *Transport Phenomena*. Wiley, New York.

Carnahan, B., H.A. Luther, and J.O. Wilkes (1969). *Applied Numerical Methods*. Wiley, New York.

Coughanowr, D.A. and L.B. Koppel (1965). *Process Systems Analysis and Control*. McGraw-Hill, New York.

Davis, J., G. Blau, and G.V. Reklaitis (1993). Computers in undergraduate chemical engineering education: A perspective on training and applications. Technical report, CACHE Corporation. Draft 3.1.

Eaton, J.W. (1994) *Octave: A high level interactive language for numerical computations*. Center for Control and Systems Research, Report 93-003, University of Texas, Austin.

Edgar, T.F. and D.M. Himmelblau (1988). *Optimization of Chemical Processes*. McGraw-Hill, New York.

Engineering Accreditation Commission (1993). *Criteria for accrediting programs in Engineering in the United States*. Technical report, Accreditation Board for Engineering and Technology, Inc.

Fogler, H.S. (1986). *The Elements of Chemical Reaction Engineering*. Prentice-Hall, Englewood Cliffs, NJ.

Henley, E.J. and E.N. Rosen (1969). *Material and Energy Balance Computations*. Wiley, New York.

Hougen, O.A., K.M. Watson, and R.A. Ragatz (1943). *Chemical Process Principles, Part I, Material and Energy Balances*. Wiley, New York.

Lapidus, L. (1962). *Digital Computation for Chemical Engineers*. McGraw-Hill, New York.

Levenspiel, O. (1962). *Chemical Reaction Engineering*. Wiley, New York.

McCabe, W.L., J.C. Smith, and P. Harriott (1993). *Unit Operations of Chemical Engineering*. McGraw-Hill, New York, fifth edition.

Reklaitis, G.V. (1983). *Introduction to Material and Energy Balances*. Wiley, New York.

Sandler, S.I. (1989). *Chemical and Engineering Thermodynamics*. Wiley, New York, second edition.

Seader, J.D. (1989). Education and training in chemical engineering related to the use of computers. *Comp. Chem. Engr.*, **13**, 377.

Seader, J.D., W.D. Seider, and A.C. Pauls (1974). *FLOWTRAN Simulation - An Introduction*. CACHE, Ann Arbor, Michigan.

Seborg, D.E., T.F. Edgar, and D.A. Mellichamp (1989). *Process Dynamics and Control*. Wiley, New York.

Septenary Committee (1985). *Chemical engineering education for the future*. Technical report, University of Texas, Austin, TX. Also *Chem. Engr. Prog.*, **81**, 9.

Smith, C.A. and A.B. Corripio (1985). *Principles and Practice of Automatic Process Control*. Wiley, New York.

Stephanopoulos, G. (1984). *Chemical Process Control*. Prentice-Hall, Englewood Cliffs, NJ.

HISTORY OF CACHE AND ITS EVOLUTION

J. D. Seader
University of Utah
Salt Lake City, UT 84124

Warren D. Seider
University of Pennsylvania
Philadelphia, PA 19104

Abstract

A definitive archival account of the 25-year history of CACHE from 1969-1994 is given, including the events leading up to the formation of the CACHE Committee in 1969, followed by incorporation in 1975. The history is presented mainly in terms of the chronological development of the many CACHE products that have been distributed to educators and the parallel incorporation of "computing" into chemical engineering curricula. Emphasis is placed on the synergistic effects of many educators and representatives from industry working together to advance the use of computers in chemical engineering education.

Two significant events occurred in the mid-1950s that were to drastically alter the education of a chemical engineers. The first was the introduction of the IBM 704 digital computer with its built-in floating-point arithmetic. The second was the development of the easy-to-use, high-level, procedure-based programming language of J. W. Backus and others of IBM, called FORTRAN. The IBM 704 came with a FORTRAN language compiler and subroutines written in that language were automatically handled. Almost overnight, chemical engineers in universities sought ways to learn FORTRAN and write computer programs to solve difficult or tedious problems.

Quickly, it became apparent that educators needed assistance in making the transition from precomputer techniques to computer-aided methods, which often involved numerical rather than analytical means to obtain a solution. Starting in 1958, the year after the introduction of FORTRAN, the University of Michigan initiated a project under the direction of the late Professor Donald L. Katz, with support from the Ford Foundation, to study the use of computers in engineering education. This project was extended to engineering design, with funding from the National Science Foundation (NSF) to: (1) train engineering design teachers in computer-aided design, (2) study the role of the computer in design, and (3) generate representative computer-oriented design problems.

The final report of the Michigan project, which was published in 1966 under the title *Computers in Engineering Design Education*, made the following recommendations to engineering schools:

1. Give introductory courses on computer programming and digital computation.
2. Teach model building, numerical methods, and optimization.
3. Integrate computer work into engineering science and design courses.
4. Stress open-ended problems.
5. Provide time-shared interactive computing to all students.
6. Train engineering teachers to use computers.

These recommendations still hold true today. Chemical engineering education is indebted to Donald L. Katz for his vision and his success in acquiring sufficient funding and recruiting qualified faculty to conduct such an important project.

Soon after the completion of the Michigan project, three textbooks were published that provided the three teaching tools needed for Recommendation (2): the first on model building entitled *Process Analysis and Simulation*, by Professors David M. Himmelblau and Kenneth B. Bischoff, the second entitled *Applied Numerical Methods*, by Professors Brice Carnahan, H. A. Luther, and James O. Wilkes, and the third entitled, *Optimization: Theory and Practice*, by Professors Gordon S. G. Beveridge and Robert S. Schechter. The numerical methods book, which was dedicated to Donald Katz, covered both theory and applications for interpolation, differentiation, integration, use of polynomials, linear algebraic equations, nonlinear algebraic equations, ordinary differential equations, partial differential equations, and statistics. The book was complete with comprehensive explanations and solutions, including listings of stand-alone FORTRAN programs for 40 example problems. A book entitled *A FORTRAN IV Primer*, by Professor Elliott I. Organick, provided a complete discussion on programming in FORTRAN.

Two other important events took place in the late 1960s. First, a widely publicized computer-aided chemical process design program called PACER (Process Assembly Case Evaluator Routine), was developed by Professor Paul T. Shannon at Purdue University, and reported in 1963. It was made available to chemical engineering departments at universities, following an instructional PACER workshop at Dartmouth College, April 23-28, 1967 that was attended by 20 chemical engineering professors from the United States and Canada. PACER was mainly a modular executive program for directing steady-state material and energy balance calculations for flowsheets that could include recycle streams. PACER had a limited library of unit-operation models and lacked a physical property estimation package. However, it was open-ended so that users could add operation and property models.

On November 22-23, 1968, Professors Rodolphe L. Motard and Ernest J. Henley, at a workshop at the University of Houston that was attended by 42 professors, introduced a more advanced simulation program called CHESS (CHemical Engineering Simulation System). It was apparent at this workshop that exciting computer-aided tools suitable for use in teaching chemical engineers could be developed and would be welcomed by educators. However, there was also concern that programs developed independently by educators might not be well documented and debugged. Also, most chemical engineering departments were neither equipped for nor disposed toward maintaining program libraries and providing consultation services. What was needed were standards and inter-university cooperation, which could be achieved by the formation of a committee.

Also in 1968, Lewis G. Mayfield, of the National Academy of Engineering (NAE), visited the University of Pennsylvania and suggested to Professor Warren D. Seider (who had a joint appointment there in chemical engineering and electrical engineering) the formation of a national committee of chemical engineering educators to be patterned after a similar committee for computer science in electrical engineering, called COSINE, which was sponsored by the NAE. As a result of this suggestion, Warren Seider, together with Professor Brice Carnahan of the University of Michigan and Professor Rodolphe (Rudy) L. Motard of the University of Houston organized a meeting, held on April 11, 1969 at the University of Michigan. This meeting was attended also by Professor William Surber of Princeton University and a member of the COSINE Committee, and the following 11 chemical engineering educators who were selected because they were engaged in the preparation of computer programs at their respective colleges and universities:

James Christiansen, University of Oklahoma

Eugene Elzy, Oregon State University

Lawrence B. Evans, Massachusetts Institute of Technology

Edward A. Grens, University of California at Berkeley

A. I. Johnson, McMaster University

Ernest J. Henley, University of Houston

Richard R. Hughes, University of Wisconsin at Madison

Matthew J. Reilly, Carnegie-Mellon University

Paul T. Shannon, Dartmouth College

Samuel L. Sullivan (for Robert E. C. Weaver), Tulane University

Arthur W. Westerberg, University of Florida

Brice Carnahan acted as chairman for the meeting and Warren Seider was acting secretary. The attendees agreed that there were numerous incentives for inter-university cooperation and that a committee should be formed, with the goal "to accelerate the integration of digital computation into the chemical engineering curriculum by sustained inter-university cooperation in the preparation of recommendations for curriculum and course outlines and new computing systems." Following a review of computing activities being conducted by each attendee, working sub-groups were formed in the areas of curriculum, standards, physical properties, new projects, and proposals. The group agreed to meet again in the Fall of that year. Immediately following the meeting, Rudy Motard suggested the acronym SEED for the name of the committee. However, SEED was already copyrighted and so a second choice, CACHE (Computer Aids for Chemical Engineering Education), was selected.

The committee needed a central base for its operations and a source of funds to support meetings, workshops, and the preparation and distribution of reports. With the encouragement of Dr. Newman A. Hall, Executive Director of the Commission on Education (COE) of the NAE, the proposal sub-group, consisting of Warren Seider, Brice Carnahan, and Richard R. Hughes, prepared a 53-page proposal, which, after minor changes by the NAE, was submitted on July 14, 1969 to the NSF as a proposal from the COE of the NAE, under the signatures of Gordon S. Brown, Chairman of the COE, Newman A. Hall, B. L. Kropp, Deputy Business Manager of NAE, and James H. Mulligan, Jr., Executive Secretary of NAE. A total of $120,812 was requested to start on January 1970 for a two-year period.

The first official meeting of the CACHE Committee was held on Tuesday, November 18, 1969 in Washington, D.C., during the week of the AIChE Annual Meeting, from 10:00 AM to 12:15 PM in the Columbian Room of the Mayflower Hotel. The founding members, all of whom were appointed by COE of the NAE for the interim until the conclusion of the initial forthcoming (hopefully) NSF funding period, consisted of the above attendees at the April 11, 1969 meeting, plus:

Robert V. Jelinek, Syracuse University

J. D. Seader, University of Utah

Imre Zwiebel, Worcester Polytechnic Institute

Professors Cameron M. Crowe of McMaster University, C. Judson King of the University of California at Berkeley, and Dale F. Rudd of the University of Wisconsin at Madison had been invited to join the committee, but declined because the committee included other faculty from their universities. Pending NSF funding, Newman Hall offered the services of Mrs. Jean P. Moore, Administrative Assistant of the COE/NAE, for distribution of mailings and support of subcommittee activities. Much of the meeting was spent in discussing efforts of members of the committee in areas of steady-state and dynamic simulation. An additional sub-committee was formed for the latter area. In addition, an ad-hoc subcommittee was established to draft governing rules and by-laws.

At the first official meeting, the CACHE Committee voted in favor of maintaining Brice Carnahan as Acting Chairman and Warren Seider as Acting Secretary until funds had been received from the NSF. During the 25-year period from 1969-1994, CACHE has been fortunate to have had outstanding and dedicated leadership and membership. Complete lists of CACHE officers and members are given in Appendices I, II, and III of this history.

The inevitable delays in obtaining federal funding occurred. Although meetings of subcommittees were held in the interim period, the next full meeting of the CACHE Committee was not held until April 2-3, 1971 in Ann Arbor, Michigan, following notice on February 18, 1971 that the NSF had finally approved the proposal, but had reduced the budget to $80,060 for the two-year period of January 1, 1971-December 31, 1972. This funding allowed $29,650 per year for all CACHE Committee activities, exclusive of overhead charges. The Ann Arbor meeting was attended by 16 of the 17 founding members of the CACHE Committee plus four observers, Dr. Newman Hall, and Jean Moore. Warren Seider was elected Chairman, with Lawrence Evans as Vice-Chairman, and Arthur Westerberg as Secretary.

Of an approved two-year budget of $63,650, $28,450 was allocated to the five subcommittees (Curriculum, Standards, Physical Properties, Dynamic Systems, and New Projects), each of which presented a report on their progress since the Washington, D.C. meeting held 17 months earlier. The most impressive report came from Professor Ernest Henley, Chairman of the Curriculum Committee, who reported on plans for the collection, review, and publication of small stand-alone FORTRAN computer programs in seven curriculum areas: stoichiometry, kinetics, control, transport, thermodynamics, stagewise computations, and design. The programs would be solicited from faculty by a letter dated April 10, 1971. The success of this project may be judged by the fact that proposals for almost 500 FORTRAN programs were submitted to CACHE from all over the world in 1971. Of these, 97, representing the efforts of almost 100 faculty members, were selected for publication. Each selected program had to be thoroughly documented and tested by the curriculum area coordinator.

By March 1973, using camera-ready copy supplied by program authors, the first two volumes (Stoichiometry with 13 programs and Kinetics with 21 programs) of a seven-volume set of programs had been distributed to all departments of chemical engineering in the United States and Canada, as well as interested departments in foreign countries. In conjunction with the assembly of the seven volumes of computer programs, the Standards Subcommittee of CACHE, chaired by Prof. Paul Shannon, issued in May, 1972 an important CACHE publication entitled *Standards for CACHE FORTRAN Computer Programs*, which included a detailed and widely referred to example of recommended programming style and documentation, prepared by Prof. Brice Carnahan for a program called GOLDEN, which performs a single-variable optimization by Golden-Section search.

Two other decisions were made at the Ann Arbor meeting. The first was to periodically publish and distribute a newsletter to all interested departments of chemical engineering. The first newsletter, edited by Prof. Lawrence Evans, was distributed to approximately 2000 faculty members in June, 1971, shortly after the Ann Arbor meeting. The cover page of the newsletter displayed, for the first time, the CACHE logo, which was provided by Jean Moore and has been used ever since. Until 1981, the newsletter was published approximately once each year. However, starting with the March 1981 issue (No. 12), it has been a biannual publication. Larry Evans (1971-73 and 1977-80), Cecil Smith (1975-77), J. D. Seader (1981-84), and David Himmelblau (1984-date) have served as Editor of the *CACHE News*. For some unknown reason, issue No. 10 was never published. A special 25th Anniversary Issue for Fall 1994 (No. 39), with a silver-colored cover was published in November 1994.

The second decision made in Ann Arbor was to designate a local CACHE representative in each of the departments in the United States and Canada. Each representative was to serve as a focal point for communication between CACHE and the institution of the representative. By March 1973, 123 local CACHE representatives from as many universities had been designated, with their names and schools listed in *CACHE News* No. 3. Starting with the 1977 AIChE Annual meeting, held in New York City, CACHE began holding an annual reception for CACHE representatives and interested faculty. At these receptions, CACHE members, task force members, or invited speakers discuss and demonstrate new products and receive feedback from faculty. At the November 1978 and 1979 AIChE meetings in Miami Beach and San Francisco, there were luncheon meetings of the CACHE representatives, with 70 and 66 in attendance, respectively. Another luncheon for 44 representatives was held at the New York City

AIChE meeting in November 1987. Today, almost every chemical engineering department in the United States and Canada, as well as many foreign countries, has a CACHE representative. On November 16, 1994 at the AIChE Annual Meeting in San Francisco, the CACHE 25th Anniversary Reception was held for representatives, associates, and former CACHE trustees. Approximately 400 attended this very successful reception.

Although voting membership in the CACHE Committee was limited from the beginning, the committee made every attempt to involve as many faculty members as possible in the work of the CACHE subcommittees. For example, in mid-1971 the Physical Properties Subcommittee, chaired by Prof. Rudy Motard, had 11 members, 9 of whom were not members of the CACHE Committee. The Dynamic Simulation Subcommittee, chaired by A. I. Johnson, had 8 members, 6 of whom were not members of CACHE and 3 of whom were graduate students.

One of the main reasons for forming the CACHE Committee was the possibility of coordinating efforts to introduce computer-aided process simulation into the chemical engineering curriculum. Indeed, almost every one of the 17 founding members of CACHE had been involved in the development of this new and exciting area of computer-aided design and analysis. Paul Shannon had developed the PACER program. Rudy Motard had developed the CHESS program, which included a physical property program called TAP by Ernest Henley. Motard had developed the MAD version of PACER and was preparing a dynamic simulation program called PRODYC. A. I. Johnson had developed a modified PACER-like program called MACSIM, and an interactive, time-sharing version called GEMECS, and was preparing DYNSYS, a dynamic simulation program, LINSYS, a material balancing system for linear equations, and GEMOPT for optimization. Warren Seider had developed a physical properties system for simulators, an automatic translator for converting FORTRAN programs into PACER routines, and a dynamic simulator called REMUS. Eugene Elzy had written the DISCOSSA simulator, which was tailored to each run by a compiling and linking step. Richard Hughes, while at Shell Development, had helped develop CHEOPS, a large simulation and optimization program. J. D. Seader, while at Chevron Research in 1959, had initiated and supervised the development of the Chevron system for generalized heat and material balancing, which used an equation-solving approach. Art Westerberg assisted Prof. Roger Sargent at Imperial College in the development of SPEEDUP, an equation-based, steady-state and dynamic simulator. Jim Christensen had developed programs in SNOBOL for recycle analysis and design variable selection. Imre Zwiebel used the PACER executive routine together with his own FORTRAN subroutines for a pyrolysis reactor, distillation columns, and a partial condenser and quench system to complete the detailed design and analysis of an ethylene manufacturing plant. Brice Carnahan was completing the development of an interactive dynamic simulation program called DYSCO. Larry Evans had just completed, with partial support from the NSF, an extensive survey on the current status of and future prospects for computer-aided chemical process design, which was published in the April 1968 issue of *Chemical Engineering Process*.

The first efforts by CACHE in the simulation area were taken by the Dynamic Simulation Subcommittee, which had been formed at the first meeting of CACHE. In December, 1971 that subcommittee, chaired by A. I. Johnson, recommended the use of the DYFLO dynamic simulation computer program of Roger G. E. Franks, a Senior Consultant for Engineering Computation and Analysis at E. I. du Pont de Nemours & Co., Inc. in Wilmington, Delaware. The FORTRAN code for DYFLO was published by Franks in a 1972 textbook entitled *Modeling and Simulation in Chemical Engineering* by Wiley-Interscience. To provide instruction on the use of DYFLO, CACHE held a three-day workshop at Purdue University on October 26-28, 1972 for 25 faculty members. Roger Franks was the principal lecturer, with assistance from Purdue Prof. John Woods. Dynamic simulation represented the first major encounter, for many chemical engineers, with the problems of solving "stiff" systems of ordinary differential equations. During the 20-year period from 1972 to 1992, DYFLO was used by many universities, but is now being displaced by more comprehensive programs developed by Aspen Technology, Hyprotech, and Simulation Sciences. In 1985 and 1989, CACHE attempted to obtain use by universities of Shell's DYMODS and Imperial College's SPEEDUP dynamic simulators, respectively, but without success. However, the latter is now available to universities through Aspen Technology.

At the Buck Hill, Pennsylvania meeting of CACHE on September 2-3, 1971, the Large-Scale Systems Task Force was formed, with J. D. Seader as Chairman. With the dynamic simulation program of Roger Franks soon to be available, this new task force was given the charge of investigating the possibility of acquiring the use of an industrial steady-state process simulation computer program for educational use at universities. The academic programs, PACER and CHESS, were already being used at a number of universities, but the 1968 survey by Larry Evans showed that much more useful simulation programs, with large data banks for pure chemicals and large libraries of models for processing equipment and mixture physical properties had been and were being developed by industry. Hopefully, one or more of these simulators might be made available to chemical engineering students.

By the following CACHE meeting, held at the Sonoma Mission Inn on December 3-4, 1971, following the annual AIChE meeting in San Francisco, the Large-Scale Systems Task Force had sent questionnaires to 15 potential suppliers of these industrial steady-state process simulators and had formulated a sample problem that would test their ease of use, robustness, and efficiency. By the next CACHE meeting, held at the Lake of the Ozarks, Missouri on May 24-26, 1972, the task force had received four favorable responses and four industrial simulation systems had been tested: GEPDS of General Electric, FLOWTRAN of Monsanto Company, PACER-245 (a commercial version of PACER) of Digital Systems Corp., and PDA of McDonnell-Douglas Automation, Inc. Of these four, the task force determined that FLOWTRAN was the most desirable and the task force was instructed to seek an agreement with Monsanto for the use of FLOWTRAN by universities over a national computer network. Furthermore, this system met the guidelines listed in the CACHE publication, *CACHE Guidelines for Large-Scale Computer Programs*, which was issued by the task force in February, 1973.

By December 1, 1972 at the CACHE Committee meeting at Grossinger's following the New York AIChE annual meeting, the Large Scale Systems Task Force reported that a preliminary oral agreement had been reached between CACHE and Monsanto on the use of FLOWTRAN, but details still needed to be worked out before a final approval could be obtained. The original request was for the use of FLOWTRAN through the service bureau of Monsanto Enviro-Chem Systems, Inc., established for outside commercial users of FLOWTRAN. However, in March 1973 Monsanto terminated the service bureau, thus eliminating that means of accessing FLOWTRAN. University of Texas' Professor John J. McKetta of the CACHE Advisory Committee called Mr. F. E. Reese, Vice President and General Manager of Monsanto, to discuss other means of acquiring FLOWTRAN. Mr. Reese's response was encouraging. Therefore, on June 13, 1973, in a formal letter to Mr. John W. Hanley, President of Monsanto Company, CACHE requested that Monsanto consider other means for allowing chemical engineering faculty and students to use FLOWTRAN for both course work and research. With the additional support and assistance of M. C. Throdahl, J. R. Fair, and S. I. Proctor of Monsanto, approval for the use of FLOWTRAN via a national computer network, was granted by Monsanto Company in a letter from Jim Fair dated November 9, 1973. This approval included assistance in implementing the system by providing a grant and loaning specialists from the Monsanto Corporate Engineering Department.

Starting on May 10, 1974, Dr. Allen C. Pauls of Monsanto directed a three-day training course on FLOWTRAN, which was attended by Richard R. Hughes, H. Peter Hutchison, J. D. Seader, Warren D. Seider, and Arthur W. Westerberg. On June 5, 1974, at a meeting of the CACHE Committee in Seven Springs, Pennsylvania, Dr. Proctor of Monsanto formally presented CACHE with a Monsanto grant, in the amount of $7,000 in cash and $21,000 in services, and conditions for making FLOWTRAN available to universities. For a department to use FLOWTRAN, a three-party agreement, drawn up by Monsanto, had to be signed by the department, CACHE, and Monsanto, so as to make sure that the program would only be used for educational purposes.

The FLOWTRAN project marked the first attempt by CACHE to distribute computing services via a computer network. In connection with this effort, the Program Distribution Task Force, chaired by Warren Seider, completed the document *CACHE Guidelines for Computer Networks* in June 1974. As a result of that effort, CACHE began a cooperative venture with EDUCOM (The Inter-university Communications Council) to help develop an educational computer network. In 1976, an article by the Program Distribution Task Force entitled, "Aspects of Software Dissemination in Chemical Engineering," was published in the proceedings of the EDUCOM Fall Conference. On September 28-29, 1978 at Washington, D.C., CACHE, under the direction of Seider, Westerberg, and EDUCOM representatives, sponsored an NSF-funded networking conference entitled "How Can the Chemical Engineering Discipline Best Utilize Networks for the Sharing of Computer-Based Resources in Research and Teaching"; it was attended by 25 educators.

By the end of 1977, CACHE had: (1) held five 4-day FLOWTRAN workshops, under the administration of the Continuing Education Department of the AIChE, in Evanston, Houston, Boston, Madison, and Philadelphia to instruct a total of 90 faculty in the use of the program; (2) written and distributed three books, *FLOWTRAN Simulation - An Introduction, CACHE Use of FLOWTRAN on UCS* (a national network), and *Exercises in Process Simulation Using FLOWTRAN;* (3) formed a FLOWTRAN User's Group; and (4) began to issue a FLOWTRAN Newsletter edited by Professors J. Peter Clark and Jude T. Sommerfeld.

Until 1982, FLOWTRAN could only be accessed from the UCS network. After that, Monsanto released the source code to CACHE and approved the preparation of load modules for 14 different computer/operating system combinations so that universities could run FLOWTRAN on their own computers. The conversion of the FLOWTRAN source code to load modules for various systems was greatly facilitated by a very careful conversion carried out by the University of Michigan under the direction of Brice Carnahan, which was the subject of two detailed reports that were distributed to other converters. The cost to license the load module was $175 to universities supporting CACHE and $250 to others. In May 1987 Professor Lorenz T. Biegler of Carnegie-Mellon University, and a member of CACHE, completed the preparation of an SQP optimization routine as an add-in to FLOWTRAN.

On April 4, 1994, after almost 20 years of providing the use of FLOWTRAN to universities and when a number of more advanced commercial simulators had entered the market and were being licensed to universities at low cost, Monsanto announced that they would discontinue licensing FLOWTRAN to universities. During the 20-year period of the CACHE FLOWTRAN project, 59 universities used the program on the UCS network, 190 FLOWTRAN load modules were distributed to universities (141 in the United States, 11 in Canada, and 38 in 21 other foreign countries), and 15,000 copies of three editions of *FLOWTRAN Simulation - An Introduction*, which had been printed in Ann Arbor, Michigan under the direction of Brice Carnahan, were sold. Today, almost all departments of chemical engineering teach their undergraduates the use of computer-aided, steady-state chemical process simulation.

Although the biannual *CACHE News* and the annual receptions for CACHE representatives have been and continue to be the main means of communicating CACHE activities and products to faculty, CACHE officers, trustees, and committee members have, from time to time, published general and specific articles about CACHE in magazines, journals, and special reports. The first such article appeared as a CACHE report on January 1, 1972, and was entitled *Origins and Organization of the CACHE Committee*, by Seider, Evans, and Westerberg. Other articles have followed, including:

"Computers in Education: How Chemical Engineers Organized the CACHE Committee," by Seider, Evans, and Westerberg, in *EDUCOM Bulletin*, Vol. 8, No. 2, pp 10-17, Summer 1973.

"Use of FLOWTRAN Simulation in Education," by J. Peter Clark and Jude T. Sommerfeld," in *Chemical Engineering Education*, p 90, Spring 1976.

"What is CACHE?," by Himmelblau and Hughes, in *Chemical Engineering Education*, pp 84-87, Spring 1980.

"Computer Graphics in Chemical Engineering Education," by Edgar, in *Chemical Engineering Progress*, pp 55-59, March 1981.

"Computer Graphics in ChE Education," by Reklaitis, Mah, and Edgar, in *Engineering Education*, pp 147-151, December 1983.

"The Impact of Computers on Undergraduate Education," by Finlayson, in *Chemical Engineering Progress*, pp 70-74, February 1984.

"Computer Aids in Chemical Education," by Edgar, Mah, Reklaitis, and Himmelblau, in *ChemTech*, pp 277-283, 1988.

"Education and Training in Chemical Engineering Related to the Use of Computers," by Seader, in *Computers and Chemical Engineering*, Vol. 13, pp 377-384, 1989.

"Computer Aids for Chemical Engineering Education: An Assessment of CACHE — 1971-1992," by Seider, in *Computer Applications in Engineering Education*. Vol.1, No. 1, pp 3-10, 1992.

Since its inception, CACHE has sought ways to cooperate with the ASEE in providing summer education in computing for chemical engineering faculty. In August 1972, Warren Seider made a presentation on the mission and activities of CACHE to an audience of nearly 100 educators, on a rainy afternoon, at the ASEE Summer School for Chemical Engineering Faculty in Boulder, Colorado. A similar presentation was made by Paul Shannon on June 27, 1973 to another 40-50 educators. In August 1977, at the ASEE Summer School for Chemical Engineering at Snowmass, Colorado, CACHE held demonstrations and workshops on interactive computing. At subsequent ASEE summer schools in 1982 at the University of California at Santa Barbara, in 1987 at Southeastern Massachusetts University, and in 1992 at Montana State University, CACHE typically organized and assumed responsibility for seven or eight computing sessions. These summer sessions have provided an exceptional opportunity for CACHE to inform chemical engineering educators of new educational computing tools.

During its first quarter century, CACHE also maintained close ties with the AIChE. One trustee (Hughes) served as Vice-President and then President of AIChE from 1981-1982, eight trustees (Hughes, Himmelblau, Evans, Seader, Seider, Seinfeld, Edgar, Finlayson and Fogler) served 3-year terms as Director, and another (Edgar) was elected Vice President of the Institute in 1995 and assumed the Presidency in 1996. Many CACHE trustees have held offices in the CAST Division, where they have been responsible for computing sessions at AIChE meetings. A number of trustees have received Institute awards: (1) Seinfeld, Edgar, Stephanopoulos, Morari, and Kim for the Allan P. Colburn award; (2) Sandler and Denn for the Professional Progress award; (3) Finlayson, Denn, Seinfeld, and Westerberg for the William H. Walker award; and (4) Douglas and Fogler for the Warren K. Lewis award. The Founders award was given to Himmelblau in 1992. Since its inception in 1979, the Computing in Chemical Engineering Award of the CAST Division has been given to a CACHE trustee (Hughes, Carnahan, Mah, Evans, Westerberg, Reklaitis, Himmelblau, Douglas, Seader, Motard, Seider, Stephanopoulos, and Grossmann) every year but three. The CAST Computing Practice Award was given to a CACHE industrial trustee, Siirola, in 1991. Seinfeld, Smith, Morari, and Arkun have received the Donald P. Eckman Award and Edgar has received the AACC Education Award. The annual Institute Lecture has been given by Seinfeld (1980), Seader (1983), Westerberg (1989), and Douglas (1993).

At the February 20-23, 1972 AIChE Annual Meeting in Dallas, the CACHE Committee sponsored its first symposium, which consisted of two sessions on Computer-Aided Process Synthesis. Over 100 persons attended the sessions, which were chaired by Seader and Elzy. A review article on that topic, co-authored by Seader, appeared in the January 1973 issue of the *AIChE Journal*. Since then, CACHE has sponsored symposia at AIChE meetings in St. Louis in May 1972, in Washington, D.C. in November 1983, and in Anaheim in May 1984. On November 5, 1989, Biegler presented an AIChE seminar on optimization at the San Francisco meeting. On November 11, 1990, Cutlip presented a seminar on the use of POLYMATH for numerical calculations at the AIChE Annual meeting. A special symposium entitled, "Computers in Chemical Engineering Education - 25 Years of CACHE," was held on November 16, 1994 at the AIChE Annual meeting in San Francisco, chaired by Cutlip and Himmelblau. Four papers were presented: (1) Edgar, "Process Control Education: Present, Past, Future," (2) Fogler, "Ghosts of Interactive Computing: Past, Present, Future," (3) Stephanopoulos, "Knowledge, Computers, and Process Engineering: A Critical Synthesis," and (4) Carnahan, "2001 -

A Computing Odyssey."

At the Buck Hill meeting on September 2-3, 1971, operating policies were established for the CACHE Committee. Some of the important policies were: (1) a maximum of 18 members, (2) a task force structure in which the chairman must be a CACHE member, but other members of the task force could be non-members, including participants from industry and government, as well as faculty, (3) election of new members each year by the present members, (4) a three-year term of membership, but with provision for re-election, and (5) election of a new vice-chairman and secretary each year, with the current vice-chairman automatically becoming the next chairman. These operating policies remained in effect, with only minor changes, until March 25, 1974, when CACHE was incorporated as a non-profit corporation in the state of Michigan, with Brice Carnahan as the Registered Agent, but with a business address in Cambridge, Massachusetts. The Articles of Incorporation were prepared by Betty Ann Weaver, Attorney at Law in Glen Arbor, Michigan and sister of Bob Weaver, newly elected CACHE secretary.

At that time, CACHE had a part-time secretary, Cindy Driear, a lawyer, William Thedinga of the firm of Bingham, Dana & Gould, and an accountant, Joseph Cullinan, all residing in the Boston area. CACHE had also established checking and savings accounts with the Cambridge Trust Company and Cambridge Savings Bank, respectively. The first balance statement showed assets of $5,519 and liabilities of $2,994. On June 6, 1974, at Seven Springs Inn in Champion, Pennsylvania, the first meeting of the trustees (made up from current CACHE Committee officers and members) of the newly formed CACHE Corporation was held. New Bylaws were presented by Bob Weaver and ratified after a few minor revisions. These Bylaws added an Executive Officer, who would also act as Treasurer, and an Executive Committee. Meetings of the trustees would be held twice each year; new officers would be elected each year, except that the Vice-President would automatically succeed to the presidency; and the number of trustees would be limited to 21 chemical engineering educators, who would serve for three-year terms, but with the possibility of re-election.

On October 2, 1974 Solomon Watson of the law firm of Bingham, Dana & Gould in Boston suggested that CACHE could enhance its attempt to gain tax-exempt status with the IRS if CACHE would form a Massachusetts not-for-profit corporation of the same name, into which the recently formed CACHE Corporation of Michigan would merge. On December 6, 1974 at a trustee's meeting at Boar's Head Inn in Charlottesville, Virginia, the CACHE membership voted to accept the merger plan. On February 26, 1975 the CACHE Committee became The CACHE Corporation in The Commonwealth of Massachusetts. Early in 1975, the IRS approved an exemption from Federal income tax and on March 15, 1976 the corporation in Michigan was dissolved.

When the CACHE Corporation Bylaws were approved on June 6, 1974, Ernie Henley questioned the restriction of membership to those in academia. At the December 2-4, 1976 meeting at The Abbey, Lake Geneva, Wisconsin, the trustees approved a motion by Duncan Mellichamp and David Himmelblau that CACHE elect industrial affiliates as non-voting members to serve for one year. At the August 5-7, 1977 meeting at Snowmass Village, Colorado, the trustees approved a motion to extend the term for industrial members to three years. The first three industrial members elected were Theodore Leininger of DuPont, Edward Rosen of

Monsanto, and Louis Tichacek of Shell Oil. At the May 17-19, 1979 meeting in St. Louis, the term of office for President, Vice-President, and Secretary was increased to two years. At the November 12-13, 1981 meeting at Cancun, Mexico, an extensive revision of the Bylaws was approved. The number of (voting) trustees was increased to 28, but the number of academic trustees was limited to 21 and the number of industrial (including government employees and consultants) was limited to 7. The Executive Officer-Treasurer would be appointed by the President for an unspecified number of years. To date, the Executive Officers of CACHE have been Larry Evans (1974-1980), J. D. Seader (1980-1984), and David Himmelblau (1984-date). At the November 5-7, 1992 meeting in Coral Gables following the AIChE meeting in Miami Beach, CACHE approved an extensive revision of the Bylaws prepared by Jeffrey Siirola, an industrial trustee. The revision was made to clarify certain statements, add consistency, improve organization, and delete obsolete or irrelevant material.

By 1974, the impact of CACHE on the use of computers in chemical engineering education began to be recognized by industry and, as mentioned above, CACHE began to recognize the importance of working closely with industry. Starting in 1974 with the Monsanto grant to CACHE for the FLOWTRAN project, CACHE began to receive some industrial grants annually. The following companies have contributed financial support to CACHE during the 20-year period from 1974 to 1995: Chevron Research Corporation, Chiyoda Chemical Engineering and Construction Company, Digital Equipment Company, Dow Elanco, DuPont Committee on Educational Aid, EXXON Educational Foundation, The Halcon SD Group, Monsanto Chemical Company, Olin Chemicals Corporation, Pfizer Foundation, Process Simulation International, Rust International Corporation, Simulation Sciences Incorporated, Shell Companies Foundation, Tektronix, Tennessee Eastman Company, Weyerhaeuser Company, and the Xerox Foundation.

The first of the three most memorable CACHE meetings was the one held on December 3-4, 1971 at the Sonoma Mission Inn, following the annual AIChE meeting in San Francisco. No one present has forgotten it. The Inn is located in the heart of the wine country north of San Francisco. California is noted for the many beautiful Catholic Missions located along most of the length of the state. But the Sonoma Mission Inn turned out to be a retirement home that also served as an overflow motel for the region. The rooms were spacious, but old and poorly furnished with door locks that used latch keys, many of which didn't work. Meals were served mostly family style on paper table cloths. Except for the retired people living there, we were the only guests. To make matters worse, it rained both days we were there. Following this meeting, at the suggestion of Ernest Henley, the Ed Grens Memorial Prize was established for acknowledging never-to-be forgotten meeting arrangements. To date that cherished prize has gone to: (1) Edward Grens for the 1971 Sonoma meeting, (2) Warren Seider for the 1978 Spring Lake meeting, and (3) Thomas Edgar for the 1986 Biloxi meeting. On the positive side at the Sonoma Meeting, Larry Evans presented the outline of a proposal to the NSF for continued support of CACHE for the three-year period of 1973-75 to: (1) develop and evaluate new computer-based courses and curricula, (2) create special computer aids for education, and (3) determine effective mechanisms for distributing computing materials. Without additional funding from NSF, the future operation of CACHE would have been in peril because no other sources of funding had been forthcoming.

Like the first proposal to the NSF for support of CACHE, the second proposal would be sponsored and submitted by the COE of the NAE. It was determined that the best chance for funding the proposal was with the Division of Undergraduate Education in Science, where Dr. Gregg Edwards was the Assistant Program Director for the Science Course Improvement Program. In January 1972, the CACHE officers discussed plans for the proposal with Dr. Edwards, who gave little encouragement. However, on a return visit by the CACHE Executive Committee in May 1972, in which some proposal changes were discussed, Dr. Edwards expressed some optimism. The 54-page formal proposal, written largely by Larry Evans, CACHE Vice-Chairman, was sent on September 27, 1972 with a request for $94,014 for the 3-year period. The proposal emphasized the exploitation of new computing technology to carry out projects that would lead to self-sufficiency for CACHE. Accordingly, the proposal called for the establishment of three new task forces: (1) the Modularized Instruction Task Force to coordinate the development of 100 teaching modules, (2) the Real-Time Laboratory Task Force to coordinate the preparation of instructional material for undergraduate laboratories, and (3) the Program Distribution Task Force to identify workable mechanisms, including computer networks, to distribute small and large-scale computer programs and modules. Prior to the meeting of June 6-8, 1973 at Ann Arbor, Michigan, the NSF funded the three-year proposal for a total of $93,884. CACHE now had until the end of 1975 to find new ways to support its activities.

At the November 30 - December 2, 1972 meeting at Grossingers in New York, David M. Himmelblau was elected to membership in CACHE. He quickly developed a concern for the future financial survival of CACHE and determined that a reliable and steady source of income was essential. Because it was becoming apparent that CACHE could provide continuing and worthwhile educational computing services to departments of chemical engineering in universities, he suggested at the December 5-7, 1974 meeting at Charlottesville, Virginia that CACHE should consider a periodic solicitation of all departments in the United States and Canada. In return, the departments would receive discounts on CACHE products. At the following meeting, held on April 2, 1975 in Houston, his idea was put into effect. At the August 21-23, 1975 meeting in Andover, Massachusetts, Himmelblau was appointed Chair of a Standing Committee for Development. At the November 20-22, 1975 meeting at San Diego, Himmelblau announced that approximately 65 schools had pledged or contributed $200. A second solicitation was initiated in the Spring of 1978 which resulted in 64 contributions from schools in the United States and Canada. Since then, schools have been solicitated for two-years of support every other year and more recently on a yearly basis. Starting with the September 1982 issue (No. 15) of *CACHE News*, departments supporting CACHE have been listed. That list has continually grown and has expanded to include departments outside of the United States and Canada. In the Spring 1994 issue (No. 38) of *CACHE News*, the list included 110 departments in the United States, 16 departments in Canada, and 22 in other countries. During the past 3 years, the solicitation of departments has brought an average annual income to CACHE of $15,500.

The first CACHE proposal to a Federal agency for a specific project came from the Modularized Instruction Task Force, chaired by Ernest Henley, who became the President of CACHE in 1975. Funded by the NSF as Grant No. HES75-03911 on July 1, 1975 through the office of Dr. Gregg Edwards for a total of $145,790, the CHEMI (Chemical Engineering Modular Instruction) Project, had as its goal the development and distribution of from 50 to 80 self-study, single concept, text modules covering the entire chemical engineering undergraduate curriculum, including the seven areas covered in the earlier CACHE computer program volumes, which had been developed under the leadership of Ernest Henley. Each module was to be from 7 to 15 pages in length, containing theory and examples suitable for a one-hour lecture and with homework exercises. The modules would be solicitated from the worldwide chemical engineering community. William Heenan of the University of Puerto Rico agreed to be the Executive Director for the project, which was announced in the December 8, 1975 issue of *Chemical and Engineering News*.

This project was a model of organization. By June 1976, the preparation of about 70% of the modules had been commissioned. By June 1977, 67 modules had been completed. Although the 3-year project was moving along quite well, it became apparent early in 1978 that it could not be completed by mid-1978. Accordingly, CACHE requested and was granted, on June 2, 1978, a no-cost contract extension until May 31, 1979. At the May 17-19, 1979 meeting of CACHE in St. Louis, Ernest Henley reported that 111 modules had been completed in six areas, the Design area having been delayed. He also reported that an agreement had been reached with Hal Abramson of the AIChE to publish the modules, which ultimately numbered 230.

CACHE donated its copyrights to all the modules developed under the CHEMI Project and provide $16,000 toward printing the volumes. The AIChE set aside $100,000 towards the printing and distribution of all six volumes. Later, The AIChE made the decision to cluster the modules into groups of 3-11 and publish them in six series of 3-7 volumes each under the title *AIChEMI Modular Instruction*. In the October 1980 issue of *Chemical Engineering Progress*, the AIChE announced the publication of the first six volumes of the AIChEMI Modular Instruction Series available for $15 each. In January 1981, all chemical engineering departments in the United States and Canada received a copy of the 93-page Vol. 1: "Analysis of Dynamic Systems" of Series A: *Process Control*, edited by Thomas F. Edgar, who had been elected a CACHE Trustee at the December 2-4 meeting at Lake Geneva, Wisconsin. By 1987, the Design series had been added and the AIChE had published a total of 36 volumes. Solutions to the homework exercises were available under separate cover.

At the May 17-19, 1979 meeting of CACHE in St. Louis, Missouri, David M. Himmelblau, who became President of CACHE on June 7, 1978 at the meeting at Spring Lake, New Jersey, announced that he had submitted to the NSF a follow-up proposal to the CHEMI Project entitled, "Assessment of Alternative Distribution Systems for Modular Instructional Materials." The purposes of this project were to complete additional undergraduate-level teaching modules, create 80 graduate-level modules, prepare 500 abstracts of topics not covered in the modules, test the modules that had been and were being produced, and experiment with ways to disseminate and encourage use of the materials through an on-line system on a computer network. On September 12, 1979 CACHE received notice from the NSF for funding of the proposal under Grant No. SED-79-13021 for a total of $298,500 for the 4-year period from September 12, 1979 to September 30, 1983. An additional $36,384 under Grant No. SED-81-16698 was added on August 19, 1981 and the expiration date was extended to September 30, 1984 on March 9, 1983. Funding actually ran out in March 1984 and the final report to the NSF was sent on March 30, 1984. This project was one of the first to demonstrate that a large educational data base could be computerized. During the period of the CHEMI Project, a Conference on Software Portability was held at the University of Texas at Austin on November 23,

1981.

At the September 2-3, 1971 meeting at Buck Hill Falls, Pennsylvania, Lawrence Evans reported that, based on a survey of 153 universities, interest in real-time computing was growing rapidly. By 1972, it was estimated that approximately 50 departments would have equipment for such computing. Accordingly, a task force was formed, chaired by Eugene Elzy, to assess the role of real-time computing in the chemical engineering curriculum. In October 1973, the task force published an interim report which listed: (1) specifications for real-time digital computing systems, (2) experiments utilizing real-time digital computers, and (3) sample outlines for courses in real-time computing. A second survey, completed in January 1974, indicated continued growing interest. Therefore, a CACHE Workshop, arranged by Cecil Smith (who had become a Trustee in 1972) as part of the AIChE Continuing Education program, was held on December 1-2, 1974 in Washington, DC. The 13 paid attendees received instruction on establishing and operating real-time computing systems. At the following CACHE Trustee's meeting held on December 5-7, 1974, Duncan Mellichamp, who had become a Trustee just a year before, became the new Chairman of the Real-Time Task Force and announced that plans were well underway for 12 members of the task force to prepare the *CACHE Monograph Series in Real-Time Computing*. The first of eight monographs, which eventually totaled more than 700 pages, was published by CACHE in the summer of 1977. Through generous grants from EXXON and Shell, CACHE distributed a full set of monographs to each CACHE-supporting department. The project was completed in 1978 and the monographs were available for purchase until 1985. Approximately 500 sets of the monographs were sold by CACHE at prices ranging from $15 to $28.

Another task force formed early in the history of CACHE was the Physical Properties Task Force, chaired by Rudy Motard, whose CHESS program included TAP, the first widely available computer program, developed by Henley, for estimating the properties of gas and liquid mixtures. In August, 1972 the task force published a very useful 35-page *CACHE Physical Properties Data Book* by Michael R. Samuels, which included extensive listings of sources of physical property data, including 19 handbooks, 69 general data tabulations, and 24 reference sources, indexed by property type. This book was distributed to all chemical engineering departments in the United States and Canada.

At the AIChE National Meeting at St. Louis in 1972, Ronald L. Klaus and Rudy Motard presented a paper entitled, "Design of a Physical Property Information System for Undergraduate Education." A detailed report on a proposed project for developing such a system was presented by the task force in an October 1974 final report. The system would be modular so that a university could assemble its own system structure. The project was abandoned when it was learned that many such systems were being developed by industry. One such system, developed jointly by the Institution of Chemical Engineers and the Engineering Sciences Data Unit in England, is PPDS (Physical Property Data Service). In 1981, Rudy Motard obtained grants from DuPont and Simulation Sciences, Inc. to enable the installation of an academic version of PPDS on the TELENET communication network at Carnegie Mellon University. His booklet, *Introduction to CACHE Version of Physical Property Data Service*, described the use of the system on the network. The CACHE version of PPDS contained data records of 18 constants for 50 common chemicals, and methods for calculating 15 different vapor and liquid mixture properties. Unfortunately, the service was little used because the biggest need for properties was in simulation, and programs like FLOWTRAN that had built-in physical property estimation systems. Nevertheless, the project demonstrated the use of a computer network to share a proprietary program and stimulated interest in large physical property data systems.

In the early 1970s, little interest was shown by academia in the United States in teaching plant safety to undergraduate students in chemical engineering. The 1968 second edition of *Plant Design and Economics for Chemical Engineers* by Peters and Timmerhaus devoted only one page to the subject. The 1980 edition increased this to about three pages. Finally in the 1990 edition, following the tragic December 1984 Bhopal, India accident in a pesticides plant that released a lethal cloud of methyl isocyanate, killing more than 6,000 people, injuring 200,000 or more, and leaving tens of thousands permanently impaired, almost 30 pages were devoted to health and safety in chemical plants. This early neglect had not been the case in England, where a guide entitled, *Flowsheeting for Safety* was published by the Institution of Chemical Engineers in 1976 and widely used thereafter. At the November 30 - December 2, 1972 meeting at Grossinger's in New York, Gary J. Powers was elected a Trustee. Powers, with Dale F. Rudd and Jeffrey J. Siirola, had developed in the early 1970s the first process synthesis computer program, AIDES, and the first textbook on process synthesis. Powers was now turning his attention to the related topics of reliability and safety. At the December, 1974 meeting at Charlottesville, a Safety and Reliability Task Force was organized, with Powers as Chair, to assemble material on safety and reliability technology for the chemical engineering curriculum. On April 17, 1975 Powers and Henley submitted a CACHE proposal to the Sloane Foundation for funding a safety and reliability project. However, because CACHE did not have a tax-exempt status at that time, the Sloane Foundation would not entertain the proposal. Subsequently, a revised proposal for about $100,000 was submitted to the Exxon Foundation to install safety and reliability programs on a network for use by faculty and students. That proposal also failed to be funded and activity by CACHE in this area was suspended until 1982 when Himmelblau obtained a process troubleshooting computer program developed by Professor Ian D. Doig of Australia. That program, which simulates malfunctions in two different plants and was described in *CACHE News* No. 17 (September 1983), was distributed by CACHE in the Spring of 1983. Today, material for teaching safety to chemical engineers is widely available through the Center for Chemical Process Safety of the AIChE and in recent textbooks such as *Chemical Process Safety* by D. A. Crowl and J. F. Louvar (1990).

Also at the December 1974 meeting at Charlottesville, Weaver proposed a new CACHE effort for assembling teaching materials in resource management, with particular attention to the data bases. The Resource Management Task Force was organized with Weaver as Chair. By the August 1975 meeting in Andover, Massachusetts, the task force had submitted a CACHE proposal to NSF for funds to produce educational materials for undergraduate courses. At the suggestion of the NSF, the proposal was expanded to include macroeconomics, allocation studies, and the creation of data bases. A third revision was submitted in 1976 for about $200,000. At the November 16-18, 1978 meeting at Nassau, following the annual AIChE meeting in Miami, several unsuccessful attempts at funding with the NSF, the Dreyfus Foundation, and the Ford Foundation were reported. The primary difficulty had been in trying to define the problem from a chemical engineering viewpoint. Although a number of trustees felt that there was a need for instruction in resource management, the task force was dissolved at the May 17-19, 1979 meeting in St. Louis. A textbook for chemical engineers on the subject of resource management, entitled *The Structure of the Chemical Processing Industries*, was published in 1978 by Professors Wei, Russell, and Swartzlander of the University of Delaware.

At the December 5-7, 1974 meeting in Charlottesville, Virginia, the Computer Graphics Ad-Hoc Task Force was formed with Richard S. H. Mah as Chairman. Dick Mah had been elected a Trustee the previous year and, along with Brice Carnahan and H. Scott Fogler, who had been a elected Trustee in November 1975, believed that computer graphics could have a tremendous impact on education. However, at that time extraordinary resources were required to store graphical images and standards were lacking. On August 3-4, 1977, the task force held an "Interactive Computing Workshop" in conjunction with the ASEE Summer School for Chemical Engineering Educators at Snowmass Village, Colorado. At that time, they also conducted a survey on graphics usage in education. In 1978, the task force published a 66-page CACHE report entitled *Computer Graphics in Chemical Engineering Education*, in which software and hardware were discussed, along with potential applications in chemical engineering. Included were the results of the survey, which showed that although 90% of the 31 respondents had graphics terminals, only 15% were using computer graphics in coursework. At the November 17-19, 1977 meeting in Buck Hill Falls, Pennsylvania, the ad-hoc task force was made a full task force, with Thomas Edgar as Chairman. By August 1, 1979, as reported in an article entitled *Computer Graphics in Chemical Engineering Education* by Edgar in the March 1981 issue of *Chemical Engineering Progress,* a new survey showed that 67 of 87 reporting departments now had graphics devices. Thirty-five percent of the respondents were using graphics to help in the teaching of such topics as dynamics and control, design, kinetics, heat transfer, thermodynamics, distillation, and stoichiometry. Many were using graphics for simple plotting.

In the Spring of 1983, a third CACHE report, entitled *Computer Graphics in the ChE Curriculum, ASEE/NSF Position Paper*, by G. (Rex) V. Reklaitis, Mah, and Edgar was published. Rex Reklaitis had been elected a Trustee on November 30, 1980 at the meeting in Hershey, Pennsylvania and had become the Chairman of the Computer Graphics Task Force at the Fontana, Wisconsin meeting, November 20-22, 1980. The position paper presented a five-year plan for the revitalization of undergraduate education by the focused introduction of interactive computer graphics technology and courseware. The plan called for an equipment grant program, a computer-aided instruction courseware development program, a software clearinghouse, and a faculty training program. CACHE would be successful later in obtaining grants for courseware development. Edgar and Reklaitis of the Computer Graphics Task Force were guest editors for a special CACHE-sponsored 1981 issue (No. 4) of *Computers and Chemical Engineering*, which included 11 refereed articles on the application of computer graphics in chemical engineering.

At the August 21-23, 1975 meeting held at Andover, Maine, a standing Committee for Publications was organized, with Brice Carnahan as Chairman. In 1974, Carnahan had brought to the attention of CACHE that books could be printed from camera-ready copy and bound with a long-lasting soft-cover binding at a low cost by several printers in Ann Arbor, Michigan. Furthermore, Ulrich's Bookstore in Ann Arbor could be used as the distributor. Carnahan became the CACHE publisher. The products were either sold from Ulrich's Book Store or distributed from the Carnahan warehouse (his garage). CACHE publications were advertised in the *CACHE News*. By 1989, so many publications were available that the first *CACHE Catalog of*

Products was published after preparation by Michael F. Doherty (who had been elected a Trustee at the November 6-8, 1986 meeting in Key West, Florida). The 1994 issue of the catalog, edited by Margaret Beam of the CACHE office, lists 75 different CACHE products.

On August 12, 1981, IBM announced its personal computer (PC), based on the Intel 8088 microprocessor. This unit had 10 times the internal memory of the first microcomputers, and could process 16 bits at a time internally; an Intel 8087 coprocessor chip could be added to greatly speed-up floating-point scientific and business calculations. By the end of 1983, IBM had shipped an estimated 500,000 machines and a wide variety of software had become available, including word processing, spreadsheets, database management, communications, data processing, and graphics. Computing was ready to migrate from the Computer Center to the desktop. CACHE had already been made aware of the impending development of microcomputers at the April 14-16, 1976 meeting at the Lake of the Ozarks in Missouri because of information provided by Joseph D. Wright (industrial trustee from XEROX), who had became a Trustee in November 1975 and Chairman of the Real-Time Task Force at the Ozarks meeting.

At that same meeting, H. Scott Fogler, who had a great interest in the improvement of teaching methods, became a Trustee. At the June 7-9, 1978 meeting at Spring Lake, New Jersey, Fogler was appointed Chair of the Personal Computers Task Force. At the Nov. 29-30, 1979 meeting at Carmel, California, following the AIChE Annual meeting in San Francisco. Fogler announced that he and Carnahan had submitted a CACHE proposal to NSF on Personal Computing. Although the initial version of the proposal considered the preparation of programs for hand-held calculators with magnetic strips, that task was deleted and the proposal was focused on desk-top microcomputers of the PET and Apple type. However, in 1982 the task force did publish *Hand-Held Programmable Calculators: A Review of Available Programs for Chemical Engineering Education*, edited by Professor F. William Kroesser; this volume was followed by a more extensive listing of published calculator programs in the March 7, 1983 issue of *Chemical Engineering* by another source.

On June 23, 1980, NSF awarded a two-year grant of $128,000 to CACHE, to fund the proposal by Carnahan and Fogler. This project, which would become known as the MicroCACHE project, involved: (1) the development, by Carnahan, of a microcomputer-based delivery or authoring system for educational materials and programs and (2) the production of a small number of educational modules to test the system and demonstrate its effectiveness. The project was completed in April 1983. The final authoring system included a graphics package, a database management package, a numerical analysis package written in FORTRAN, and 13 prototype educational modules to test and demonstrate the system. In addition, two stand-alone modules, written under Fogler's direction, involved the simulation of a packed-bed reactor and the design of a multicomponent distillation column. Although the Apple II Plus microcomputer was selected as the hardware vehicle, an IBM PC version of MicroCACHE, prepared with other sources of funding was the initial system distributed by CACHE, starting late in 1984. Early in 1986, a second IBM PC version of the MicroCACHE software was announced. By April 1987, 30 copies of the software had been purchased from CACHE by departments. The most recent version of MicroCACHE is called MicroMENTOR, which consists of system management software for distribution of departmental computing resources to networked personal computers, lesson authoring software, and 11 instructional modules prepared with the authoring system for PCs running under DOS or OS/2.

Another venture by CACHE into the use of microcomputers began in 1981 with the election of Peter R. Rony as a Trustee. At that time many chemical engineering faculty recognized that computer interfacing in the laboratory was about to shift from relatively expensive minicomputers to much less expensive microcomputers, as predicted by Wright. Starting in 1974, Rony had written and edited monthly columns on computers and, more particularly, on microcomputer interfacing, in *American Laboratory and Computer Design*. He had authored nine popular books (called the "Bug" books) on integrated circuits, programming, and interfacing of microprocessors, and had conducted several short courses on those topics. At the April 8-9, 1981 meeting in Houston, CACHE approved a five-day course on microcomputer interfacing/ programming to be given by Rony for 20 interested faculty, at the lowest possible cost, at Purdue University on February 11-15, 1982. The great success of the course led to an additional five-day course on October 21-25, 1982 at the University of Pennsylvania. One participating faculty member commented that this course was the most important thing that CACHE had done to date.

On February 12, 1984, Rony and Wright published the results of a survey entitled, "Microcomputers and Personal Computers in American and Canadian Departments of Chemical Engineering." They received 46 responses and concluded that interest in the use of microcomputers was growing rapidly and that the introduction of the IBM PC with a 16-bit Intel chip promised a very bright future for the personal computer. In November 1984, Wright arranged for appearance of the survey results in the April 1980 issue (No. 14) of *CACHE News*, and signaled the importance of faculty members having their own personal computers. In 1982, he began compiling lists of programs for microcomputers and arranged a visit by the CACHE Trustees to the Xerox Palo Alto Research Center (PARC) where much of the early research that culminated in the graphical interface of the Apple Macintosh computer was carried out.

At the November 20-22, 1980 meeting at Fontana, Wisconsin, Bruce A. Finlayson was elected a Trustee. Finlayson had authored two important books in numerical mathematics for chemical engineers and was one of the first faculty to have his own personal computer and to recognize its advantages over mainframe computers, particularly in the graphics area. Beginning with the April 1984 issue (No. 18) of *CACHE News*, he wrote a series of articles entitled "Programs for PCs," which described in detail available software of interest to chemical engineers. By the Spring 1993 issue, Finlayson had described 24 programs.

Starting in the 1970s, and especially in 1976 with funding by the NSF and the DOE for the LINPACK project of the Applied Mathematics Division of Argonne National Laboratory, high-quality public-domain software for numerical methods began to become available. At the November 12-13, 1981 meeting in Cancun, William E. Schiesser, an important contributor to the development of software for the numerical solution of ordinary and partial differential equations, became a Trustee of CACHE. Beginning in 1975, he had served as Chairman for the biannual AICA/IMACS International Symposia on Computer Methods for Partial Differential Equations. Starting in the September 1982 issue (No. 15) of *CACHE News*, he wrote a series of articles on available high-quality software for the solution of systems of linear, nonlinear, ordinary differential, partial differential, and differential-algebraic equations. Most importantly, he became a distributor for several of the more widely used packages, so that copies could be obtained by faculty from him. By the Spring 1989 issue of *CACHE News*, Schiesser had written seven articles on these programs. Much of this type of software is now readily available over the Internet by accessing netlib@ornl.gov, from which an index of all available programs can be obtained by E-mail request. While this software was almost exclusively used on mainframes, minicomputers, and workstations when the software first became available, much of the software runs at an adequate speed on today's PCs.

Another early thrust by CACHE in computer-based instruction using desktop computers was made by Stanley I. Sandler and Michael B. Cutlip, who were also elected Trustees at the November 1981 meeting. At that time, the Control Data Corporation had developed a worldwide educational computer system called PLATO, which used special touch-screen terminals connected to a dedicated mainframe computer. Sandler and Cutlip were among the chemical engineers who were leaders in developing PLATO software. At the July 30 - August 1, 1982 meeting at Ojai, California, they were appointed Co-Chairs of a new Computer-based Instruction Task Force. The PLATO system had already been evaluated by Cameron M. Crowe, based on his experiences with second-year chemical engineering students in a material balance, energy balance, and thermodynamics class. The PLATO lessons had been prepared by Sandler, Charles A. Eckert, N. A. Ashby, and S. C. Miller. Crowe's evaluation, which appeared in the April 1982 issue (No. 14) of *CACHE News*, stressed the importance of PLATO in practicing problem solving and reviewing concepts. Unfortunately, the cost of using the PLATO system was too high for it to receive widespread use at universities. However, the potential use of a desktop computer with a graphical user interface to assist computer-based instruction was well established by the task force.

On July 6-11, 1980 at New England College, Henniker, New Hampshire, Professors Richard S. H. Mah, who had been elected Trustee in 1973, and Warren D. Seider conducted the first International Conference on the Foundations of Computer-Aided Chemical Process Design (FOCAPD), which was organized by the CAST Division of the AIChE, supported by the NSF, and sponsored by the Engineering Foundation and the AIChE, with the proceedings being published by the AIChE. The conference brought together 146 participants from industrial and governmental laboratories and universities of 16 countries to listen to, discuss, and critique 30 papers. Because design methods and tools were being improved at a rapid rate due to the ready availability of computers of increasing capability, the success of the Henniker conference suggested that such a conference should be held every three or four years. Just prior to the conference, Mah wondered why CACHE couldn't provide the management and arrangements for such a conference. Of course, that would include setting the fee for the conference, providing the advertising, assuming the financial risk in the event that the conference did not at least break even, and publishing the proceedings. At the June 12-14, 1980 meeting at Hershey, Pennsylvania, Mah presented a proposal for CACHE to provide management and arrangements for specialized research conferences and an Ad Hoc Committee on Conferences was formed with Mah as the Chair. At the November 20-22, 1980 meeting at Fontana, Wisconsin, Mah's proposal was approved and in 1982, the CACHE Conferences Task Force was formed with Mah as the Chair. Prior to the formation of this task force, CACHE had only engaged in projects involving teaching. Now CACHE would also involve itself in sponsoring conferences related to research. Periodic conferences managed by CACHE could provide a much-needed steady source of additional revenue through conference fees and sales of proceedings. CACHE could provide financial assistance to young faculty to attend such conferences. The main goal of the conferences would be to provide a mechanism for engineers from industry, government, and academia to share their viewpoints and directions.

The first CACHE-arranged research conference, called FOCAPD-83, was the Second International Conference on Foundations on Computer-Aided Process Design held at Snowmass Village, Colorado on June 19-24, 1983. Art Westerberg and Dr. Henry H. Chien of Monsanto Company served as conference chair and co-chair, respectively. The conference, whose management by CACHE was approved by the CAST Division of AIChE, was sponsored by the NSF, nine companies, and CACHE, and had 162 attendees from 12 countries divided almost equally between industry and academia. The conference proceedings, which included 22 papers, were edited, printed from camera-ready copy, and distributed by CACHE, under the direction of Carnahan. Registration and conference arrangements were handled by Seader and Vickie S. Jones. Total income to CACHE for the conference was $104,000, with $21,500 coming from the NSF, $11,000 from industrial sponsors, and $71,500 from registration fees at an average of $440 each. The NSF funds were the result of a proposal of May 1, 1982 to the NSF by Westerberg. All of the NSF and industrial-support funds were used to partially defray conference fees and travel expenses for invited speakers and session chairs. In addition, CACHE provided a total of $2,600 for the partial support of young faculty who were invited to attend the conference. Total expenses for the conference were $83,400. By May 30, 1984, following sales of the proceedings to libraries and individuals not attending the conference, the net income to CACHE for the conference was about $25,000. Of this, $15,000 was set aside to back-up future conferences. The first CACHE conference was considered a huge success both technically and financially. However, with respect to the latter, it was recognized that the profit-attendance curve is very steep. Had the attendance been only 100, CACHE would have lost money.

At the November 18-20, 1982 meeting at Newport Beach, California, Thomas F. Edgar, who had become President of CACHE at the April 8-9 meeting in Houston, suggested that CACHE should offer to manage the Third International Conference on Chemical Process Control. These so-called CPC conferences were initiated in 1976, with the second conference in 1981. The offer was accepted, and CPC III, chaired by Manfred Morari and Thomas J. McAvoy, was held from January 12-17, 1986 at Asilomar Conference Grounds, California, with 145 participants from 16 different countries. The ability of CACHE to manage research conferences was now firmly established and had become an important source of income for the organization. Since the FOCAPD-83 and CPC III conferences, the following additional CACHE conferences have been held:

July 5-10, 1987 – "Foundations of Computer-Aided Process Operations (FOCAPO I) – Chaired by G. V. Reklaitis and H. Dennis Spriggs – Park City, Utah – 135 participants.

July 10-14, 1989 – FOCAPD-89 – Chaired by J. J. Siirola, I. E. Grossmann, and G. Stephanopoulos - Snowmass Village, Colorado – 170 participants.

February 17-22, 1991 – CPC IV– Chaired by Y. Arkun and Harmon Ray – Padre Island, Texas – 158 participants.

July 18-23, 1993 – FOCAPO II – Chaired by David W. T. Rippin, J. C. Hale, and J. F. Davis – Mount Crested Butte, Colorado – more than 100 participants.

July 10-15, 1994 – FOCAPD-94 – Chaired by M. F. Doherty and L. T. Biegler – Snowmass Village, Colorado – 143 participants.

July 9-14, 1995 – ISPE-95 – "Intelligent Systems in Process Engineering" – Chaired by G. Stephanopoulos, J.F. Davis, and V. Venkatasubramanian – Snowmass Village, Colorado – Snowmass Village, Colorado – Snowmass Village, Colorado – 145 participants.

January 7-12, 1996 – CPC-V – Chaired by Jeffrey Kantor and Carlos Garcia– Snowmass Village, Colorado – Tahoe City, California – 143 participants.

To facilitate the management of conferences, especially with regard to the sharing of duties between CACHE management and conference chairs, Mah, with assistance from the task force, prepared the *CACHE Conference Chairman's Manual* that was issued on March 15, 1982. At the November 11-13, 1993 meeting in St. Louis, Siirola became the new chair of the Standing Committee on Conferences and, with Mah, prepared a major revision of the conference chairman's manual. Plans are underway for future conferences to be held in 1997, 1998, and 2000.

Starting in August of 1966 and finishing in 1977, Professor Buford D. Smith of Washington University, with partial funding from the Exxon Education Foundation, supervised the development of a series of 22 design case studies prepared by chemical engineers at Monsanto, Amoco, DuPont, Phillips Petroleum, Foster Wheeler, Washington University, Newark College of Engineering, Notre Dame University, University of Delaware, University of Toledo, Purdue University, and University of Arkansas. During the 1970s, these design case studies, together with AIChE Student Contest Problems, had been become the major sources of design problems for senior design courses in chemical engineering.

At the November 20-21, 1980 meeting at Fontana, Wisconsin, George Stephanopoulos, who had been elected Trustee in 1977, was appointed Chair of the newly organized Process Design Case Studies Task Force whose assignment was to determine whether a new set of design case studies that made more use of computer-aided design tools should be prepared. At the November 12-13, 1981 meeting, Manfred Morari, who had been elected Trustee at the Fontana meeting and who became the new Chair of the Process Design Case Studies Task Force at that meeting, presented the following findings of the task force concerning the available design case studies: (1) they did not exercise an integrated approach, (2) they usually did not stress the synthesis of alternatives, and (3) they did not take advantage of computer-aided design tools. Therefore, Morari proposed that the task force supervise the development and publication of a new series of process design case studies based on current computer-aided design philosophy.

The first CACHE process design case study, completed in September 1983 and entitled *Separation System for Recovery of Ethylene and Light Products from a Naphtha-Pyrolysis Gas Stream*, was based on a problem statement posed by EXXON and developed into a case study by A. Michael Lincoff of Carnegie Mellon University under the supervision of Ignacio E. Grossmann, who was elected a Trustee at the November 4-5, 1983 meeting in Williamsburg, Virginia, and Gary E. Blau who was later elected as an industrial trustee (Dow Chemical Co. and later Dow-Elanco) on December 1-3, 1988 at the Leesberg, Virginia meeting. The design study involved the synthesis of a separation sequence, optimization of column pressures, and synthesis of a heat-exchanger network. The design calculations were made with an interactive process simulation program called DESPAC.

Since the first CACHE Process Design Case Study, six additional case studies have been edited by Morari and Grossmann and published by CACHE for use by universities; the most popular is Vol. 6, with more than 90 copies sold as of early 1995:

Vol. 2 - *Design of an Ammonia Synthesis Plant* - Problem posed by Phillip A. Ruziska of Exxon Chemicals - Solution by Stacy G. Bike under the supervision of Grossmann - May 1985.

Vol. 3 - *Preliminary Design of an Ethanol Dehydrogenation Plant* - Problem posed by Union Carbide - Solution under the supervision of Biegler and Hughes - May 1985.

Vol. 4 - *Alternative Fermentation Processes for Ethanol Production* - Solution by Samer F. Naser under the supervision of Steven E. LeBlanc and Ronald L. Fournier of the University of Toledo - April 1988.

Vol. 5 - *Retrofit of a Heat Exchanger Network and Design of a Multiproduct Batch Plant* - Solutions by Richard D. Koehler and Brenda A. Raich under the supervision of Grossmann - May 1990.

Vol. 6 - *Chemical Engineering Optimization Models with GAMS* - Problems prepared by faculty and students at Carnegie Mellon, Northwestern, and Princeton with coordination by Grossmann - October 1991.

Vol. 7 - *Process Integration of an Ethylene Plant* - Problem posed by EXXON and solution prepared by Gert-Jan A. F. Fien and Y. A. Liu of Virginia Tech - 1994.

In the Spring 1989 (No. 28) issue of *CACHE News*, Jeffrey C. Kantor, who became a CACHE Trustee in November 1992, published an article entitled, "Matrix Oriented Computation Using Matlab (Matrix Laboratory)," which had been developed in FORTRAN by Cleve Moler in 1981 and included most of the numerical algorithms from LINPACK and EISPACK, of which Moler also was a co-author. Initially, Matlab was a public-domain program, but with the advent of PCs in the early 1980s, Moler and others formed The MathWorks, Inc. to market PC-Matlab, written in C for portability and efficiency. By 1991, Matlab had grown considerably in popularity. Toolboxes had been added, including the Control System Toolbox, and the program was being used by students at more than 80 universities worldwide. In his article, Kantor showed how Matlab could be used to increase the productivity of a course in process control. At the April 6-7, 1991 meeting in Woodlands, Texas, The Process Design Case Studies Task Force announced that Manfred Morari and N. Lawrence Ricker, with Douglas B. Raven and a number of other contributors, were developing a CACHE Model Predictive Control Toolbox (CACHE-Tools) based on the use of Matlab functions for the analysis and design of model predictive control (MPC) systems. MPC had been conceived by industry in the 1970s and had steadily gained in popularity until by the 1990s it had become the most widely used multivariable control algorithm in the chemical process industries. The first version of CACHE-Tools was distributed on October 30, 1991. A presentation and demonstration of CACHE-Tools was made at the November 20, 1991 CACHE Open House at the Los Angeles AIChE meeting by Yaman Arkun, who had been elected a Trustee in November 1986. An article on CACHE-Tools by Morari and Ricker appeared in the Spring 1992 issue (No. 34) of *CACHE News*. By early 1995, more than 40 copies of CACHE-Tools had been sold to departments.

At the November 4-5, 1983 meeting in Williamsburg, Virginia, a suggestion was made for CACHE to focus more on the ChE curriculum and its needs and less on computer technology. Accordingly, an Ad Hoc Committee on Curriculum was formed with Morton M. Denn, who had been elected a Trustee in November 1981, as Chair, with the goal of assisting chemical engineering departments to effectively integrate computing and computer technology into undergraduate education. One of the factors considered by the committee was the expectations of industry for chemical engineers entering the work force. At the November 29-December 1, 1984 meeting at Carmel, California, the Curriculum Task Force was formed with Denn and Seider as Co-Chairs. At the following meeting on March 28-29, 1985 at Woodlands, Texas, the task force presented a position paper entitled, "Expectations of the Competence of Chemical Engineering Graduates in the Use of Computing Technology." This paper, which was published in the Winter 1986 issue of *Chemical Engineering Education* with an introduction by Denn, was approved by CACHE for distribution to the AIChE Education and Accreditation Committee. The paper listed the following expectations:

1. The graduate must be familiar with at least one computer operating system.
2. The graduate must be competent in at least one scientific programming language.
3. The graduate must be experienced in computer-aided acquisition and processing of information.
4. The graduate should have conducted at least one information retrieval search from an electronic data base.
5. The graduate should have experience in the use of a word processor and a graphics program for the generation of reports.
6. In the near future, the graduate should have experience with electronic mail and external data bases.
7. The graduate should have an appreciation of the concepts of numerical analysis, including convergence and stability.
8. The graduate should be familiar with the use of spreadsheets.
9. Most importantly, computing should be integrated throughout the curriculum and more use should be made of open-ended problems.

The March 28, 1985 position paper was transmitted by letter of May 6, 1985 from Mah, then President of CACHE, to Professor Bryce Andersen, then Chairman of the AIChE Education and Accreditation Committee. Also, all chairs of chemical engineering departments in the United States were sent a copy of the position paper and informed of the letter to Andersen by another letter from Mah of May 16, 1985. Concurrent with the development of the CACHE position paper, the Chemical Engineering Department of the University of Texas at Austin commissioned a group of industrial leaders to recommend changes in chemical engineering education. Their report, which was sent to all chemical engineering departments in the United States and is summarized in the October, 1985 issue of *Chemical Engineering Progress,* included the following corroborative statement about computing:

> "Educators must fully recognize the major position that the computer and professionally written software play in the modern practice of the profession. The computer should become the dominant calculational tool early in the curriculum."

The impact of CACHE and the University of Texas on computing in the undergraduate chemical engineering curriculum may be judged by the following quote, which did not appear in the early 1980s but does now appear, in the 1994-95 "Criteria for Accrediting Programs in Engineering in the United States" of ABET in cooperation with the E & A (Education and Accreditation) Committee of AIChE:

> "IV.C.3.g. Appropriate computer-based experience must be included in the program of each student. Students must demonstrate knowledge of the application and use of digital computation techniques for specific engineering problems. The program should include, for example, the use of computers for technical calculations, problem solving, data acquisition and processing, process control, computer-assisted design, computer graphics, and other functions and applications appropriate to the engineering discipline. Access to computational facilities must be sufficient to permit students and faculty to integrate computer work into course work whenever appropriate throughout the academic program."

The recommendations of CACHE for computing skills were amended at the December 1-3 meeting in Leesberg, Virginia and transmitted in a letter of December 20, 1988 from Seider to John W. Prados of ABET. These recommendations now are included as an appendix to the ABET accreditation materials distributed by the E & A Committee of AIChE to chemical engineering program evaluators for accreditation.

To assist professors in making use of computing technology in under-graduate courses other than design and control, the Curriculum Task Force, at the Woodlands, Texas meeting in 1985, proposed that CACHE publish a series of open-ended problems with an emphasis on problem formulation and the seeking of a solution, rather than the development of a computer program. In the Fall of 1987, CACHE published and distributed *CACHE IBM PC Lessons for Chemical Engineering Courses Other than Design and Control*, with the following six lessons in color graphics format prepared largely by students under the direction of the listed professors:

1. "Slurry Flow in Channels" - Denn
2. "Supercritical Fluid Extraction" - Seider
3. "Gas Absorption with Chemical Reaction" - Seinfeld
4. "Design of Flash Vessels and Distillation Towers" - Finlayson
5. "Heterogeneous Reaction Kinetics" - Fogler
6. "CSTR Dynamics and Stability" - David T. Allen

At the November 14-16, 1985 meeting in St. Charles, Illinois, George Stephanopoulos, who was re-elected as Trustee in December 1984 after his return from Greece, informed the trustees that considerable progress had been made recently in artificial intelligence (AI) and that important contributions could now be made in process engineering by knowledge-based (or expert) systems. He proposed that CACHE introduce knowledge-based, computer-aided tools into the mainstream of chemical engineering education. Accordingly, an Ad Hoc Committee on Expert Systems was formed with Stephanopoulos as Chair. At the November 6-8, 1986 meeting in Key West, the committee became the Expert Systems Task Force, with work under way to publish case studies and monographs. Three volumes of case studies, written by Venkat Venkatasubramanian who was elected Trustee on November 7, 1992, edited by Stephanopoulos, and entitled *Knowledge-Based Systems in Process Engineering*, were published and distributed in the Fall of 1988. The volumes, which presented the following four subjects, provided an excellent introduction to the subject of expert systems for chemical engineers:

1. *A General Introduction to Knowledge-Based Systems.*
2. *CATDEX: An Expert System for Diagnosing a Fluidized Catalytic Cracking Unit.*
3. *PASS: An Expert System for Pump Selection.*
4. *CAPS: An Expert System for Plastics Selection.*

The next step taken by the Expert Systems Task Force was to prepare an in-depth series of monographs covering all aspects of artificial intelligence as they might apply to chemical engineering. James F. Davis, who had been elected Trustee on November 14, 1987, became a co-editor of the volumes with Stephanopoulos. The first three volumes of the series, entitled *Artificial Intelligence in Process Systems Engineering*, with the following titles, were published and distributed in the Fall of 1990, followed by a fourth volume in the Spring of 1994:

1. *Knowledge-Based Systems in Process Engineering: An Overview* by Stephanopoulos.
2. *Rule-Based Systems in Chemical Engineering* by Davis and Murthy S. Gandikota.
3. *Knowledge Representation* by Lyle H. Ungar and Venkat Venkatasubramanian.
4. *An Introduction to Object-Oriented Programming in Process Engineering* by Ronald G. Forsythe, Jr., Suzanne E. Prickett, and Michael L. Mavrovouniotis.

In addition, Stephanopoulos and Mavrovouniotis were guest editors for a special September/October 1988 issue of *Computers and Chemical Engineering*, which featured 15 refereed articles on research and development in artificial intelligence in chemical engineering and was distributed with CACHE funds to all chemical engineering departments in the United States and Canada.

At the July 23-24, 1993 meeting at Mt. Crested Butte, Colorado, the Expert Systems Task Force reported that the case studies and monographs were attracting interest in AI, and process systems engineers were now beginning to come to a good appreciation of what AI could and could not do. AI techniques, such as knowledge-based systems, heuristic search, neural networks, and object-oriented programming were being successfully integrated with applied mathematics and operations research for process engineering applications. They proposed that CACHE manage an international conference on intelligent systems in process engineering to be held in 1995. The conference was approved by the CACHE Conferences Task Force.

Although E-mail is widely used and taken for granted today by chemical engineers, this was not the case in 1983 when CACHE organized the Ad Hoc Committee on Communications (E-mail) at the Williamsburg, Virginia meeting with Rony as Chair. Although E-mail, together with TELNET to log on to a remote computer and FTP to transfer files between a local and a remote computer, had been available at universities via the ARPANET national network, conceived in 1968 and in service by 1971, these services had been used mainly by computer scientists and other engineering professionals with government contracts. In *CACHE News* No. 19 of September 1984, Rony announced that in a one-year experiment, the committee would: (1) gather information on existing networks, (2) coordinate an E-mail experiment with a small group of interested chemical engineers, (3) determine potential chemical engineering applications, and (4) determine whether a ChE E-mail service should be established.

At the March 28-29, 1985 meeting at The Woodlands, Texas, the ad hoc committee became the Electronic Mail Task Force. In the April 1986 issue of *CACHE News* (No. 22), Rony announced, in a two-part article, some of the results of the CACHE national electronic mail experiment. Despite much publicity, interest in E-mail among chemical engineers remained low at that time. Although CACHE authorized 150 mailboxes on the COMPMAIL+ service, only 10-15 accounts were used actively. However, the use of E-mail to transfer files at a relatively low cost was amply demonstrated by Rony, Editor of *CAST Communications*, who transmitted by COMPMAIL+ the articles for the April 1985 issue to Associate Editor Wright, who supervised the final layout of the issue for printing. Some article files were also sent by COMPMAIL+ to Himmelblau for *CACHE News*.

By 1987, a number of wide-area networks had been established, including ARPANET, CSNET, BITNET, USENET, and NSFNET. In particular, BITNET (Because It's Time), which had been established as a financially self-supporting network with no central administration or paid staff by CUNY in 1981 to link university computers for inter-university communications and with only 65 hosts, had, by 1985, reached 1,000 host computers. Besides E-mail, BITNET offered other services including a bulletin board and file transfer. In the April 1986 issue of *CACHE News* (No. 22), Rony recommended that chemical engineers use BITNET for E-mail. In the Fall 1989 issue of *CACHE News* (No. 29), the Netlib system for distributing public domain mathematical software via electronic mail was announced. The index can be obtained by sending the message Send Index to netlibl@epm.ornl.gov. In that same issue and the following one, Rony presented an extensive user's guide to electronic mail, including the Internet, which had now become the major gateway for connecting networks. By January of 1994, 2,217,000 host computers were on the Internet. Today, almost all chemical engineers in academia, industry, and government have E-mail addresses.

Thanks to the Internet, E-mail addresses are relatively simple, being of the general form: username@machine.dept.inst.domain. For example, seader@uuserv.cc.utah.edu. A large list of E-mail addresses for chemical engineers is maintained on a gateway machine at the University of Texas at Austin by Rony, James B. Rawlings, and John W. Eaton. The list can be obtained by sending the message Send Index to chelib@che.utexas.edu.

At the November 14-16, 1985 meeting in St. Charles, Illinois, an Ad Hoc Committee on Simulated Laboratory Experiments was organized with Reklaitis as Chair. He told CACHE of plans by Robert Squires and himself to pursue funding from industry and the NSF through CACHE to prepare a series of simulated process engineering laboratory/pilot plant experiments with an industrial process orientation for use by students at universities to fill an important gap in their education. The approach would be for the student to solve a process engineering problem as if he/she were in industry. The student would: (1) view a video tape of an actual industrial process showing the equipment and the control panel, along with a discussion of operating conditions and safety features, (2) be given a description of a process engineering problem for a section of the process that requires a solution, (3) carry out some experimental work to obtain necessary data, (4) carry out simulations with the data to solve the problem using both steady-state and dynamic models, and (5) prepare a report with recommendations. A problem involving a new resid hydrotreater of Amoco was already formulated.

At the December 1-3, 1988 meeting at Leesberg, Virginia, the ad hoc committee was replaced by the Simulated Laboratory Modules Task Force and Reklaitis reported that NSF had agreed to fund the development of five simulated laboratory modules over three years for $270,000, with additional industrial support for the formulation of the problems from Amoco (hydro-desulfurization), Dow Chemical (latex emulsion polymerization), Mobil (catalytic reforming), Air Products (process heat transfer), and Tennessee-Eastman (methyl acetate from coal). On August 10, 1989 a workshop for interested faculty was held at Purdue University on use of the Amoco problem. By 1990, the Amoco, Dow, and Mobil modules were in beta-test, and a workshop for 23 faculty on the use of the modules was held at Purdue University on July 27-29, 1990. An article on the Amoco module appeared in the Spring 1991 issue of *Chemical Engineering Education*. A third workshop for representatives from 34 departments was held on July 26-28, 1991 at Purdue. By the Spring of 1993, the Amoco, Dow, Mobil, and Tennessee-Eastman modules were ready for distribution by CACHE. All five modules are discussed by Squires, P. K. Andersen, Reklaitis, S. Jayakumar, and D. S. Carmichael in an article entitled, "Multimedia-Based Educational Applications of Computer Simulations of Chemical Engineering Processes," which appeared in *Computer Applications in Engineering Education*, Vol. 1, No. 1, pages 25-32, 1992.

At the St. Charles, Illinois meeting of November 1985, an Ad Hoc Committee on Laboratory Applications of Microcomputers was formed with Mellichamp as Chair. With the approval of the Trustees, Mellichamp reported that he and Ali Cinar would prepare a survey of the use of computers in undergraduate laboratory experiments. That survey, which resulted in 116 replies to a questionnaire from departments in 10 countries, was published by CACHE in September 1986. From that survey, Cinar and Mellichamp selected 21 experiments and included them in the CACHE publication, *On-Line Computer Applications in the Undergraduate Chemical Engineering Laboratory: A CACHE Anthology*, which was published in June 1988. Construction and instrumentation details are given for each experiment. The authors show how a PC can be used to take the data, and analyze and display the results.

By the mid-1980s, the use of steady-state process simulation was well established in undergraduate education. However, a growing number of chemical engineers were becoming involved in batch processing. GPSS (General Purpose Simulation System) had been created by IBM in 1959 to solve such discrete-event simulation problems, by allowing the user to use time as the basic variable, with varying parameters and process layouts. To assist chemical engineering educators in introducing students to discrete-event simulation, CACHE, in 1988, published a book, written by Daniel J. Schultheisz and Jude T. Sommerfeld, entitled, *Exercises in Chemical Engineering using GPSS*, which included the educational version of the GPSS/PC program and 18 solved problems, covering a wide range of complexity from the unloading of oil tankers to the batchwise manufacture of PVC.

Since 1975, CACHE has had a New Projects and Long-Range Planning Committee, which from time-to-time has been requested by the CACHE Executive Committee to make recommendations. The first major report was issued by the committee, with Seider as Chair, in May 1984. Some major recommendations included the following: (1) consider projects other than computer-related educational aids, (2) keep abreast with the latest changes in computing and systems technology, (3) carefully consider new projects, (4) consider new means of financing activities and projects, and (5) find ways to help young faculty. At the March 5-6, 1988 meeting at New Orleans, Siirola became the first industrial trustee elected to Vice-President, and assumed the task of long-range planning. At the next meeting, in Leesberg, Virginia, he led a soul-searching discussion, during which the following questions were raised: (1) Why don't more faculty use CACHE products? (2) What new products are needed? and (3) How do we achieve better feedback from our customers? At the July 1989 meeting at Snowmass, Siirola arranged for the first of two Future Planning Exercises, the first of which was led by David Stump of Eastman Kodak, to consider the topics of suppliers, inputs, processes, outputs, and customers. At the March 17-19, 1990 meeting at Orlando, Siirola became President of CACHE and reported the results of the Future Planning Exercises, which were slightly modified and issued in final form at the following meeting in Chicago. The accepted mission and vision statements were:

> "The mission of CACHE is to provide leadership and resource materials for advancing the use of computer-based methods and technology in chemical engineering education and practice."
>
> "The goal of CACHE is to be the recognized facilitator in the identification, creation, distribution, understanding, and use of timely computer-based products and services which have a substantial impact on undergraduate, graduate, and postgraduate chemical engineering education."

The key results of the exercises were the following goals:

1. Identify our customers and understand their needs.
2. Gain customer acceptance.
3. Identify existing or facilitate the development of superior products.
4. Facilitate the distribution and understanding of our products and services.
5. Recognize our limits and resources.
6. Operate CACHE in a fiscally sound manner.

In 1974, Fogler published the text-book, *The Elements of Chemical Kinetics and Reactor Calculations,* which included information on and stressed the importance of programmed learning. In his subsequent 1986 book, *Elements of Chemical Reaction Engineering*, he further elucidated that approach and illustrated it throughout the book. This led to an interest in how to develop innovation in problem solving, with the concern that if we are to be economically competitive, our engineers must be able to generate new ideas, processes, and products. In 1988, with the help of six undergraduate engineering students, he assembled a booklet entitled, *A Focus on Developing Innovative Engineers*, which formed the basis of an NSF proposal. The booklet included an approach to teaching innovation and the application of the approach to a large number of industrial problems.

At the December 1988 meeting at Leesberg, Virginia, CACHE organized the Innovative Engineering Task Force chaired by Fogler, who announced that his University of Michigan proposal to NSF, entitled *A Focus on Developing Innovative Engineers*, had been funded for two years at $250,000, with a subcontract to CACHE of $26,000. With Susan M. Montgomery as Project Manager, the project was extended to 1993, by which time (August), four sets of interactive computer modules for PCs running MS-DOS were completed for distribution by CACHE. These four sets, containing a total of 24 modules, were developed to enhance the teaching of material and energy balances (5 modules), fluid mechanics and transport phenomena (5 modules), separations (5 modules), and chemical reaction kinetics (9 modules). Some of the modules include new technologies, while all utilize graphical animations and entertaining motivators, whose design was facilitated by use of the QUEST authoring system. The modules are discussed by Fogler, Montgomery, and Robert P. Zipp in the article, "Interactive Computer Modules for Undergraduate Chemical Engineering Instruction," which appeared in *Computer Applications in Engineering Education,* Vol. 1, No. 1, pages 11-24, 1992.

At the November 9-11, 1989 meeting at St. Helena, California, Finlayson announced that he had received an NSF grant through the University of Washington with support for CACHE assistance in a project for the development, testing, and evaluation of computer programs for reactor design using state-of-the-art numerical methods on UNIX workstations running X-Window. By 1994, the project was completed. Known as the Chemical Reactor Design Toolbox, the program simulates batch reactors, CSTRs, plug-flow reactors, plug-flow reactors with axial dispersion, and 2-D reactors with axial dispersion, including multiple reactions and heterogeneous reactions, with and without heat transfer and pressure drop.

Also at the St. Helena meeting, the Trustees were informed that Tektronix was marketing a new computational chemistry system, advertised in the May 1, 1989 issue of C&EN under the trade name CACHE. Concern was expressed about the use of the CACHE name and it was decided to have a lawyer investigate what protection CACHE might have and whether the name could be trademarked. At the November 1990 meeting in Chicago, Himmelblau reported that Tektronix had agreed to pay CACHE $10,000 plus $1,000/year for three years as a settlement. CACHE did attempt to register the CACHE trademark, but the application was rejected because of previous related applications by other companies and the potential cost of litigation.

By the November 12-14, 1987 meeting at Southampton, Bermuda, the Large-Scale Systems Task Force, which had been chaired by Jeffrey J. Siirola since November 1984 (Siirola had become an industrial Trustee at the Snowmass meeting in June 1983) announced that it was actively seeking software products in addition to FLOWTRAN. More and more, as PCs and workstations were becoming more pervasive and capable, programs were migrating from main frames to the smaller computers. Accordingly, at that meeting the name of the task force was changed to the Process Engineering Task Force. H. Dennis Spriggs, who had been elected industrial Trustee at the Snowmass meeting while with Union Carbide, was then with Linnhoff-March Process Integration Consultants, who had developed software for heat-exchanger network synthesis. Spriggs informed the task force that one of their products, TARGET II, could be made available to educators. This software was designed to determine the minimum utilities and pinch temperatures for an exchanger network and produce *composite* and *grand composite* plots with a graphical display. Following a favorable evaluation by the task force, Seader prepared a user's guide and distribution of TARGET II began in 1987.

The TARGET II program did not actually synthesize a heat-exchanger network, but only performed the targeting. At the Snowmass meeting in July, 1989, Edward M. Rosen, who had been elected an industrial Trustee at the Snowmass meeting in August, 1977, and was appointed Chair of the Process Engineering Task Force at the Leesberg, Virginia meeting in December, 1988, announced that CACHE had received a favorable evaluation for an educational program, called THEN (Teaching Heat Exchanger Networks), written by S. Pethe, R. Singh, S. Bhargava, R. Dhoopar, and F. Carl Knopf. This program locates pinch temperatures and then allows a student to use heuristics to develop a network using graphics. The program permits stream splitting and identifies heat exchanger loops, as described in an article by Pethe, Singh, and Knopf in Vol. 13, No. 7 of *Computers and Chemical Engineering*, pages 859-860. Distribution of THEN began in 1989.

The next software package that was offered to CACHE and received a favorable evaluation by the Process Engineering Task Force was PIP (Process Invention Procedure), an interactive PC code, written under the supervision of James M. Douglas, for the synthesis of petrochemical gas-liquid processes. Douglas had been elected Trustee at the 1988 Bermuda meeting and in 1988 published an innovative textbook for a senior design course entitled *Conceptual Design of Chemical Processes*. PIP became a CACHE product in the Fall of 1989. At the November 14, 1990 CACHE Open House during the Chicago AIChE meeting, Douglas gave a detailed description of PIP and its use in design. Ever since CACHE was first organized, there was considerable interest in obtaining and distributing *process synthesis* software. With TARGET II, THEN, and PIP, CACHE now had three such products.

At the July 1989 meeting in Snowmass, Michael B. Cutlip, who was elected a Trustee in November 1981, and Mordechai Shacham offered the POLYMATH software package to CACHE for distribution. This was a commercial program that had been developed with a the financial support of and marketed earlier by Control Data Corp. and had been in use at the University of Connecticut since 1983. The package included programs for curve fitting, solving nonlinear systems of equations, solving initial value problems in ordinary differential equations, linear and multiple regression, and manipulation of matrices. By the November 9-11, 1989 meeting at St. Helena, California, Rosen and Seader of the Process Engineering Task Force had evaluated the POLYMATH program and, after some minor revisions, had found it to be a very easy-to-use and useful program that should appeal greatly to undergraduate students. Fogler, who was then President of CACHE had also tested the package and, in a written "President's Perspective" of September 20, 1989 indicated that POLYMATH could be one of a growing number of very user-friendly software packages that were beginning to cause a paradigm shift in teaching toward more complex and open-ended problem assignments. Because of the growing number of undergraduate students who own their own PCs, Rosen developed a novel method of distribution for POLYMATH that was discussed at the November 15-17, 1990 meeting at Chicago. A chemical engineering department would license POLYMATH for $125 the first year and $75 per year thereafter. For that fee, the package could be copied for use during the year by any student or faculty member in the department. Thus, for the first time, a CACHE product would be put into the hands of a student at no cost to the student. Distribution of POLYMATH began in 1990, following a presentation on the package by Cutlip on November 11, 1990 at the AIChE meeting in Chicago and an article by Cutlip and Shacham that appeared in the Fall 1990 issue (No. 31) of *CACHE News*. During the first year of availability, POLYMATH was licensed by 75 departments. Version 3.0 of POLYMATH was described in the Fall 1994 issue of *CACHE News*. Version 4.0 was released in the Summer of 1996.

At the November 15-17, 1990 meeting at Chicago, Ross Taylor and Harry Kooijman offered CHEMSEP, a suite of PC programs for performing rigorous multicomponent multistage separation process calculations, for license to universities by CACHE. In 1977, Fredenslund, Gmehling, and Rasmussen had published a Newton-type code, coupled to the UNIFAC method for determining activity coefficients, for solving such problems. By the end of the 1980s, that type of method had been incorporated into most steady-state simulation programs. However, it was believed that an easy-to-use stand-alone PC-program with a built-in property data bank for about 200 components and a wide variety of thermodynamic property models would appeal to instructors and students for use in a separations course, especially if the program could produce tables and McCabe-Thiele and triangular plots of results. As discussed in an article by Kooijman and Taylor in *CACHE News* No. 35 (Fall 1992), the CHEMSEP project was initiated in February 1988 at the University of Technology in Delft, where the first version of the program was used successfully with undergraduate and graduate students. In 1991, an extended version 2.0 was completed. That version was evaluated by Rosen and Seader and approved for license by CACHE at the March 28-29, 1992 meeting in New Orleans. Following a detailed presentation on CHEMSEP by Taylor at the CACHE Open House on November 4, 1992 at the AIChE meeting in Miami, distribution of the program began. By November 1994, the program had been licensed by 40 departments. Version 3.0, which will include a rate-based model, is scheduled for distribution in the near future.

At the November 21-23, 1991 meeting at Indian Wells, California, Rosen announced that he had been approached by Douglas J. Cooper, who with his students had developed an interactive PC program called PICLES (Process Identification and Control Laboratory Experiment Simulator) that provided a student with "real-world" experience in the study of dynamics and control. The program uses colorful graphics and allows the student to follow the action as decisions are made, and includes models for tanks in series, a heat exchanger, and a distillation column, with controller algorithms ranging from P-only to PID. Following an evaluation by the Process Engineering Task Force and some program revisions, PICLES Version 3.1 was accepted as a CACHE product for license to universities in the Fall of 1993. Cooper presented a detailed description of the software at the CACHE Open House on November 10, 1993 at the St. Louis AIChE meeting. As with POLYMATH, with a department license, the software can be copied for use by students and faculty on their own computers. PICLES was described in issues

37 and 38 of *CACHE News*. By the end of 1994, the program had been licensed by 25 departments. Version 4.0 is described in the Fall 1994 issue of *CACHE News*.

The CD-ROM (Compact-Disk Read-Only Memory) appeared on the market about eight years ago. Although a bright future was predicted, particularly for the coupling of the CD-ROM with graphics, sound, and animation to achieve the multimedia PC (MPC), the promise was more than the reality, mainly because of a lack of standards. That changed late in 1991, when the Microsoft-endorsed MPC standard, ISO (International Standards Organization) 9660, for storing data was adopted. By March 1992, two MPCs and three multimedia upgrade kits were on the market, but none of them rated an Editors' Choice by PC Magazine. By mid-1992, seven companies were offering MPCs and seven different multimedia upgrade kits were available. However, at $800 to $1,900, the price of the kits was high. Finally, by the end of 1994, CD-ROMs and multimedia were here, with more than 5,000 CD-ROM titles, MPCs from almost every PC vendor, and upgrade kits priced at less than $400. Today, anyone purchasing a PC must give serious consideration to making it an MPC.

At the March 28-29, 1992 meeting at New Orleans, Cutlip became President of CACHE and announced that CACHE would celebrate its 25th anniversary in 1994. He also made a compelling case for the use of CD-ROMs for the distribution of CACHE products and documentation. On August 13, 1992 at the ASEE Summer School for Chemical Engineering Faculty at Montana State University, Rony was appointed Chair of an Ad Hoc CD-ROM Task Force with the objective of creating a demonstration CD-ROM disk. Rony's university, Virginia Tech., established a multimedia lab on September 18, 1992 and by 1993 was distributing software to students on CD-ROMs. Rony published two articles on CD-ROM technology in the Fall 1993 and Spring 1994 issues of *CACHE News*. In the latter article, he called for contributions to a 25th Anniversary CACHE CD-ROM. More than 1000 copies of that CD-ROM were distributed to chemical engineering students, faculty, and practitioners at the November 13-18, 1994 meeting of the AIChE in San Francisco. The CD-ROM contains 42 Mb of MacIntosh files and 636 Mb of Windows/DOS files, including Adobe Acrobat Reader for Windows, described in the Spring 1994 issue of *CACHE News*, and several CACHE software products. The 3M Company produced the copies and donated most of the blank disks. Yet another CD-ROM was prepared by Peter Rony during 1995; more than 1000 copies of this new CD-ROM were distributed to students and faculty in the Spring of 1996.

On November 18, CACHE held a 25th Anniversary Dinner at the Silverado resort in Napa, California for former and current Trustees and their spouses. For that event, Seider prepared a CACHE Time-Line that listed all of the significant events that had occurred during the first 25-years of the history of CACHE and displayed most of the CACHE products together with photographs that had been taken at some of the CACHE meetings. The dinner was attended by 15 former CACHE Trustees. The first 25 years of CACHE were over. Much had been accomplished by untold hours of donated time by CACHE Trustees and a large number of interested faculty members. To begin CACHE's second quarter century, the Trustees thought it appropriate to elect, as President, Warren D. Seider, who helped organize CACHE in 1969.

Appendix I — CACHE Officers

Chair:

Brice Carnahan	1969-1971
Warren D. Seider	1971-1973
Lawrence B. Evans	1973-1974
Brice Carnahan	1974-1975
Ernest J. Henley	1975-1976

President:

Robert E. C. Weaver	1976-1977
Duncan A. Mellichamp	1977-1978
David M. Himmelblau	1978-1980
Richard R. Hughes	1980-1981
Thomas F. Edgar	1981-1984
Richard S. H. Mah	1984-1986
Gintaris V. Reklaitis	1986-1988
H. Scott Fogler	1988-1990
Jeffrey J. Siirola	1990-1992
Michael B.Cutlip	1992-1994
Warren D. Seider	1994-1996
Lorenz T. Biegler	1996-

Vice Chair:

Lawrence B. Evans	1971-1973
Brice Carnahan	1973-1974
Ernest J. Henley	1974-1975
Robert E. C. Weaver	1975-1976

Vice President:

Duncan A. Mellichamp	1976-1977
David M. Himmelblau	1977-1978
Richard R. Hughes	1978-1980
Thomas F. Edgar	1980-1981
David M. Himmelblau	1981-1982
Richard S. H. Mah	1982-1984
Gintaris V. Reklaitis	1984-1986
H. Scott Fogler	1986-1988
Jeffrey J. Siirola	1988-1990
Michael B. Cutlip	1990-1992
Ignacio E. Grossmann	1992-1994
Lorenz T. Biegler	1994-1996
James F. Davis	1996-

Secretary:

Warren D. Seider	1969-1971
Arthur W. Westerberg	1971-1973
J. D. Seader	1973-1974
Robert E. C. Weaver	1974-1975
Duncan A. Mellichamp	1975-1976
Rodolphe L. Motard	1976-1978
Richard S. H. Mah	1978-1980
H. Scott Fogler	1980-1982
Gintaris V. Reklaitis	1982-1984
Edward M. Rosen	1984-1986
John C. Hale	1986-1988
Michael B. Cutlip	1988-1990
Lorenz T. Biegler	1990-1992
Yaman Arkun	1992-1994
James. F. Davis	1994-1996
Andrew N. Hrymak	1996-

Executive Officer:

Lawrence B. Evans	1974-1980
J. D. Seader	1980-1984
David M. Himmelblau	1984-

Appendix II — Former CACHE Trustees

Name of Former Trustee	Affiliation at Time of Service	Dates of Service
James H. Christensen	University of Oklahoma	1969-1973
Eugene Elzy	Oregon State University	1969-1973
Edward A. Grens	University of California, Berkeley	1969-1973
Robert V. Jelinek	Syracuse University	1969-1973
A. I. Johnson	University of Western Ontario	1969-1973
Matthew J. Reilly	Carnegie-Mellon University	1969-1973
Imre Zwiebel	Worcester Polytechnic Institute	1969-1973
Robert E. C. Weaver	Tulane University	1969-1979
Arthur W. Westerburg	University of Florida	1969-1981
Lawrence B. Evans	Massachusetts Institute of Technology	1969-1984
Rodolphe L. Motard	Washington University	1969-1984
Ernest J. Henley	University of Houston	1969-1985
Richard R. Hughes	University of Wisconsin	1969-1986
Paul T. Shannon	Dartmouth College	1973-1974
Gary J. Powers	Massachusetts Institute of Technology	1973-1977
Cecil L. Smith	Louisiana State University	1973-1979
D. Grant Fisher	University of Alberta	1975-1977
Ronald L. Klaus	University of Pennsylvania	1975-1978
W. Fred Ramirez	University of Colorado	1975-1979
Duncan A. Mellichamp	University of California, Santa Barbara	1975-1987
Lewis J. Tichacek	Shell	1977-1985
Theodore L. Leininger	DuPont	1977-1980
James White	University of Arizona	1978-1980
Bruce A. Finlayson	University of Washington	1981-1992
John C. Hale	E. I. DuPont de Nemours	1981-1995
William E. Schiesser	Lehigh University	1982-1984
Irven H. Rinard	Halcon	1982-1985
Morton M. Denn	University of California, Berkeley	1982-1986
Stanley I. Sandler	University of Delaware	1982-1987
John H. Seinfeld	California Institute of Technology	1983-1991
H. Dennis Spriggs	Union Carbide	1984-1990
John J. Haydel	Shell	1986-1991
Norman E. Rawson	IBM	1986-1994
Yaman Arkun	Georgia Institute of Technology	1987-1995
James M. Douglas	University of Massachusetts	1988-1990
Gary E. Blau	Dow Elanco	1991-1995
Carlos E. Garcia	Shell	1993-1995

Appendix III — CACHE Trustees as of June 1996

Name of Present Trustee	Affiliation	Date Service Began
Brice Carnahan	University of Michigan	1969
J. D. Seader	University of Utah	1969
Warren D. Seider	University of Pennsylvania	1969
David M. Himmelblau	University of Texas	1973
Richard S. H. Mah	Northwestern University	1975
Thomas F. Edgar	University of Texas	1977
H. Scott Fogler	University of Michigan	1977
Edward M. Rosen	Monsanto Company	1977
Joseph D. Wright	Xerox	1977
Manfred Morari	California Institute of Technology	1978
George Stephanopoulos	Massachusetts Institute of Technology	1978
Gintaris V. Reklaitis	Purdue University	1980
Peter R. Rony	Virginia Polytechnic Institute and State University	1981
Michael B. Cutlip	University of Connecticut	1982
Ignacio Grossmann	Carnegie-Mellon University	1984
Jeffrey J. Siirola	Eastman Chemical Company	1984
Lorenz T. Biegler	Carnegie-Mellon University	1986
Michael F. Doherty	University of Massachusetts	1987
James F. Davis	Ohio State University	1988
Andrew N. Hrymak	McMaster University	1991
Sangtae Kim	University of Wisconsin	1991
Jeffrey Kantor	University of Notre Dame	1993
Venkat Venkatasubramanian	Purdue University	1993
Feran Kayihan	Weyerhaeuser Corp.	1994
Ross Taylor	Clarkson University	1995
David Smith	E. I. DuPont de Nemours	1995

INTERACTIVE COMPUTER-AIDED INSTRUCTION

Susan Montgomery and H. Scott Fogler
Department of Chemical Engineering
University of Michigan
Ann Arbor, MI 48109

Abstract

This paper first describes how interactive computing can address the needs of both traditional and non-traditional learners through Soloman's Inventory of Learning Styles and also focuses on the higher level skills in Bloom's taxonomy. Next, the types of interactive computing are described and classified: Presentation, Assessment, Exploration, and Simulation. After a brief historical overview, CACHE's participation in a number of interactive computing projects and the resulting products are presented. The CACHE products described are POLYMATH, PICLES, the Michigan Modules, Purdue-Industry Computer Simulations and the Washington Chemical Reactor Design Tool. The paper closes with a discussion of applications using multimedia, hypertext, and virtual reality.

Introduction

Students learn a course's content best when exposed to the subject matter using a variety of teaching styles. A majority of undergraduate students learn best through experimentation and active involvement in the subject matter (Felder and Silverman, 1988). The standard textbook-lecture-homework triad allows for little of this type of learning. Consequently, it is necessary to enhance the curriculum by supplementing the standard teaching methods to increase the number of learning modes available to the student. This enhancement can be achieved through interactive computing that can provide students with supplementary exposure to the fundamental concepts in chemical engineering, as well as to give them an opportunity to apply and further explore these concepts. We will start this chapter by addressing the pedagogical needs and learning styles of engineering students, and discussing the types of interactive computer instruction that can address these needs and learning styles. We then provide a historical overview of the development of interactive computer materials in chemical engineering education, and a look to the future of interactive computing.

Focus on Learning Styles

Engineering education often focuses on disseminating technical information, without helping students make the connections between this information and both their own life experiences and the processes they will encounter in their careers (Griskey, 1991). However, extensive research in cognitive psychology has shown that effective learning of new principles requires explicit presentation of situations where these abstract principles are relevant (Vander-Stoep and Seifert, 1993).

Traditional teaching also ignores the needs of non-traditional learners and often results in students perceiving the material covered in these courses as foreign to any experience they may have had or will have as engineers. The need to address non-traditional learning styles is particularly important in efforts to attract and retain women and underrepresented minorities, who typically do not conform to traditional learning styles. Multimedia and interactive computing materials address these needs, by allowing students to interact with information using a variety of active mechanisms, rather than being passively exposed to information without goals (Qasem and Mohamadian, 1992). Kulik and Kulik (1986, 1987) reported that most studies found that computer-based instruction had positive effects on students. Specifically, students:

- learn more.
- learn faster (the average reduction in instructional time in 23 studies was 32%).
- like classes more when they receive computer help.
- develop more positive attitudes toward computers when they receive help from them in school.

One of the key factors to successfully developing and using interactive computing in courses is to first identify the activities that cannot be accomplished by other means (e.g., pencil and paper, calculator). By activities, we mean those exercises that are used to practice certain skills, learn new material, or test comprehension of previously learned material. In deciding which activities to include in educational software, one can take advantage of Bloom's Taxonomy of Educational Objectives (Bloom, 1956), shown in Table 1,which classifies the intellectual skill levels of various activities. A typical undergraduate course usually focuses on the first three levels only. Computer-based materials are one way to allow students to exercise their higher level thinking skills.

Once one has identified the skills one wants the user to practice, one needs to determine how to best reach the student to ensure that these skills are indeed exercised. Myers-Briggs Type Indicators have been widely used to classify student learning styles (McCaulley et al., 1983, Felder et al., 1993a). They can be used to classify people's personalities according to four dimensions: Extroversion vs. Introversion, Sensing vs. Intuition, Thinking vs. Feeling and Judging vs. Perceptive. This test's focus on personality as a whole is too broad for an identification of learning styles, however. An assessment tool that combines the simplicity of the Myers Briggs inventory with an emphasis on teaching is the Inventory of Learning Styles, by Barbara Soloman (1992). After answering 28 simple questions, a student is classified along four learning-style dimensions, shown in Table 2.

In combination, Bloom's Taxonomy and Soloman's Inventory of Learning Styles provide a means of determining both the content of the computer package and the presentation to the student, as shown in Fig. 1.

Keeping these considerations in mind, we can address the types of interactive computer software packages that can be used to assist in the training of chemical engineering student.

Table 1. Bloom's Taxonomy of Educational Objectives (Bloom, 1956).

1. *Knowledge*: The remembering of previously learned material. Can the problem be solved simply by defining terms and by recalling specific facts, trends, criteria, sequences, or procedures? This is the lowest intellectual skill level. Examples of knowledge-level assignments and questions are: *Write* the equations for a batch reactor and *list* its characteristics. *Which* reactors operate at steady state?
Other words used in posing knowledge questions: *Who . . . , When . . . , Where . . . , Identify . . . , What formula . . . .*

2. *Comprehension*: This is the first level of understanding and skill level two. Given a familiar piece of information, such as a scientific principle, can the problem be solved by recalling the appropriate information and using it in conjunction with manipulation, translation, or interpretation? Can you *manipulate* the design equation to find the effluent concentration or extrapolate the results to find the reactor volume if the flow rate were doubled? *Compare* and *contrast* the advantages and uses of a CSTR and a PFR. *Construct* a plot of N_A as a function of t.
Other words: . . . Relate . . . , Show . . . , Distinguish . . . , Reconstruct . . . , Extrapolate . . . .

3. *Application*: The next higher level of understanding is recognizing *which set* of principles ideas, rules, equations, or methods should be applied, given all the pertinent data. Once the principle is identified, the necessary knowledge is recalled and the problem is solved as if it were a comprehension problem (skill level 2). An application level question might be: *Make use* of the mole balance to solve for the concentration exiting a PFR.
Other words: . . . Apply . . . , Demonstrate . . . , Determine . . . , Illustrate . . . .

4. *Analysis*: This is the process of breaking the problem into parts such that a hierarchy of sub problems or ideas is made clear and the relationships between these ideas are made explicit. In analysis, one identifies missing, redundant, *and* contradictory information. Once the analysis of a problem is completed, the various subproblems are then reduced to problems requiring the use of skill level 3 (application). An example of an analysis question is: What conclusions did you come to after reviewing the experimental data?
Other words: . . . Organize . . . , Arrange . . . , What are the causes . . . , What are the components . . . .

5. *Synthesis*: This is the putting together of parts to form a new whole. A synthesis problem would be one requiring the type, size, and arrangement of equipment necessary to make styrene from ethyl benzene. Given a fuzzy situation, synthesis is the ability to formulate (synthesize) a problem statement and/or the ability to propose a method of testing hypotheses. Once the various parts are synthesized, each part (problem) now uses the intellectual skill described in level 4 (analysis) to continue toward the complete solution. Examples of synthesis level questions are: *Find a way* to explain the unexpected results of your experiment. *Propose* a research program that will elucidate the reaction mechanism?
Other words: . . . Speculate . . . , Devise . . . , Design . . . , Develop . . . , What

alternative . . . , Suppose . . . , Create . . . , What would it be like . . . , Imagine . . . , What might you see . . . .

6. *Evaluation*: Once the solution to the problem has been synthesized, the solution must be evaluated. Qualitative and quantitative judgments about the extent to which the materials and methods satisfy the external and internal criteria should be made. An example of an evaluation question is: *Is the author justified* in concluding that the reaction rate is the slowest step in the mechanism?
Other words: . . . Was it wrong . . . , Will it work . . . , Does it solve the ***real*** *problem . . . , Argue both sides . . . , Which do you like best . . . , Judge . . . ,*

Table 2. Dimensions of the Inventory of Learning Styles (Soloman, 1992).

Dimension	Range	Comments
Processing	Active/Reflective	Active learners learn best by doing something physical with the information, while reflective learners do the processing in their heads (Felder, 1994).
Perception	Sensing/Intuitive	Sensors prefer data and facts, intuitors prefer theories and interpretations of factual information (Felder, 1989).
Input	Visual/Verbal	Visual learners prefer charts, diagrams and pictures, while verbal learners prefer the spoken or written word.
Understanding	Sequential/Global	Sequential learners make linear connections between individual steps easily, while global learners must get the "big picture" before the individual pieces fall into place.

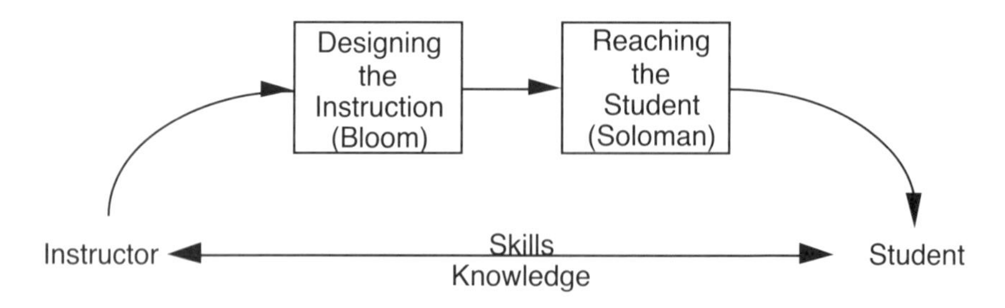

Figure 1. Incorporating Bloom's Taxonomy and Soloman's Inventory to impart skills and knowledge to students.

Types of Interactive Computing

Interactive computer-aided instruction in chemical engineering can be divided into four categories:

- Presentation
- Assessment
- Exploration
- Simulation

Presentation focuses on the delivery of technical material, which can occur in a number of ways. The following list is representative of these ways and the corresponding Soloman learning style dimension.

1. Display of text material (verbal).
2. Access to expanded explanation of text material through hot keys (active, sequential).
3. Visual and graphical representation of material (visual).
4. Use of animation to display phenomena (global) and manipulate equations (active).
5. Use of video clips to display industrial situations and situations where motion is involved (global, visual).

In the *presentation* phase the primary focus is on the knowledge, comprehension, and application levels of Bloom's Taxonomy. In the *assessment* category the student is tested on mastery of the material. The use of multiple-choice questions coupled with interactive simulations which the student must run to answer questions is one of the most effective testing methods. The simulations are closed-ended and focus on the first four levels of Bloom's Taxonomy (knowledge, comprehension, application, and analysis). The correct solutions to the questions are displayed immediately after the student's solution is entered. These types of assessments are particularly suited to the active, sequential and sensing learners.

The third category, *exploration*, allows users to better understand the role of various parameters on the performance of a given process through exploration of the process. These are exploratory simulations within a confined parameter space (simulations where equations are given and last terms may be dropped and the parameters take on any value). Instructional modules can also provide for the planning of experiments by allowing students to choose experimental systems, to take simulated "real" data, to modify experiments to obtain data in different parameter ranges, to manipulate data so as to discriminate among mechanisms, and to design a piece of equipment or process. These interactive computer modules can provide students with a variety of problem definition alternatives and solution pathways to follow, thereby exercising their divergent-thinking skills. Active learners enjoy the chance to manipulate information, and sensors and global learners get to experience a real process, or at least a simulation of it. This module type focuses on levels 3 and 4 of Bloom's Taxonomy (application and analysis). We will limit our discussion to those materials used in the core technical content courses, not the complex process simulations available for use in process design or laboratory courses. Examples of these exploratory materials are shown in Table 3, and are discussed in more detail in the historical overview.

The *simulation* category includes those activities in which the equations and parameter values are not given but can be easily entered in an ODE solver, such as POLYMATH, Maple, Mathematica, MATLAB, MathCAD, and spreadsheets. These tools allow users to better understand the role of various parameters on the performance of a given process through exploration of the process. This type gives the student practice of the higher levels of Bloom's Taxonomy (synthesis and evaluation).

Table 3. Exploratory Computer-Aided Instruction Materials.

PLATO Materials	Various
UM Interactive Modules (Drug Patch, Heat Effects)	Scott Fogler, Susan Montgomery et al. (U. of Michigan)
Lab Modules	Robert Squires, G.V. Reklaitis, S. Jayakumar et al. (Purdue U.)
Reactor Tool Kit	Bruce Finlayson (U. of Washington)
PICLES™	Doug Cooper (U. of Connecticut)

Historical Overview of Interactive Computing

PLATO

The early materials for interactive computer-aided instruction were developed for mainframe computers. While these materials proved very useful within individual universities, they were not easily distributable to other universities. They also suffered from the lack of availability of the materials to the students: most universities did not have adequate computer facilities to allow interactive access to lessons by large numbers of students. With the advent of personal computers such as IBM-PCs, the opportunities for interactive computer-aided instruction grew enormously. Interactive computer materials benefited greatly from the development of Control Data Corporation's PLATO (Programming Logic for Automated Teaching Operations) educational computer system. This system, at the University of Illinois in 1959, featured the use of terminals with touch-sensitive screens, as well as a highly efficient management and recording system (Smith, 1970, Smith and Sherwood, 1976). With the PLATO system, each terminal had a screen and keyboard that the students used to carry out their self-paced instructional lessons. The first interactive lessons used yes/no and keyword responses (e.g., does the unknown react with C_6H_5COCl in pyridine?). Later lessons simulated actual laboratory experiments. For example, students could explore the reaction of an olefin with bromine in methanol by varying the initial concentrations. Using known kinetic data, the computer immediately calculated and displayed the product composition. By experimenting with several sets of reaction conditions, students could understand the degree of completion involved in the reaction. Other lessons included the construction of NMR spectra, synthesis pathways, and determination of the unknowns. PLATO opened up the possibilities for truly interactive instruction.

POLYMATH

One of the earliest applications available to chemical engineers was POLYMATH, whose developers, Mordechai Shacham and Michael Cutlip, took advantage of the features of PLATO (Shacham and Cutlip, 1981a, 1981b, 1982, 1983). Using POLYMATH, students are able easily to set up a system of equations, to obtain an intuitive feeling of the problem being studied. This feeling and understanding is obtained because the student is able to take a significant amount of time to explore complex problems by varying the systems parameters and operating conditions rather than spending tedious time writing programs to enter and to solve the model equations for the physical system, including the required numerical methods. As a result, the student not only learns through discovery from the results of parameter variation, he or she has the opportunity to be creative in the solution to the problem. Thanks to POLYMATH, virtually every problem or homework assignment in chemical engineering can be turned into an open-ended problem that provides students with the extra time to practice their creative and synthesis skills.

University of Michigan Interactive Computer Modules

In the early 1970s, a number of interactive computer modules on reaction engineering were developed by Scott Fogler at the University of Michigan for use on a mainframe time-sharing computer. These modules focused on problem solving, freeing the student from the hassle of computation and mere formula plugging, so that the focus could be on process exploration and analysis. One popular example from this set is the Columbo module, in which students had to use their knowledge of chemical kinetics to solve a murder mystery. These simulations were enthusiastically received at a national meeting. Unfortunately, the programs were not easily transportable to computing environments at other universities. In the early 1980s, most of the six original reaction engineering modules were translated to IBM-PC format, using BASICA. These modules, especially one involving a styrene micro-plant, were distributed to and used at a limited number of universities.

The computing scene has changed tremendously from the early days described above. Students now have access to rooms full of computers at their universities, including IBM personal computers, Macintoshes, and UNIX workstations. In addition, many students have computers of their own, often connected to their university's computing environments, and through them, to world-wide networks such as the Internet. With the advent of sophisticated LCD panels, professors can display the contents of the computer screen using an overhead projector, easily bringing computers directly into the classroom. In the early 1990's, twenty-four interactive computer modules were developed at the University of Michigan and distributed by CACHE to every chemical engineering department in the United Sates and Canada. They were developed by Fogler, Montgomery, and Zipp and are described in detail elsewhere (Fogler, Montgomery and Zipp, 1992).

These modules employ presentation, assessment and exploration to enhance the students' mastery of the material. In addition, the sophistication required to satisfy a more computer-oriented student body has resulted in the need for more "polished" modules than in the past. The resulting interactive computer module should provide a learning experience that supplements the typical classroom and standard homework activities, reaching those students, particularly active, sensing, and global learners, who are not reached through traditional activities.

There are many advantages to using computer-based learning tools. There are also some pitfalls that one must be aware of and avoid. In addition to ensuring the technical accuracy of the material and simulations in the module, there are other considerations we have become aware of through our student testing and the comments of the external faculty testers. Some of the aspects of the use of interactive computer modules in engineering education that should be considered by all computer module developers are:

- Ease of use
- Introduction of new technologies
- Maintaining the focus on the concepts
- Eliminating tediousness!
- Promoting learning
- Individual guidance

The interactive computer modules for chemical engineering instruction developed at the University of Michigan typically consist of the following components:

Introduction
Review of pertinent fundamentals
Demonstration
Interactive exercises
A branching component
Solution to the exercise
Evaluation

The review section makes extensive use of animation in the derivation of equations. The problem-solving session often includes a scenario that captures the student's interest. For example, in the SHOOT module (see Fig. 2), the objective is to master simplification of the equations of motion. The student must determine which terms in the equation should be dropped for a given situation. To make the learning more interesting, these decisions are made in an amusement park setting.

Students greatly appreciate this type of interaction. While they must still master the technical material, the experience itself is made more pleasant by the use of these external motivators.

The problem-solving session often includes a scenario that captures the student's interest. For example, in Kinetics Challenge 2 (see Fig. 3), the objective is to master some of the basic principles of stoichiometry. The student must answer a large number of multiple-choice questions. To make the learning more interesting, these questions are asked in a setting similar to the quiz show Jeopardy.

These modules were developed under sponsorship of the National Science Foundation and the University of Michigan; they are available to chemical engineering departments through CACHE.

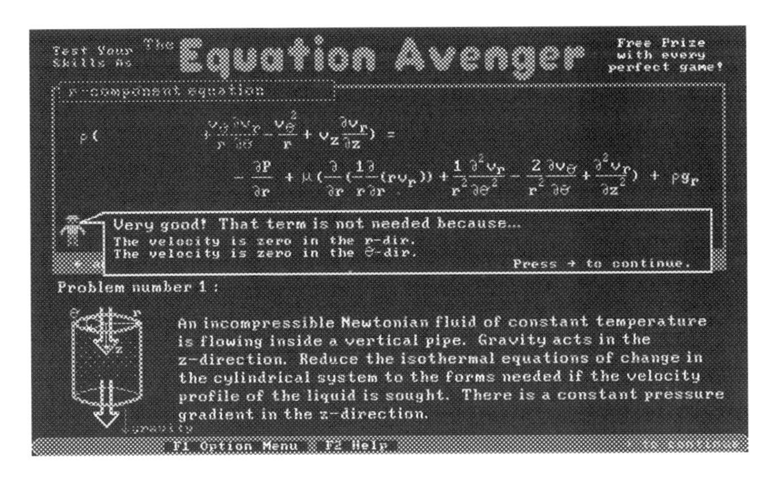

Figure 2. Interaction in SHOOT module, University of Michigan interactive computer modules.

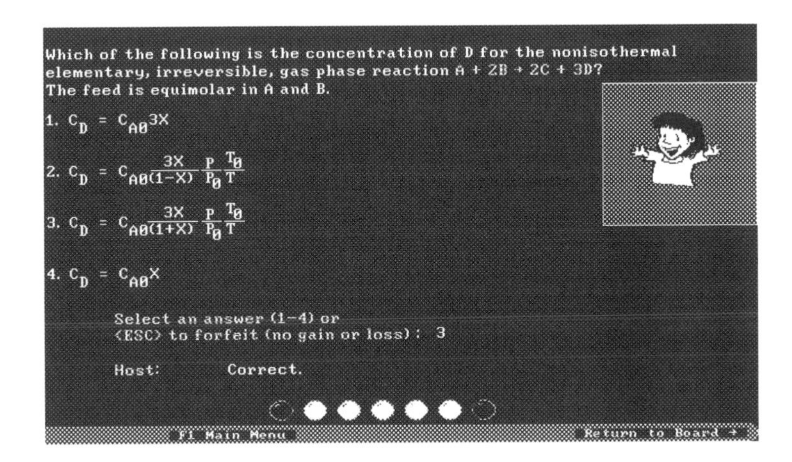

Figure 3. Interaction in KINETICS CHALLENGE 2 module, University of Michigan interactive computer modules.

Chemical Reactor Design Tool - University of Washington

The Chemical Reactor Design Tool developed at the University of Washington is primarily a set of exploration tools. This textbook supplement allows students to vary parameters and observe trends in complex chemical reaction engineering problems, including CSTR's, batch reactors, and plug flow reactors with axial and radial dispersion. The exploration of parameters allows users to very easily make their own deductions about the importance of physical phenomena, as well as to make design decisions that may revolve around conflicting constraints, none of which can be easily handled if one writes the program oneself. The interface is written in X-windows, providing the user with 3D perspective views, 2D contour plots, and solution variables, making it easy for the student to make comparison studies. In addition, the program automatically uses the correct, robust tools to solve the problem, so that the focus is on experimentation. This is ideal for active and global learners. The programs were developed under sponsorship of the National Science Foundation and the University of Washington, and are available to chemical engineering departments through CACHE.

Purdue-Industry Chemical Engineering Computer Simulation Modules

The Purdue-Industry Chemical Engineering Computer Simulation Modules are examples of the educational benefits that can result from collaborations with industry (Squires et al., 1992, Jayakumar et al.,1993). Their materials combine videotaped tours of portions of chemical plants with computer simulations of the systems, and allow students to perform "real world" design experiments to solve open-ended problems. These simulations have seen extensive use both in undergraduate laboratories, and in reactor design and process control courses. The key to the success of these modules is two fold: students can (1) see the actual plant through the videotape, and (2) use the computer simulation to study the effects of changes of system parameters on the operation of the system. These two features make the modules ideal for global and active learners.

Each module is written as an industrial problem caused by a change of conditions in an existing process, requiring an experimental study to re-evaluate the characteristic constants of the process. These might include, for example, reaction rate constants, equilibrium constants, heat transfer and mass transfer coefficients, and phase equilibrium constants. The student teams are expected to design experiments that will enable them to evaluate the needed constants. This is referred to as the *measurements* section of the problem.

After the constants have been determined, the student must validate them by using an existing computer model of the process, and comparing the simulated and experimental results. When they are convinced that their constants are reliable, the students must use these constants to predict some other specific process performance characteristics in the *applications* sections.

The Purdue modules are meant to supplement, not to replace, traditional laboratory experiments. Computer-simulated experiments have a number of advantages over traditional experiments:

- Processes that are too large, complex, or hazardous for the university laboratory can be simulated with ease on the computer.
- Realistic time and budget constraints can be built into the simulation, giving the students a taste of "real world" engineering problems.
- The emphasis of the laboratory can be shifted from the details of operating a particular piece of laboratory equipment to more general considerations of proper experimental design and data analysis.
- Computer simulation is relatively inexpensive compared to the cost of building and maintaining complex experimental equipment.

- Simulated experiments take up no laboratory space and are able to serve large classes because the same computer can run many different simulations.

These modules were developed under sponsorship of the National Science Foundation, Amoco Chemicals Corporation, Dow Chemical Company, Mobile Corporation, Tennessee Eastman Corporation, and Air Products and Chemicals, Inc., and are made available to universities through CACHE.

PICLES

One of the great opportunities afforded by the use of computers is the chance to simulate a real-time interaction with process equipment. One such simulator, focusing on process dynamics and control, is PICLES^{TM} (Process Identification and Control Laboratory Experiment Simulator), developed by Doug Cooper at the University of Connecticut (Cooper, 1993). PICLES is an IBM-PC training simulator that provides hands-on experience to students of process dynamics and control (see Fig. 4). It is used by over 60 chemical engineering departments around the world, and is very intuitive. Students using PICLES get experience in real-time use of P, PI, and PID control as well as using the Smith predictor, feed forward and cascade control, decouplers, and digital control of various systems, including: fluid level in gravity draining tanks, exit temperature in a heat exchanger, liquid level in a surge tank, and distillate and bottom composition in a distillation column (see Fig. 4).

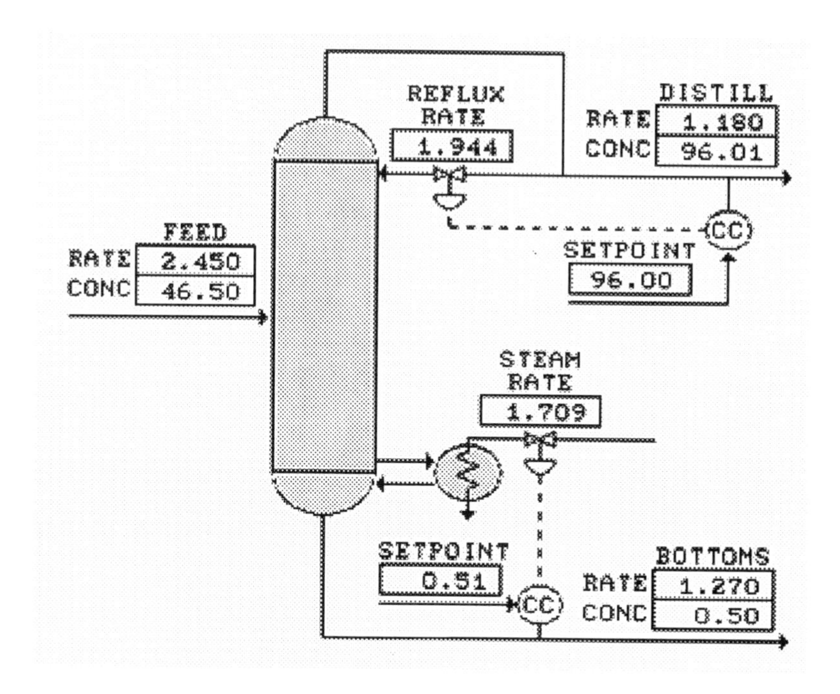

Figure 4. Process control of distillation columns using PICLES.

Current Practice and Future Directions

With the onset of CD-ROM technology, increasing computing speed and expanding computer memory, as well as more robust world-wide computer networks, we are at the threshold of a new revolution in interactive computer-aided instruction. Presentation of materials, which until now has been limited to the type of theoretical material presented in textbooks, is starting to incorporate investigation of "real world" processes, making the theoretical material come alive. Investigation, limited in the past to numerical simulations of processes, now is starting to include visual explorations into the processes themselves. Some examples are shown below.

Multimedia and Hypertext

Multimedia presentations allow students to interact with information in a variety of ways. Qasem and Mohamadian (1992) found that multimedia allows the student to take an active role in the educational process, freeing the student from being a passive information target. Pohjola and Myllyla (1990) discuss an object-oriented hypertext approach to organizing educational chemical engineering information, setting the groundwork for future efforts, and highlighting the use of animation and other techniques to assist students in creating mind pictures of the steps occurring at the molecular level in chemical engineering processes. The next level of sophistication is the integration of still and moving images into these educational projects. Coburn et al. (1992), for example, includes video images, animation, sound and full-motion video in their modules for introductory thermodynamics. The materials are well presented, but incorporation of video images and full-motion video appear to be restricted to the motivators in the derivations of the concepts and in the historical perspective.

Computer-based educational materials that take full advantage of multimedia are starting to emerge. Susan Montgomery at the University of Michigan has developed some prototype multimedia materials for use in the material and energy balances course as well as in the chemical engineering undergraduate laboratory. These computer-based instructional materials integrate graphics, animation, video images and video clips into multimedia packages that allow students to learn the basic concepts in chemical engineering through exploration of actual situations ranging in scope from simple bench-scale experiments and day-to-day experiences to industrial chemical plants. For example, for an open-ended problem on mass balances, a multimedia module allows students to tour the phosphate coating system (see Fig. 5) of Ford Motor Company's Wixom Assembly Plant. The module includes a description of each stage in the system, chemical usage, tank size and dump schedule information, and a short video clip of each stage.

In a module on multiphase systems (see Fig. 6), students can apply their expertise using T-xy diagrams to actual industrial equipment, a valuable experience for global learners

CACHE's role in the implementation of multimedia materials for chemical engineering has been invaluable. The successful completion and delivery of the 25th anniversary CD-ROM was the culmination of uncounted hours of work for Peter Rony of Virginia Tech, head of the CACHE CD-ROM Task Force, and made possible by the vision of Michael Cutlip of the University of Connecticut, current past-president of the CACHE Corporation. The exploration of CD-ROM as an avenue for distribution before most universities had CD-ROM drives available is an example of the pro-active role CACHE has taken in fostering the development and implementation of interactive computer-based materials for chemical engineering courses.

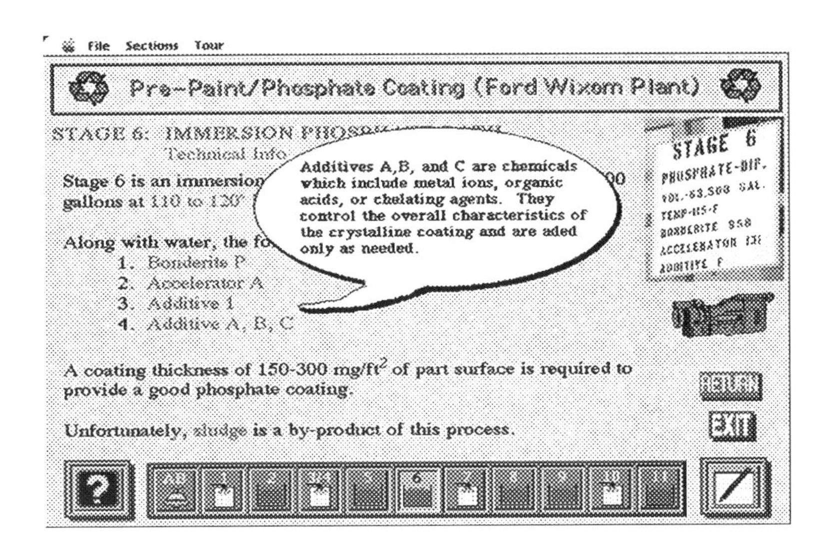

Figure 5. Analysis of phosphate coating system.

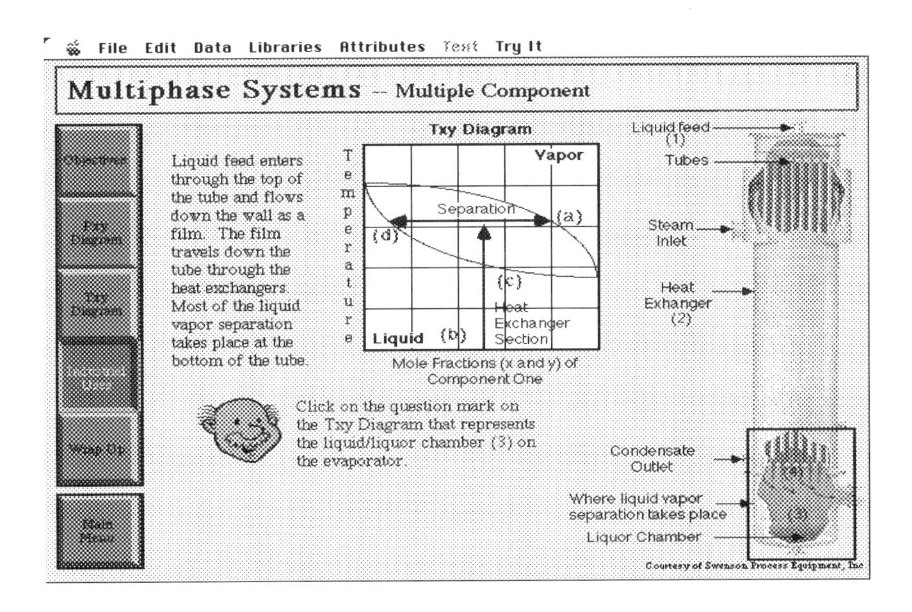

Figure 6. Real world applications of liquid-vapor separation principles.

Virtual Reality

Another exciting area on the horizon is virtual reality (VR). CACHE has formed a Virtual Reality Task Force and within the next five years we should see a significant number of modules developed using virtual reality. One module currently under development at the University of Michigan by John Bell and Scott Fogler is the prototype of a chemical plant that uses a straight-through transport reactor with a coking catalyst (see Fig. 7). Here, the student uses VR to enter the plant lobby where he or she is given an overview of the process and is free to explore various parts of the room and video tapes at will, simply by moving a joy stick. After this introduction, the student enters the reactor room where he or she can change the operating parameters and see their effect on the reaction variables such as degree of coking and conversion. The student can travel *inside* the reactor to observe the coking and catalyst transport, and can even enter the catalyst pellet to view the pore space inside the pellet and see molecules reacting on the surface.

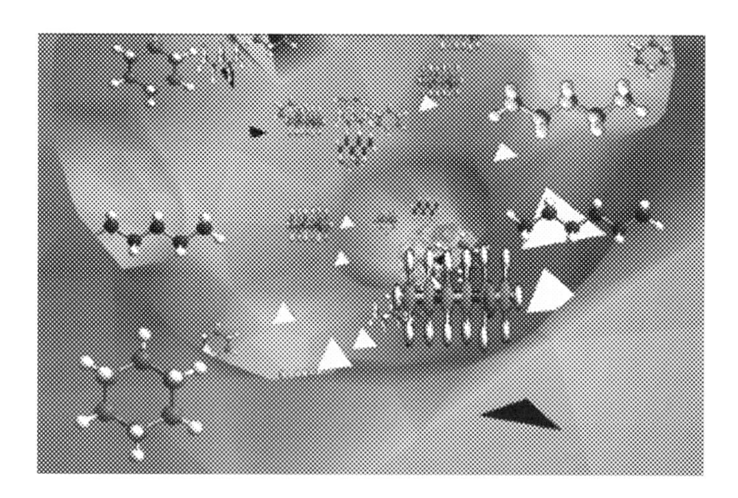

Figure 7. Exploration of the reactions taking place within a catalyst pellet - University of Michigan.

This visualization of the process and reaction mechanisms will greatly enhance the students' understanding and appreciation of this reaction engineering process. In general, the advent of virtual reality tools opens the door wide for all types of exploration.

Postscript

In surveying the literature prior to writing this article, the authors were surprised to see how few articles there are in the literature about instructional software that we know to exist. Our guess is that most early software developers were lone pioneers, with little time to devote to writing articles about their work, given the lack on importance placed on the development of educational materials by tenure and review committees. Recently, however, research focusing on the development of interactive computer-based educational materials seems to have garnered increasing respect in the academic community, as shown by the increasing number of peer-reviewed journals that can serve as outlets for dissemination about computer-aided instruction. These now include, among others, *Computer Applications in Engineering Education, The International Journal of Engineering Education, Computer Applications in Chemical Engineering, the Journal of Engineering Education,* and *Computers and Chemical Engineering.*

References

Bloom, Benjamin S. (1956). *Taxonomy of Educational Objectives; the Classification of Educational Goals, Handbook I: Cognitive Domain.* New York: David McKay Company.

Coburn, W.G. G.C. Lindauer, R.L. Collins, T.E. Mullin, and W.P. Hnat (1992). Development of multifaceted instructional modules for introductory thermodynamics. *Proceedings of IEEE Conference on Frontiers in Education.* 150-155.

Cooper, D.J. (1993). PICLES: The process identification and control laboratory experiment simulator. *CACHE News* **37**, 6-12.

Felder, R.M. (1989). Meet your students. I. Stan and Nathan. *Chemical Engineering Education*, Spring, 68-69.

Felder, R.M. (1994) Meet your students: V. Edward and Irving. *Chemical Engineering Education*, Winter, 36-37.

Felder, R.M., K.D. Forrest, L. Baker-Ward, E.J. Dietz, and P.H. Mohr (1993). A longitudinal study of engineering student performance and retention: I. Success and failure in the introductory course. *Journal of Engineering Education*, Jan, 15-21.

Felder, R.M. and L.K. Silverman (1988). Learning and teaching styles. *Engineering Education*, 674-681.

Fogler, H.S., S.M. Montgomery, and R.P. Zipp (1992). Interactive computer modules for chemical engineering instruction. *Computer Applications in Engineering Education,* **1**(1), 11-24.

Griskey, R.G. (1991). Undergraduate education: Where do we go from here? *Chemical Engineering Education, Spring*, 96-97.

Jayakumar, S., R.G. Squires, G.V. Reklaitis, P.K. Andersen, K.R. Graziani, B.C. Choi (1993). The use of computer simulations in engineering capstone courses: A chemical engineering example - the Mobil catalytic reforming process simulation. *International Journal of Engineering Education,* **9**(3), 243-50.

Kolb, D.A. (1984). *Experiential Learning: Experience as the Source of Learning and Development*. Prentice-Hall, Englewood Cliffs, N.J.

Kulik, C.C. and J. A. Kulik (1986). *AEDS Journal*, **19**, 81.

Kulik, J.A. and C. C. Kulik (1987). *Contemporary Education Psychology*, **12**, 222.

McCaulley, M.H., E.S. Godleski, C.F. Yokomoto, L. Harrisberger and E.D. Sloan (1983). Applications of psychological type in engineering education. *Engineering Education*, 394-400.

Pohjola, V.J., and I. Myllyla (1990). Object-oriented hypermedia as a teaching aid in chemical engineering education. In TH. Bussemaker and P.D. Iedema (Eds.). *Computer Applications in Chemical Engineering*, Elsevier, Amsterdam. 199-202.

Qasem, I. and H. Mohamadian (1992). Multimedia technology in engineering education. *Proceedings. IEEE Southeastcon'92 (Cat. No. 92CH3094-0) vol. 1.* IEEE, New York. 46-49.

Shacham, M. and M.B. Cutlip (1981a). Educational Utilization of PLATO in chemical reaction engineering. *Computers & Chemical Engineering*, **5**, 215-224.

Shacham, M. and M.B. Cutlip (1981b). Computer-based instruction: is there a future in ChE education? *Chemical Engineering Education*, 78.

Shacham, M. and M.B. Cutlip (1982). A simulation package for the PLATO educational computer system. *Computers & Chemical Engineering*, **6**, 209-218.

Shacham, M. and M.B. Cutlip (1983). Chemical reactor simulation and analysis at an interactive graphical terminal. *Modeling and Simulation in Engineering*, 27.

Smith, S.G. (1970). The use of computers in teaching organic chemistry. *J. of Chemical Educatio*n, **47**, 608.

Smith, S.G. and D.A. Sherwood (1976). Educational uses of the PLATO computer system. *Science*, **192**, 344.

Soloman, B.S. (1992). *Inventory of Learning Styles*. North Carolina State University.

Squires, R.G., P.K. Andersen, G.V. Reklaitis, S. Jayakumar, and D.S. Carmichael (1992). Multimedia-based applications of computer simulations of chemical engineering processes. *Computer Applications in Engineering Education,* **1**(1), 25-30.

VanderStoep, S.W. and C.M. Seifert (1993). Learning 'How' vs. learning 'When': improving transfer of problem-solving principles. *Journal of the Learning Sciences,* **3**(1), 93-111.

GENERAL-PURPOSE SOFTWARE FOR EQUATION SOLVING AND MODELING OF DATA

Mordechai Shacham
Ben-Gurion University of the Negev
Beer-Sheva 84105, Israel

Michael B. Cutlip
University of Connecticut
Storrs, CT 06269

N. Brauner
Tel-Aviv University
Tel-Aviv 69978, Israel

Abstract

Much of the educational emphasis in numerical computing has shifted from FORTRAN or other source code programming languages to the use of general-purpose packages which typically solve Nonlinear Algebraic Equations (NLE), Ordinary Differential Equations (ODE), and carry out the computations required for data modeling and correlation. The educational advantage in using these programs is that they require the user to develop and input the model equations, but carry out the technical numerical details of the solution method without user intervention. In this paper comparisons of four such packages (MAPLE, MATLAB, MATHEMATICA and POLYMATH) are made for solving NLE's, ODE's, and for regression of data. References for test problems are identified, and the performances of the packages with the test problems are compared. Improvements are suggested in the solution and result storage algorithms, as well as in some "user friendly" features.

Introduction

The role of computers for numerical solution of chemical engineering problems was recognized early on over thirty years ago. The first textbook to address this subject was that by Lapidus (1962). This textbook included chapters on polynomial approximation, solution of ODE's, partial differential equations (PDE's), linear equations and NLE's, and the least squares error approach. There was also a chapter on optimization and control.

Seven years after the publication of the textbook by Lapidus (1962), the CACHE Corporation was founded. This paper concentrates on the solution of NLE's, ODE's and on data modeling and correlation. Optimization and control are covered in two additional separate papers in this monograph. The numerical solution of PDE's is still considered too difficult to be included in the undergraduate curriculum and is not covered here.

The history of the use of computers in chemical engineering education has been documented by Seader (1989). The publication of the textbook by Carnahan, Luther and Wilkes (1969) on numerical methods and the textbook by Henley and Rosen (1969) on material and energy balances formulated for digital computer use are mentioned as important developments in computer applications in chemical engineering. The CACHE Corporation published a seven volume set of books entitled *Computer Programs for Chemical Engineering Education* in 1972 with examples in seven curriculum areas. The examples included problem statements with solutions that were mostly via programs in FORTRAN. A few CSMP (Continuous System Modeling Package, available on mainframe computers) programs were also included. A typical computer assignment in that era would require the student to carry out the following tasks: (1) derive the model equations for the problem at hand, (2) find an appropriate numerical method to solve the model (mostly NLE's or ODE's), (3) write and debug a FORTRAN program to solve the problem using the selected numerical algorithm, and (4) analyze the results for validity and precision.

It was soon recognized that the second and third tasks of the solution were minor contributions to the learning of the subject matter in most chemical engineering courses, but they were actually the most time consuming and frustrating parts of a computer assignment. The computer indeed enabled the students to solve realistic problems, but the time spent on technical details which were of minor relevancy to the subject matter was much too long.

In order to solve this dilemma there was a growing tendency to provide the students with computer programs that can solve one particular type of problem. Listings, or even disks containing small programs, were included in textbooks, and large scale process simulators, such as FLOWTRAN (Seader, Seider and Pauls, 1979) were made available to departments through CACHE or from commercial vendors. The provision of a complete program to students for a particular problem has the disadvantage that the connection between the mathematical model and the problem is obscured. Students can only provide the input data, and then can only observe the results. The very important step of converting physical phenomena to a mathematical model is missing. Furthermore, each class of problems requires learning use of a different software tool and the input format of a particular program does not contribute to the learning of the subject matter.

The latest approach in chemical engineering education is to use general-purpose packages for problem solving. These packages require the student to input the mathematical model and the numerical data, but the programs carry out all the technical steps of the numerical solution. This approach is demonstrated by Fogler (1993), who discusses his method of teaching the chemical reaction engineering course. For example, his proposed approach for isothermal reactor design is to use mole balance, rate laws and stoichiometry in setting up the mathematical model and then to use a user friendly ODE solver such as POLYMATH (Shacham and Cutlip, 1994a) for combining and solving the equations. In this way, the use of the computer becomes just a routine, efficient and natural step in problem solution.

A general-purpose program for educational use must be versatile enough so that it can be used in most courses and laboratories during a four year educational program. It must be inexpensive and capable of running on inexpensive computers so that students can install and run it on their own computers. It must be user friendly so that it is easy to learn and easy to relearn. The technical details of the solution should take up only a minimum amount of the student's time. While most programs claim to be user friendly, there are actually objective measures to compare user friendliness of numerical software for equation solving and regression programs which will be presented later in this paper.

A recent survey by CACHE (Davis, Blau and Reklaitis, 1994) has indicated that only a disappointingly small percentage of engineers use numerical methods. Only 3% of the engineers use numerical libraries frequently and only 1% use mathematical packages. In comparison, 74% use spreadsheet programs frequently and 23% use them occasionally. There is a growing gap between the specialized users who, according to an article by Boston, Britt and Tayyabkhan (1993), will be able to solve "systems involving several hundreds of thousands of equations with several hundreds of degrees of freedom" by the year 2000 and the more typical engineers whose most sophisticated computational tool will remain the spreadsheet. The reasons for the limited use of numerical methods can be best described by observations made by Seader (1989) and deNevers and Seader (1992). Seader (1989) notes: "One problem is the lack of chemical engineering textbooks which include examples and exercises that presume that the student has a working knowledge of numerical techniques for solving sparse and dense systems of linear and nonlinear equations. Even if the textbooks were available, they might not be widely used because of the second problem, which is that many chemical engineering educators are either not familiar with or not comfortable with the use of computers to solve equations." A similar observation is made by deNevers and Seader (1992): "Since the advent of digital computers, textbooks have slowly migrated toward computer solutions of examples and homework problems, but in many cases the nature of the examples and problems has been retained so that they can be solved with or without a computer."

In short, the problem is that the available software in recent years has not been "user friendly" enough to entice most chemical engineering educators to use numerical methods. Many faculty have practiced "computer avoidance" and consequently many efficient new design techniques have not been introduced or used in classes. Furthermore, most examples in newer textbooks remained the same as those used before the advent of the computer. Often, the easiest faculty approach is to "computerize" by using spreadsheets alone, while avoiding the more realistic educational problem solving which numerical methods can provide.

Available Software

Extensive listings of the available chemical engineering software can be found in the annual CEP software directory which is published by the AIChE as a supplement to the December issue of *Chemical Engineering Progress*. The relevant software is listed under the category of "Mathematics and Statistics." This software directory gives information on what computer configuration is needed to run the software and often also the price of the software.

The software for solving equations and modeling data can be divided into three categories: (1) subroutine libraries, (2) single- or double-purpose programs, and (3) multi-purpose mathematical packages.

The 1995-96 edition of the CEP software directory lists several subroutine libraries written in FORTRAN, BASIC, Assembler and C. Our work has extensively used the mainframe version of the IMSL (1982) library as well as many of the subroutines from the library associated with the book entitled *Numerical Recipes* (Press, et al., 1986). These subroutines are all very well debugged and are dependable. The use of subroutine libraries is a must for someone who does his or her own programming, but these libraries lack the friendly user interface which is needed for a program to be effectively utilized in undergraduate engineering education.

Single-purpose programs are available for solving NLE's (TK Solver, Slaughter et al, 1990), for solving ODE's (ACSL, SimuSolv (Steiner, Rey and McGroskey, 1993), Tutsim (p. 532 in Coughanour, 1991)) and many programs for modeling of data. These programs are usually targeted toward specialized users, and they are typically too expensive and overly complicated for use in undergraduate chemical engineering education. Use of a single-purpose program also means that the student must learn many different programs to deal with a variety of problems.

The multi-purpose general mathematical packages such as POLYMATH (Shacham and Cutlip, 1994a), MAPLE (Ellis et al., 1992), MATLAB (MathWorks, 1991), and MATHEMATICA (Wolfram, 1991) are the most appropriate general equation solving and data modeling tools in chemical engineering education. These packages have inexpensive educational site licenses or student versions for most personal computers. User interfaces are provided to minimize programming; however, the effort required by the user to learn to use the program and to prepare the input data differs from one package to another. These differences among the packages will be discussed in more detail in the next two sections.

Spreadsheet programs for equation solving are not considered in this paper. They can be used for solving NLE's, ODE's and even PDE's, as was shown by Rosen and Adams (1987), but such use does require programming. In this respect, spreadsheets do not have advantages over programming languages. They can however be very helpful in simple calculations such as required in a "Mass and Energy Balance" course (Carnahan, 1993; Misovich and Biasca, 1990). Spreadsheets can also be used for data modeling and analysis. This aspect will be further discussed in a later section. In our opinion, the undergraduate chemical engineering computational toolkit for students must contain both a spreadsheet program and a package for equation solving and data modeling.

Solution of Nonlinear Algebraic Equations

Solution of NLE's are required in all chemical engineering courses. Typical examples include calculation of compressibility factor using an equation of state, adiabatic flame temperature, phase equilibrium calculations in ideal and nonideal mixtures, chemical equilibrium calculations, flowrate and pressure drop in a pipeline and pipeline networks, and steady state material and energy balances in chemical reactors. Many such examples are discussed in detail by Shacham and Cutlip (1994b). Equations of state, adiabatic flame temperature, and phase equilibrium calculations for ideal mixtures are representative of problems that are easily solved. That is, available programs will converge to the solution from reasonable initial estimates. Chemical equilibrium calculations, steady state material and energy balances on chemical reactors, and nonideal phase equilibrium problems are often very difficult to solve.

Problems in this category require accurate initial estimates that are very close to the solution, or some manipulation of the equations in order to enable the NLE solver to achieve a solution.

Recently, Shacham, Brauner and Pozin (1994) have compared four general-purpose packages (MAPLE, MATLAB, MATHEMATICA and POLYMATH) for their ability to solve sets of test problems taken from chemical engineering applications. The test problems were taken from Shacham (1984, 1989) and Shacham and Cutlip (1994b). None of the NLE packages could solve all the test problems from all sets of initial estimates. Thus it can be expected that any one of the packages will fail occasionally; therefore, it is important that the package should clearly signal when no solution is achieved. Some of the packages converged to a local minimum without indicating that solution had not been reached, and one package went into an infinite loop necessitating the rebooting of the computer.

This study concluded that the convergence intervals of both MAPLE and POLYMATH for difficult single nonlinear equations are the widest, and that multiple solutions are usually detected. For systems of NLE's, the largest number of cases were solved by MATHEMATICA and POLYMATH. All the programs except POLYMATH converged occasionally to a local minimum without giving a warning message. The largest number of such incidents were observed with MATLAB.

Shacham, Brauner and Pozin (1994) have recommended several "user-friendly" features to be included in an interactive NLE solver in relation to the four packages:

1. Menu-based program control instead of command-based control.

The program is easiest to use when all the available options are presented on the screen. The advantage of the menu based control becomes less obvious for complicated programs because of the need to search through several menu levels. POLYMATH was the only package which was completely menu-based.

2. Notation and format used in equation entry.

For the user it is most convenient when the equations can be entered into the package using almost the same notation and format as used in the mathematical model. Obviously there must be some rules for equation input, but there were major differences between the very flexible structure used by MAPLE and POLYMATH and the subroutine like structure required by MATLAB.

3. Equation ordering and detection of implicit relationships among variables.

When formulating a mathematical model, the general rule is usually considered first and the connections among variables are defined afterwards. The order of calculation is often the opposite, and it is much more convenient for the user when the reordering is done by the package. Often implicit relationships exist among variables that are difficult to detect by inspection. The package should find such relationships and inform the user. POLYMATH, MAPLE and MATHEMATICA reorder equations in the correct computational order; MATLAB does not perform any reordering. Only POLYMATH detects and issues appropriate error messages for existence of implicit relationship among variables.

4. Debugging aids.

All the packages can detect syntax errors after an equation has been entered. Only POLYMATH keeps a list of undefined variables. Such a list can be very helpful in detecting the misspelling of the name of a variable or in reminding the user of constants or variables that have not yet been defined.

5. Root verification and multiple solutions.

Depending on the solution method used, programs may converge to a local minimum instead of the root of the system of equations. The user must be made aware that the true solution has not been found. POLYMATH always displays the norm of the function values at the solution, whereas, the user must explicitly request calculation and display of the function values in the other packages. If there are several roots for a system of equations, all packages will find one solution when started from one set of initial estimates. The only exception is that POLYMATH finds up to five roots for a single equation inside an interval specified by the user.

6. Importance of initial estimates.

All the packages require the user to provide initial estimates for the unknowns. Shacham, Brauner and Pozin (1994) have shown that selection of initial estimates can be critical for highly nonlinear problems (such as chemical equilibrium, nonideal phase equilibrium, etc.) regardless of the program used. Often, transformations can be used to reduce the nonlinearity of the problem.

Solution of Nonlinear Ordinary Differential Equations

Numerical solution of ODE's is extensively utilized in Process Dynamics and Control and Chemical Reaction Engineering courses. In fact, the first documented use of a specific ODE solver language was in control courses (CSMP was used in the "Control" volume of "Computer Programs for Chemical Engineering Education", published by CACHE in 1972). Solution of ODE's is also useful in simulating batch processes, such as batch distillation and solving steady state heat and mass transfer problems. Many examples requiring numerical solution of ODE's discussed in detail by Shacham and Cutlip (1994b).

From the numerical point of view, many types of ODE's are relatively easy to solve. They can be solved with high accuracy with some variation of the 4th-order Runge-Kutta (RK4) algorithm with error estimation and step size control (Press et al., 1986, pp. 554-562). Unfortunately, there are certain classes of problems which are difficult to solve. Carnahan and Wilkes (1981) mention reaction kinetics and distillation column modeling as being of the more difficult type. One common source of difficulty is that the system of equations is "stiff." Stiff system of equations contain variables that change (decay) in widely varying time scale. In such cases, an explicit integration method, such as the RK4 method, is forced to use extremely small step sizes; consequently, a very large number of steps is required to integrate over the desired time interval. There are implicit and semi-implicit methods that can handle stiff equations effectively as discussed by Carnahan and Wilkes (1981), but these methods are much more complex than the RK4 method.

Another source of difficulty is singular (or more precisely, almost singular) points in the region of the integration. If, for example, one of the variables approaches zero and there is division by this variable, reduction of the step size to infinitesimally small values is required. The step adjusting algorithm may often "overstep" the singular point, causing the zero-bounded variable to become negative. From this point on, the integration is continued without any difficulty, giving incorrect results.

Equations with periodic solution may present difficulties, not with the precision of the solution itself, but in the presentation of the solution. Graphical or tabular presentation of the results is done in discrete time intervals. If the time intervals used are not small enough, details of the solution may be omitted, and instead of a continuous curve, broken lines are obtained.

The consequences of all these difficulties are that the user often cannot be certain of the accuracy or even validity of the numerical solution obtained. Unlike the solution of nonlinear algebraic equations, there is no simple way to verify the accuracy of the solution. Test problems that can be handled with RK4 algorithms can be found, for example, in Shacham and Cutlip (1994), while a moderately difficult stirred tank reactor problem is presented by Shacham, Brauner and Cutlip (1994a). Very stiff test problems are listed by Enright and Hull (1976) and Johnson and Varney (1976).

A number of test problems from Brauner, Shacham and Cutlip(1994) and a few stiff systems for mainly reactor design problems from Fogler (1992) have been solved using POLYMATH, MATLAB, MATHEMATICA and MAPLE. The solution algorithms used by the different programs are shown in Table 1.

Table 1. Integration Algorithms in the Four Packages.

	Nonstiff Algorithms	Stiff Algorithm
POLYMATH	RK3-RK4	Semi-implicit Euler
MATLAB	RK2-RK3, RK4-RK5, Adams	Gear
MATHEMATICA	No documentation for algorithm used	
MAPLE	RK4-RK5	---

MATLAB stores information on all integration steps for variables for which information is requested. POLYMATH and MATHEMATICA store results at even time intervals, MAPLE does not store the results for plotting, but plots during integration using "even" time intervals for adding a new section to the plot. In MATHEMATICA the user can change the frequency of the data storage and in MAPLE the frequency of the plot update. During the solution, only POLYMATH provides a plot of step size and local truncation error history. The other packages do not interact with the user during solution time and do not provide any integration error or step size information.

The test problems were intended to detect and demonstrate weaknesses of the packages. For the nearly singular test problem where the concentration of one component approaches zero, all packages (except MATHEMATICA with a very small tolerance) "overstepped" the nearly singular point and continued with negative concentration of this component. The use of MATLAB with very high precision caused the program to stop the integration with error message "singularity likely."

For the stiff test problems, POLYMATH usually detected that the problem was stiff and recommended the use of the implicit Euler method. With the implicit Euler method, the solution obtained was unstable, oscillatory and inaccurate. When MATLAB was used with error tolerance not small enough, it did not detect the existence of stiffness and simply returned with incorrect results. With smaller error tolerance, the solution obtained was fairly accurate, but small oscillations of the stiff variable were noticeable in the results. The behavior of MAPLE for stiff systems was inconsistent. The program often returned with error messages or without any results or messages. Other times, the solutions were very accurate. For stiff problems, the performance of MATHEMATICA was the best, as it gave correct results using the default error tolerance values.

The basic quality of the plots was best with MATLAB where the results of all the integration steps are stored and can be displayed. In POLYMATH and MAPLE where the results are stored or displayed at fixed intervals, details of the solution are sometimes poorly plotted. If the solution is periodic, only two cycles could be displayed in full details with POLYMATH or the default setting of MAPLE. Attempts to increase the final time (adding more cycles) caused the graph to be displayed as broken lines instead of a continuous curve. In MAPLE, the display interval could be changed by the user; in POLYMATH there was the need to divide the integration interval to smaller parts in order to get a continuous curve in such cases. The Unix version of MATHEMATICA was used in this work, and this version did not provide plots of comparable quality to the other packages. The results when transferred to MATLAB gave plots as detailed as did the MATLAB package alone.

Solution of stiff ODE's may take considerable amount of time. For one particular example containing 12 differential equations the time required was: POLYMATH - 1 min. (including restart of the solution twice), MATHEMATICA - 1.5 min., MATLAB - 0.5 min. In MAPLE, the integration interval had to be shortened, because use of the full interval led to continued execution for 10 minutes and the graph was not plotted. Defining a shorter integration interval resulted in 5 minute run time and the results were plotted. In any case, the integration time for stiff problems may be long. Therefore, it is important to keep the user informed that calculations are proceeding and how far the integration has gone. Only POLYMATH provides such information.

The user-friendliness of the different packages was similar to that for NLE's. The format required by MAPLE and MATHEMATICA for entering the ODE's was even more rigid, and the rules for the modifications required are less obvious than for entering NLE's.

Data Modeling and Correlation

Data modeling and correlation is useful in most chemical engineering courses and laboratories. Typical examples are correlation of physical and thermodynamic properties, phase equilibrium, heat transfer coefficients, and reaction rate for homogeneous and heterogeneous reactions. Often the need arises to integrate or differentiate tabular data, such as when calculating mean heat capacity, vapor liquid equilibrium from total pressure measurements or using the differential method of reaction rate data analysis. Many detailed examples are discussed by Shacham and Cutlip (1994b).

Most of the statistical techniques for analyzing the accuracy of correlations have been known for several decades (Himmelblau, 1969), but these were rarely used on mainframe systems because of the limitations of the computational and graphical tools when using batch computations. Today, many interactive statistical analysis programs are available which can model and correlate data. The best known among these programs are SPSS (1983) and SAS (Cary, 1988). These and other similar programs are intended to be used by statisticians; they are unnecessarily complicated for routine use by engineers and engineering students to correlate engineering data. Fortunately, there are programs which focus on data modeling and analysis, and many of them are listed in the CEP software directory.

In order to select the most appropriate program for data modeling and analysis, the program requirements must be clearly defined. Some of the requirements are discussed in the book by Noggle (1993); however, the discussion is limited to models containing a single independent variable. The results of our investigations in this area are discussed in more detail in Shacham, Brauner and Cutlip (1994b).

A regression package appropriate for undergraduate engineering students or for practicing engineers must be able to perform linear, polynomial, multiple linear and multiple nonlinear regressions. It also should be able to determine cubic splines for relationships that cannot be adequately represented by known models. The package should provide options to extrapolate, differentiate and integrate tabulated data using the regression curve. Tools for transforming the original data (for models containing expressions such as log(P) or 1/T, etc.) should also be provided.

All regression programs provide some statistical information. In our experience, the reliance on one, or even several statistical variables indicating the quality of the fit (such as the variance or linear correlation coefficient) can be misleading, especially when the underlying assumptions in calculating these variables are not well understood (Shacham, Wisniak and Brauner, 1993; Shacham, Brauner and Cutlip, 1994b).

The quality of the experimental data can be checked by plotting the independent variables one versus another. This way, linear dependence between the independent variables can be detected. If the model is comprised of sums of terms of different order of magnitude (as in rate expressions for reversible reactions), plots can be derived which can indicate whether the data are precise enough to represent the contributions of the less significant terms.

When there are several candidate models for representing the same data, models that contain an insufficient number of parameters can be eliminated using residual plots. Models with excessive parameters can be eliminated using confidence intervals and the mean sum of squares to select from among candidate models. Practical selection of the best method for estimation of the parameters (i.e., linear versus nonlinear regression) should utilize error analysis of the transformation functions, residual plots and comparison of mean sum of squares. The regression package must provide options for plotting one variable versus another one, construction of residual plots, calculation of confidence intervals on the parameter values, and calculation of the mean sum of squares in order to facilitate the statistical analysis.

All of the above capabilities are available in the POLYMATH package, and we have used them successfully with undergraduate students. Most of the desirable options are also available or can be generated with little effort with the more advanced spreadsheets programs (an example is Quattro Pro 4.0). All the above capabilities (and some more) except for multiple linear and nonlinear regression, are available in the EZFIT program (Noggle, 1993).

Conclusions

There are several software packages that can be used as general-purpose tools for solving NLE's, ODE's and for modeling and correlating data that are useful within the undergraduate chemical engineering curriculum. These programs require the user to input the model and the data, but many of the technical details of the problem solution need not be considered during normal use. It is expected that the use of such packages will increase as educators realize their potential value and develop new design and calculation techniques that enable students to efficiently solve more realistic problems.

In general, existing packages can be improved by making the user interface more "friendly" by making the notation and the model entry format more flexible, carrying out equation ordering, detecting implicit connections among variables, and by providing the user with more information regarding the validity of the solution. The NLE solution algorithms should be improved to make them less sensitive to initial estimates in solving difficult problems, such as chemical or nonideal phase equilibrium problems. The ODE integration algorithms can be improved by making them monitor more closely the condition of the equations so that they can detect stiffness, singularity or other conditions which can make the solution inaccurate or incorrect and inform the user or change solution method accordingly. The ODE solvers also need an adaptive result storage scheme, where the results are stored at variable time intervals so as to closely follow the curvature of the solution.

In data modeling and correlation, the available software is adequate, but in order to use this software most effectively the study of statistical concepts must be reinforced, so that the chemical engineer can select and use the most significant statistical tests for a particular problem and not rely on tests when underlying assumptions are violated.

The use of the new interactive software tools will probably be the most beneficial for students and practicing engineers who did not learn to program or do not program well in languages such as FORTRAN. Over fifteen years of experience has reinforced our opinion that interactive programs on personal computers for the numerical solution of NLE's, ODE's, and data correlation will indeed become increasingly important tools in chemical engineering education.

References

Boston, J. F., H. I. Britt, and M. T. Tayyabkhan (1993). Software: Tackling tougher tasks. *Chem. Eng. Progr.*, **89**(11), 38.

Brauner, N., M. Shacham, and M. B. Cutlip (1994). Application of an interactive ODE simulation program in process control education. *Chem. Engr. Ed.*, **28**(2), 130-135.

Carnahan, B., H. A. Luther, and J. O. Wilkes (1969). *Applied Numerical Methods*. Wiley, New York.

Carnahan, B. and T. O Wilkes (1981). Numerical solution of differential equations - an overview. In R.S.H. Mah and W. D. Seider (Eds.), *Foundation of Computer Aided Chemical Process Design*, Engineering Foundation, New York, 225-341.

Carnahan, B. (1993). Personal Communication.

Cary, N. C. (1988). *SAS Language and Procedures Usage*, Version 6, SAS Institute.

Coughanowr, D. R. (1991). *Process Systems Analysis and Control*. 2nd ed. McGraw-Hill, New York.

Davis, J. F., G. E. Blau, and G. V. Reklaitis (1994). Computers in undergraduate chemical engineering education: A perspective on training and application. *Chem. Engr. Ed.*, submitted for publication.

Ellis, W. Jr., E. Johnson, E. Lodi, and D. Schwalbe (1992). *MAPLE V Flight Manual*. Brooks/Cole Pub. Co., Pacific Grove, California.

Enright, W. H. and T. E. Hull (1976). Comparing numerical methods for the solution of stiff systems of ODE's arising in chemistry. In Lapidus, L. and W. E. Schiesser (Eds.), *Numerical Methods for Differential Systems*, Academic Press, New York. 45-66.

Fogler, H. S. (1992). *The Elements of Chemical Reaction Engineering*, 2nd ed., Prentice-Hall, Englewood Cliffs, New Jersey.

Fogler, H. S. (1993). An appetizing structure of chemical reaction engineering for undergraduates. *Chem. Engr. Ed.*, **27**(2), 110.

Forsythe, G. E., M. A. Malcolm, and C. B. Moler (1977). *Computer Methods for Mathematical Computation*, Prentice-Hall, Englewood Cliffs, New Jersey.

Henley, E. J. and E. M. Rosen (1969). *Material and Energy Balance Computation*. Wiley, New York.

Himmelblau, D. M. (1970). *Process Analysis by Statistical Methods*. Wiley, New York.

IMSL Library Users Manual. 9th Ed., IMSL, Houston, Texas (1982).

Johnson, A. I. and J. R. Barney (1976). Numerical solution of large systems of stiff ordinary differential equations in a modular simulation framework. In Lapidus L. and W. E. Schiesser (Eds.), *Numerical Methods for Differential Systems*, Academic Press, New York. 97-124.

Lapidus, L. (1962). *Digital Computation for Chemical Engineers*. McGraw-Hill, New York.

Math Works, Inc. (1991). *MATLAB for Unix Computer's Users' Guide*. Natick, Massachusetts.

Misovich, M. and K. Biaslea (1991). The power of spreadsheets in a mass and energy balances course. *Chem. Engr. Ed.*, **25**(1), 46.

deNevers, N. and J. D. Seader (1992). Helping students to develop a critical attitude towards chemical process calculations. *Chem. Engr. Ed.*, **26**(2), 88-93.

Noggle, J. H. (1993). *Practical Curve Fitting and Data Analysis*, Prentice-Hall, Inc., Englewood Cliffs, New Jersey.

Press, W. H., B. P. Flannery, S. A. Teukolsky and W. T. Vetterling (1986). *Numerical Recipes*. Cambridge University Press, Cambridge, UK.

Rosen, E. M., and R. N. Adams (1987). A review of spreadsheet usage in chemical engineering calculations. *Comp. & Chem. Engr.*, **11**(6), 723-736.

Seader, J. D., W. D. Seider, and A. C. Pauls (1979). *FLOWTRAN Simulation - An Introduction*. CACHE, Ann Arbor, Michigan.

Seader, J. D. (1989). Education and training in chemical engineering related to the use of computers. *Comp. & Chem. Engr.*, **13**(4/5), 377-384.

Shacham, M. (1984). Recent developments in solution techniques for systems of nonlinear equations. In *Proc. of the Second International Conference on Foundations of Computer Aided Design,* (Ed.) A. W. Westerberg and H. H. Chien, pp. 891-923. CACHE Publications, Ann Arbor, Michigan.

Shacham, M. (1989). An improved memory method for solution of a nonlinear equation. *Chem. Eng. Sci.*, **44**, 7, 1495-1501.

Shacham, M., J. Wisniak, and N. Brauner (1993). Error analysis of linearization methods in regression of data for the Van Laar and Margules equations. *Ind. Eng. Chem. Res.*, **32**, 2820-2825.

Shacham, M., N. Brauner and M. Pozin (1994). Comparing software for interactive solution of systems of nonlinear algebraic equations.

Shacham, M., N. Brauner, and M. B. Cutlip (1994a). Exothermic CSTR's: Just how stable are the multiple steady states? *Chem. Engr. Ed.*, **28**(1), 30-35.

Shacham, M., N. Brauner, and M. B. Cutlip (1994b). Critical analysis of experimental data, regression models and regressed coefficients in data correlations. Proceedings of the Conference on *Foundations of Computer-Aided Process Design*, 305-308, Snowmass, Colorado.

Shacham, M. and M. B. Cutlip (1994a). *POLYMATH 3.0 Users' Manual*. CACHE Corporation, Austin, TX.

Shacham, M. and M. B. Cutlip (1994b). *Numerical Solution of Chemical Engineering Problems Using POLYMATH*. Prentice-Hall, Englewood, New Jersey (to be published, 1997).

Slaughter, J. M., J. N. Peterson and R. L. Zollars (1991). Use of PC based mathematics software in the undergraduate curriculum. *Chem. Eng. Ed.*, **25**(1), 54-60.

SPSS User Guide, McGraw-Hill, New York (1983).

Steiner, E. C., T. D. Rey, and P. S. McGroskey (1993). *SimuSolv Reference Guide*, The Dow Chemical Company, Midland, Michigan.

Westerberg, A. W. (1993). Human aspects: Redefining the role of chemical engineers. *Chem. Eng. Prog.*, **89**(11), 60.

Wolfram, S. (1991). *MATHEMATICA, A System for Doing Mathematics by Computer*, 2nd ed. Addison Wesley, NY.

THERMODYNAMICS AND PROPERTY DATA BASES

Aage Fredenslund, Georgios M. Kontogeorgis and Rafiqul Gani
Chemical Engineering Department
Technical University of Denmark
2800 Lyngby, Denmark

Abstract

The formulation of the thermodynamic equilibrium problem and the methodologies used for its solution are presented in this chapter. The resulting equations need a number of pure component and mixture equilibrium properties, which, whenever not experimentally available at the conditions of interest, must be estimated. Several predictive techniques for preliminary estimation of a variety of properties are discussed. Special emphasis is given to the group-contribution methods (especially those related to the UNIFAC model) and the recent advances in mixing rules for cubic equations of state. Three application examples illustrating the applicability of some of the presented models to difficult systems (mixtures with polymers, electrolytes and multiphase equilibria) are given. Finally, some suggestions for future research in the area of applied thermodynamics are included.

Introduction

Phase equilibrium thermodynamics plays a particularly important role for process development in chemical, petroleum, and related industries. We mention here some examples: Separation of the products of organic synthesis and removal of by-products are often encountered in engineering practice. These separations are usually accomplished with conventional separation processes (e.g., distillation, extraction), which are based on thermodynamic equilibrium between two or more phases. However, some recent applications related to the food and biochemical industries (e.g., separation and fractionation of fish-oil related compounds) are more efficiently handled with non-conventional separation techniques, such as supercritical fluid extraction. Furthermore, since major products of interest to the chemical and petroleum industries are of complex (and often macro-) molecular structure, computer-aided product design will help to identify those materials which best satisfy specific requirements from an economic point of view (Zeck and Wolf, 1993). Satisfactory knowledge of thermodynamic, as well as physical, properties of the materials is needed for efficient product design.

Product development and process design require molecular-sound correlation models, and especially, predictive methods. Group-Contribution (GC) techniques have found wide applicability due to their successful compromise between simplicity and accuracy. The purpose here is not to give a complete review of all the successful GC methods, but rather to outline those methods and procedures which are of wide applicability and seem particularly promising for phase equilibrium computations in chemical engineering thermodynamics and thermodynamics instruction.

The rest of this chapter is organized as follows: first, the general formulation and the solution methods for thermodynamic equilibrium problem are given. Next, a number of predictive (mostly GC-based) methods for the estimation of pure component and mixture properties will be presented. Then, the recently proposed EoS/G^E mixing rules for cubic Equations of State (EoS) will be described in some detail, followed by a discussion of computational aspects in thermodynamic calculations.Three examples showing the capabilities of predictive models in the description of complex mixtures will be given. Finally, together with an overall assessment of existing GC-based models, some guidelines for future research in the area of phase equilibrium thermodynamics are provided.

Phase Equilibria: the Problem and the Solution

The type of problem that phase-equilibrium thermodynamics aims to solve can be formulated as follows: Suppose that two multicomponent phases, α and β, reach an equilibrium state and the temperature T as well as the mole fractions of phase α are given. The problem is then to determine the mole fractions of phase β and the pressure P of the system. Alternatively, the pressure may be known and then the temperature has to be calculated. The number of intensive properties that must be specified to fix the equilibrium state is given by the Gibbs phase rule (Number of independent intensive properties = number of components - number of phases + 2). For example, in the typical case of a two-component two-phase system, the number of intensive properties is two (four properties are involved and two must be specified before the other two are calculated).

The starting point for the solution of every phase-equilibrium problem is provided by classical thermodynamics which requires that when a system containing N components and two phases (α and β) reaches equilibrium (at constant T and P), the total Gibbs free energy assumes its minimum value. It may therefore be stated that the chemical potential or, equivalently, the fugacities of any component i in the two phases must be equal. Thus,

$$f_i^{\alpha} = f_i^{\beta} \qquad i = 1, \ldots, N \tag{1}$$

where f_i^{α} and f_i^{β} are the fugacities of component i in the two phases.

In order to solve Eq. (1), we need to relate the fugacities with the quantities which are directly measurable and computable (temperature, pressure, mole fractions) employing auxiliary functions closer to our physical senses instead of the rather abstract variables such as chemical potential or fugacity. In this section, we will demonstrate how Eq. (1) can be more conveniently transformed to the cases involving multicomponent Vapor-Liquid Equilibrium (VLE), Liquid-Liquid Equilibrium (LLE) and Solid-Liquid Equilibrium (SLE).

Vapor-Liquid Equilibrium

Two main approaches are often used in the solution of Eq. (1):

i. The activity coefficient or γ-ϕ approach (for low pressure calculations).

ii. The Equation of State (EoS) approach (for high pressure calculations).

The γ-ϕ approach: In the γ-ϕ approach which is employed at low pressures, the liquid phase is expressed through the activity coefficient (γ) and the standard state fugacity (f_i^o). It is possible to express the standard state fugacity in terms of the fugacity coefficient at saturation (ϕ^s), the vapor pressure at the system temperature (P^s) and the Poynting effect. Eq. (1) can thus be written as:

$$y_i \Phi_i^V P = x_i \gamma_i P_i^s \Phi_i^s \exp \frac{V_i(P-P_i^s)}{RT} \tag{2}$$

where ϕ_i^V is the fugacity coefficient of component i in the vapor phase, and x_i and y_i are the mole fractions in the liquid and vapor phases, respectively. The exponential term is the Poynting correction, which is often ignored at low to moderate pressures.

The key quantities are the vapor phase fugacity coefficient and the activity coefficient in the liquid phase (especially the latter) and this is why this approach is traditionally called the γ-ϕ or Low-Pressure approach to VLE calculations. γ is usually calculated using a model suitable only for the liquid phase. This can be a correlative model (e.g., UNIQUAC, NRTL) or a predictive model (e.g., UNIFAC, ASOG). ϕ_i^V must be calculated from another model (usually a cubic EoS or the virial equation truncated after the second term) or assumed to be equal to one (often a reasonable approximation at low to moderate pressures, except for strongly associating compounds, like organic acids). Provided that the liquid phase is adequately described, complex multicomponent equilibria can be modelled. However, due to the assumptions involved, the γ-ϕ approach is only applicable to mixtures at relatively low pressures (typically below 10-15 atm.).

The Equation of State approach: At high pressures, the EoS approach is used for VLE computations and Eq. (1) is written as:

$$y_i \Phi_i^V = x_i \Phi_i^L \tag{3}$$

The fugacity coefficients of the liquid and the vapor phases depend on temperature, pressure and composition and using established thermodynamic relationships, they may be calculated from any EoS applicable to both phases. Usually cubic EoS (e.g., SRK, PR) with classical mixing rules and one or two interaction parameters (obtained from fitting experimental VLE data or from generalized correlations) are used for this purpose. Contrary to the γ-ϕ approach, Eq. (3) applies to both low and high pressures but due to the limitations of most classical EoS and their mixing rules, especially in the description of the liquid phase, the EoS approach was (until recently) used only for mixtures with relatively nonpolar components (mainly systems with hydrocarbons and gases). The need for models which would combine the positive features of the γ-ϕ and the EoS approaches has led to the development of improved mixing rules for the cubic EoS often called combined EoS/G^E mixing rules, which are described later.

Liquid-Liquid Equilibrium

When two liquid phases α and β are in equilibrium with each other, Eq. (1) can be written as:

$$(x_i \Phi_i^L)^\alpha = (x_i \Phi_i^L)^\beta \quad i = 1,\ldots,N \tag{4}$$

or alternatively

$$(x_i \gamma_i)^\alpha = (x_i \gamma_i)^\beta \quad i = 1,\ldots,N \tag{5}$$

Equations (4) and (5) hold for every component in the mixture and are solved together with the mass balance equations, i.e., the sum of the mole fractions of all the components in each phase must be equal to unity. Equation (4) is used whenever an EoS suitable for the liquid phase is available. Since this is rarely the case, Eq. (5) is often used in conjunction with an activity coefficient model.

Solid-Liquid Equilibrium

In the case of a binary solvent/solute(solid) system and under certain simplifying assumptions, Eq. (1) yields an expression for the solubility (mole fraction) of the solute in the solvent as a function of γ (the liquid activity coefficient of the solute) and the ratio of the fugacities of pure solid and pure, sub-cooled liquid. The ratio of fugacities can be calculated as a function of several thermal properties of the solute and its melting point. In the case of nonideal solutions, the activity coefficient is usually calculated from the regular solution theory or an activity coefficient model. Cubic EoS have also been used (though not very often) for modelling SLE.

Multiphase Equilibria and Need for Properties

Multiphase equilibria problems such as VLLE can be formulated by solving Eq. (2) or (3) together with Eq. (4) or (5). Cubic EoS require accurate values for the critical temperature, the critical pressure and the acentric factor of each compound involved in the calculations. For many heavy and complex compounds, these properties are not known and/or cannot be measured experimentally and consequently, estimation methods need to be used. Other pure component properties (heat of fusion, solid and liquid heat capacities, solubility parameter, liquid volume, vapor pressure) are needed either directly in the calculations (e.g., in Eq. (2)) or indirectly through the models used (e.g., regular solution theory, free-volume models) and often they have to be estimated. Mixture properties are, of course, also needed. The fugacity coefficient (of each equilibrium phase) and, in particular, the liquid phase activity coefficient are key properties for phase equilibrium calculations.

Pure Component Property Estimation and Data Base

In computerized applications, either the experimental data in the form of property constants or parameters for regressed correlations or, parameters for appropriate estimation methods need to be supplied. Property data bases such as the Design Institute for Physical Property Data (DIPPR) data compilation system (Buck and Daubert, 1990), PPDS-2 (Scott, 1987) and the DORTMUND data-bank (Gmehling, 1985) represent one way of supplying such information. Another option is the use of property data bases in process simulators such as HYSIM

(1992), PRO-II (1991) and ASPEN (1982). Both types of data bases include property constants and, usually, temperature dependent parameters for regressed correlations for a very large number of properties and pure compounds. While in the former type of data bases, emphasis is given on information of data source and quality, in the later type of data bases, emphasis is given on estimation of properties (usually by group contribution methods). Estimation methods are needed as it is not always possible to find reliable experimental data for the required properties for the compounds of interest. Information on the data source and data quality however play an important role in the development of the estimation methods. Therefore, both types of databases have importance.

In the area of estimation methods, a recently developed group contribution approach (Constantinou and Gani, 1994 and Constantinou *et al.*, 1994) for prediction of ten pure component properties deserves a brief description. With this method, the estimation is performed at two levels. In the first level, contributions from first-order groups (for example, groups from UNIFAC) are utilized while in the second more complex level, a small set of second-order groups are employed. The second-order groups use the first order groups as building blocks. Thus, the method provides both a first order approximation (first order group contributions) and a more accurate second-order prediction. Lists of first and second order groups, parameter tables and functions for ten properties (normal boiling point, normal melting point, critical pressure, critical temperature, critical volume, standard enthalpy of vaporization at 298 K, standard enthalpy of formation at 298 K, standard Gibbs energy, liquid molar volume at 298 K and the acentric factor) are given in the above mentioned publications. Several important features need to be noted. Significant improvement in the prediction accuracy has been achieved, it is possible to distinguish between some isomers, extrapolation is very reliable and finally, all the properties are estimated only from the structural information of the pure compound. Thus, those properties which are only dependent on the structural variables are called primary properties as opposed to secondary properties, which are dependent on other primary or secondary properties. The following simple functional expression is used for all ten properties:

$$f(x) = \sum_i N_i C_i + W \sum_j M_j D_j \tag{6}$$

where C_i and D_j are the first and second order group contributions respectively and N_i, M_j are the number of their occurrences in the compound. The constant W is set equal to unity for second-level estimation.

Models for Predicting Mixture Properties with Emphasis on the GC Approach

In this section, UNIFAC (Fredenslund *et al.*, 1977) and related methods for predicting a variety of mixture properties are reviewed followed by a brief discussion on the achievements and the shortcomings of cubic EoS in calculating phase equilibria. Special emphasis is given to the recently developed combined EoS/G^E mixing rules. Finally, one of the most promising EoS/G^E models, the MHV2 model, is described and evaluated.

UNIFAC and Related Models

UNIFAC has been developed by Fredenslund *et al.* (1977) and is a GC activity-coefficient model based on UNIQUAC. It contains two contributions to the activity coefficient, a combinatorial (*comb*) and a residual (*res*) term:

$$\ln \gamma_i = \ln \gamma_i^{comb} + \ln \gamma_i^{res} \qquad (7)$$

The combinatorial term, which is a function of the van der Waals volume (R), surface area (Q) and composition, takes into account the entropy effects arising from differences in size and shape between the components of the mixture. The R and Q values are estimated using the group-increments given by Bondi, and are readily available in the UNIFAC tables. The residual term is a function of the surface area, the composition, the temperature and the group-energy parameters (a_{mn}). It accounts for the energetic interactions between the functional groups in the mixture. The group parameters of the residual term are usually estimated from experimental low-pressure phase equilibrium (normally VLE) data. Occasionally, other types of experimental data are used in the parameter estimation (e.g., excess enthalpies h^E and heat capacities, infinite dilution activity coefficients, LLE, etc.). Three major UNIFAC models can be identified:

> Original UNIFAC: In the original UNIFAC, γ^{comb} was given by the Staverman-Guggenheim (SG) term, while the group-parameters were assumed to be temperature independent. For this reason, original UNIFAC, although successful for VLE calculations, does not yield quantitative predictions for excess enthalpies and extrapolations above 425 K should be avoided. The latest revision of this model is given by Hansen et al. (1991).
>
> Modified UNIFAC: In the modified UNIFAC model developed in Lyngby by Larsen *et al.* (1987), the empirical combinatorial formula originally suggested by Kikic *et al.* (1980) was adopted. This new combinatorial term performs better than the SG term for alkane mixtures. The residual term is the same as in original UNIFAC, but the group interaction parameters are assumed to be temperature dependent through a logarithmic temperature dependency. Both VLE and h^E data have been used in the parameter estimation.

Modified UNIFAC only offers a marginal improvement over original UNIFAC regarding VLE predictions. However, due to the temperature dependent parameters employed, the prediction of h^E, and hence, the temperature dependence of the activity coefficients is much improved. Thus, unlike the case with original UNIFAC, safe (large) temperature extrapolations (often up to 550-600 K) are possible with modified UNIFAC. An alternative similar modified UNIFAC model has been developed by the Dortmund group (Gmehling, 1986). VLE, h^E and infinite dilution activity coefficients have been used successfully in the parameter estimation, thus making the Dortmund version of modified UNIFAC a very attractive model (particularly suitable for infinite dilution activity coefficient calculations).

> New UNIFAC: This is the latest revision of the UNIFAC model, developed in Lyngby by Hansen *et al.* (1992). It has been found that the Kikic combinatorial formula employed in modified UNIFAC does not extrapolate well to mixtures with polyethylene, and hence it should not be used for mixtures with polymers. On the other hand, the original SG term provides a much better representation of athermal polymer solutions and is, therefore, retained in the new UNIFAC model. The residual term is that of UNIFAC using linear temperature-dependent parameters fitted exclusively to VLE data. The logarithmic term of the temperature dependence is dropped in the new model both because it is not

needed if only VLE data are to be fitted and because it leads, in some cases, to uncertain extrapolations. The performance of new UNIFAC is at least as accurate as the previous versions of UNIFAC for VLE calculations (and often better), while it retains the correct limit for polymer systems. Furthermore, it has been used as the basis for the extension of the UNIFAC model to polymers.

Application of UNIFAC to LLE/SLE Calculations

The modified UNIFAC models, with temperature dependent parameters, predict qualitatively well UCST and closed-loop phase envelopes for some systems over an extended temperature range. However, it has been found that, in general, UNIFAC with VLE-based parameters do not yield quantitative prediction of LLE. For this reason, Magnussen *et al.* (1981) developed a UNIFAC-LLE parameter table (parameters based exclusively on LLE data), which yield quantitative estimations of LLE phase compositions around 25 °C.

Gmehling *et al.* (1978) used UNIFAC for predicting SLE for some solids (e.g., phenanthrene, anthracene) in a variety of single and mixed solvents (alcohols, acids, hydrocarbons, etc.). Eutectic temperatures and compositions were also calculated for some binary systems (benzene/phenol, benzene/acetic acid). Good agreement with experiment is found. In all cases the predictions were much better than when the ideal solution ($\gamma=1$) assumption is used.

Use of γ in Transport Properties

Under certain assumptions, the surface tension of a liquid mixture (σ) is related to the surface tension of a pure component i (σ_i), the molecular surface area A_i of a pure component i, the compositions in the surface (s) and the bulk (b) liquid phases and the corresponding activity coefficients, $\gamma_{i,s}$ and $\gamma_{i,b}$. Suarez *et al.* (1988) used the modified UNIFAC model of Larsen *et al.* to predict $\gamma_{i,s}$ and $\gamma_{i,b}$, having determined σ_i and A_i from selected experimental surface tension data for binary mixtures. They obtained excellent agreement between experimental and calculated surface tensions.

Wu (1986) developed a widely applicable GC viscosity model by combining Eyring's theory with UNIFAC, but the group interaction parameters were obtained from experimental viscosity data. Recently Cao *et al.* (1993) developed a GC viscosity model with parameters taken directly from original UNIFAC-VLE (Hansen *et al.*, 1991). The viscosity predictions for liquid mixtures can be thus carried out without any viscosity information for mixtures and very good agreement is obtained with experimental data. The average deviation is only 4.1% compared to about 13% with the method of Wu.

The GC Approach for Polymers and Electrolytes

Many GC models, both EoS and γ-models, have been proposed in the last ten years for modelling polymer solutions. Two EoS models are hereafter briefly described.

1. The GC version of the Flory EoS (GC-Flory) originally developed by Holten-Andersen *et al.* and recently revised by Bogdanic and Fredenslund (1994).
2. The Group-Contribution Lattice Fluid (GCLF) EoS proposed by High

and Danner (1990), which is a GC version of the Lattice-EoS proposed earlier by Panayiotou and Vera. GCLF was recently revised (Danner, 1994) to include a GC-based interaction parameter. Both EoS are predictive models and they successfully represent polymer/solvent VLE, but their applicability is somewhat limited by their rather small (compared to UNIFAC) parameter table. Using the same VLE-based parameter table, GC-Flory has been recently shown to predict qualitatively well LLE for binary polymer solutions, including Upper and Lower Critical Solution (UCST, LCST) behavior. In several cases, especially for polar solutions (e.g., octanol/polyethylene), the predictions are quantitatively adequate for design purposes.

Original UNIFAC alone usually underestimates the activity coefficients in polymer solutions, because it ignores the significant free-volume differences between solvents and polymers, which often exist in polymer solutions. Three attempts to remedy this deficiency have been proposed. Oishi and Prausnitz (O-P, 1978) and Iwai and Arai (1989) added a Free-Volume (FV) term in the original UNIFAC model. Kontogeorgis *et al.* (1993) developed a simple γ-model (Entropic-FV) using the residual term of new UNIFAC (with linear temperature dependent parameters) and a combined comb/FV term recently proposed by Elbro *et al.* (1990). All three models make use of existing UNIFAC tables but require the component volumes at the system temperature. When experimental data are not available, both the solvent and polymer densities can be predicted with very good accuracy (1% for most solvents below the normal boiling point and 2.7% for amorphous polymers) using a recently developed GC method for the volume (Elbro *et al.*, 1991).

The O-P UNIFAC-FV and Entropic-FV models have been extensively tested and shown to provide very good predictions for VLE and infinite dilution activity coefficients for a great variety of homopolymer and co-polymer solutions. Entropic-FV is simpler than the O-P model and, unlike the latter, it can be easily extended to multicomponent solutions. Both UNIFAC and Entropic-FV models predict UCST behavior qualitatively well using the VLE-based UNIFAC parameter table. In addition, Entropic-FV is capable of predicting qualitatively well the FV-driven LCST behavior, as well as combined UCST and LCST phase diagrams. From a quantitative point of view, the LLE predictions with Entropic-FV are often rather good, particularly for relatively nonpolar solutions (e.g., polystyrene/cyclohexane).

Simple activity coefficient models have been proposed for predicting the activity coefficient of solvent in liquid mixtures containing ions. Two models have received much attention during the last years: the Chen *et al.* (1982) model and the model proposed by Sander *et al.* (1986). The latter has been simplified recently by Nikolaisen *et al.* (1993). Both models can treat aqueous as well as mixed solvent electrolyte solutions; they use a Debye-Hückel expression to describe the long-range ion-ion interactions. In the Chen model, short-range interactions are described with the NRTL equation, while in Sander's model an extended UNIQUAC equation (with concentration-dependent parameters) is used for the same purpose. In the modification of Sander's model by Nikolaisen *et al.*, only binary (linear temperature dependent) concentration independent parameters are used. Chen and Sander models give good representation of VLE for electrolyte solutions. In particular, the modification by Nikolaisen *et al.* has been found to give good predictions of mineral solubilities in ternary and multicomponent systems.

Use of Cubic EoS for Phase Equilibrium Calculations

Cubic EoS are often used for phase equilibrium calculations for nonpolar (hydrocarbon) or slightly polar (gas/hydrocarbon) systems. Usually the classical van der Waals one-fluid (vdW1f) mixing rules are used together with the geometric mean rule for the cross energy parameter and the arithmetic mean rule for the cross co-volume parameter. In most cases (especially when asymmetric systems are involved, e.g., methane/nC_{36} or ethane/nC_{44}) one or two interaction parameters (in the cross energy terms) are needed. Several generalized correlations have been reported for specific cases (e.g., the correlation by Kordas *et al.* (1994) covering the whole CO_2/hydrocarbon series). The applicability of cubic EoS to complex equilibria (e.g., for water/hydrocarbons) present many difficulties, even if adjustable parameters are used. In these cases, more sophisticated concentration-dependent or density-dependent local composition mixing rules are used, but at the cost of extra complexity in equilibrium calculations and significant increase in computing time. An interesting review on these rules is given by Heidemann and Fredenslund (1989). The need for complex mixing and combining rules stems from the recognition that many of the limitations of cubic EoS are due to the use of vdW1f mixing rules for the EoS parameters, especially for the mixture energy parameter.

In the last years the EoS/G^E approach has become very popular, since it incorporates G^E models (e.g., UNIFAC, NRTL) in the mixing rule for the energy term of cubic EoS in a way which appears to be both computationally efficient and relatively fast and successful in the description of complex equilibria. The EoS/G^E models which use Group-Contribution (GC) G^E models are particularly worth mentioning, since the resulting EoS/G^E models are purely predictive tools for phase equilibrium calculations. The basic principles of the EoS/G^E models and one of the most successful such models (MHV2) are hereafter described.

The EoS/G^E Mixing Rules for Cubic EoS

The basic principle (or starting point) in the development of most EoS/G^E models is to demand that the excess Gibbs energy from a cubic EoS be equal to the excess Gibbs energy from an activity coefficient model (e.g., UNIFAC, Wilson, etc.), i.e.,:

$$\left[\frac{G^E}{RT}\right]_P^{EoS} = \left[\frac{G^E}{RT}\right]_P^{*} \tag{8}$$

where the superscript * refers to the activity coefficient model used. The subscript P denotes that the equality of equation (8) is valid at a certain pressure, which is called "reference pressure". Vidal (1978) developed an EoS/G^E model assuming infinite reference pressure. This match, however, renders it impossible to use G^E-parameters based on low-pressure VLE data, like those in the GC tables of UNIFAC or ASOG or the NRTL (and other liquid solution models) parameters published by Dechema.

The first combination which led to simple and purely predictive EoS/G^E models was the MHV1 and MHV2 mixing rules (Modified Huron-Vidal first and second order; Michelsen, 1990). In these mixing rules, it is assumed that equation (8) is valid at zero pressurizes making it possible to use the existing UNIFAC (and related) tables. In the MHV2 model, as usually used, the SRK EoS is combined with the modified UNIFAC model (Larsen *et al.*, 1987), but the approach is readily applied to any EoS and G^E model.

An EoS/G^E model without specific reference pressure (LCVM model; Boukouvalas *et al.*, 1994), particularly successful for phase equilibrium calculations for asymmetric mixtures (i.e., those with components differing significantly in size), has been recently developed.

A theoretical limitation of the aforementioned models is that they do not obey the quadratic composition dependence of the mixture second virial coefficient, required by statistical mechanics. Although this limitation seems not to have serious impact on the accuracy of equilibrium calculations, Wong and Sandler (1992) recently proposed an alternative combination approach (where equation (8) is valid for the excess Helmholtz free energy instead of G^E) at infinite pressure, which provides the theoretically correct composition dependence for the mixture second virial coefficient.

The MHV2 Model

Let us assume the SRK EoS. Taking the limit of EoS at zero pressure in combination with the basic equation (8), the following mixing rule can be easily derived for the reduced dimensionless energy parameter ($=a/bRT$):

$$q^{ex}(\alpha) = \left[\frac{G^E}{RT}\right]^* + \sum_i x_i q^{ex}(\alpha_i) + \sum_i x_i \ln\left(\frac{b}{b_i}\right) \tag{9}$$

where $q^{ex}(\alpha)$ is the function:

$$\ln(fo/RT) + \ln b = -1 - \ln(u_o - 1) - a\ln\left(\frac{u_o + 1}{u_o}\right) = q^{ex}(a) \tag{10}$$

where f_o is the fugacity at zero pressure. u_o ($=V_o/b$) is the reduced volume at zero pressure, obtained by solving the SRK EoS. It is defined only when $\alpha > \alpha_{lim} = 3+2\sqrt{2}$. Equation (9) is the basic equation for the exact modified Huron-Vidal mixing rule, as originally proposed by Michelsen (1990). In principle, any mixing rule can be used for b, although usually the linear rule is used. On the right-hand side of equation (9), UNIFAC or any other G^E model can be used. Equation (9) ensures that the cubic EoS reproduces exactly the G^E model it is combined with. Equation (9) has to be solved iteratively for the energy parameter. An additional problem of the exact model is that it is not valid for $\alpha < \alpha_{lim}$. Such low α values correspond to systems where the reduced temperature of one of the components is higher than approximately 0.9, for example, mixtures with a gas and a heavy hydrocarbon, where the gas is in supercritical condition. These problems of the exact model find a solution with Michelsen's observation that $q(\alpha)$ is almost a linear function of α, i.e., $q(\alpha) \approx q_o + q_1\alpha$. When this relation is inserted in equation (9), an explicit mixing rule for α, the so-called MHV1 mixing rule is obtained.

A much better reproduction of the G^E model is obtained with a quadratic $q(\alpha)$ function, i.e., $q(\alpha) \approx q_o + q_1\alpha + q_2\alpha^2$. The resulting mixing rule is called MHV2 (and is no more explicit with respect to α):

$$q_1\left(\alpha - \sum_i x_i\alpha_i\right) + q_2\left(\alpha^2 - \sum_i x_i\alpha_i^2\right) = \left(\frac{G^E}{RT}\right)^* + \sum_i x_i \ln\left(\frac{b}{b_i}\right) \tag{11}$$

q_1 and q_2 are constants depending on the EoS used and they are fitted in the interval $10 < \alpha < 13$.

Using equation (11), MHV2 model can also be used for systems with supercritical components. In order to do so in a predictive way, Dahl *et al.* (1991) added 13 new groups involving gases in the modified UNIFAC parameter table. The basic results with MHV2 can be summarized as follows:

1. Low pressure VLE is predicted as accurately as using UNIFAC and the γ-ϕ approach.
2. VLE at high pressures and temperatures is also predicted well, even for binary and ternary systems involving highly polar compounds (Dahl and Michelsen, 1990). For example, the following overall errors in Bubble Point Pressure calculations have been reported: acetone/water 0.8% up to 60 bar, water/ethanol 2.3% up to 186 bar, methanol/benzene 2.7% up to 58 bar, acetone/methanol/water 8% up to 82 bar.
3. Binary and ternary gas solubilities at high temperatures and pressures are predicted very well, even for systems containing methanol, water and gases. In particular, MHV2 predicts reasonably well the three phase equilibria for carbon dioxide/ethanol/water and water/2-butanol/butane, systems which are important in process design and for supercritical extraction processes. Furthermore, solubilities in multicomponent systems are predicted (based only on binary data) better than when more complex density-dependent mixing rules are used. This makes the EoS/G^E approach a very attractive and predictive alternative to these complex mixing rules.
4. The one limitation of MHV2 is that it cannot describe accurately phase equilibria for asymmetric systems, e.g., carbon dioxide or ethane with heavy hydrocarbons and carbon dioxide with fatty acids and esters. This deficiency may be partly explained by the fact that, in these cases, the quadratic $q(\alpha)$ function employed in MHV2 (or any other approximate $q(\alpha)$ function) cannot guarantee full reproduction of the G^E model implemented. Note that the Wong-Sandler mixing rule also fails to reproduce the G^E model used and, thus, it yields unsatisfactory results for asymmetric systems, like 1-hexanol/water (Coutsikos *et al.*, 1994). A quadratic composition dependence in the mixture co-volume parameter in conjunction with the MHV1 mixing rule (Coniglio *et al.*, 1994) or a different EoS/G^E approach, like the empirical LCVM model (Boukouvalas *et al.*, 1994), have been shown to represent successfully asymmetric systems, preserving at the same time the EoS/G^E solution to this problem. In particular, the LCVM model yields very good predictions for the activity coefficients in asymmetric alkane systems and reasonable estimates for the solubilities of solid aromatic hydrocarbons in supercritical carbon dioxide.

Computational Aspects and Examples of Application

Among the many different models and methods applicable to a particular thermodynamic property, it can be argued that some are more suitable than others (O'Connell, 1983). Even the most popular models have differing levels of accuracy, reliability, computational efficiency and parameter development. Also, purposes for which property values are used can vary.

Therefore, in this section, some aspects related to the appropriate selection of the thermodynamic model for a specific problem, the correct and efficient utilization of the selected model and finally, the validity of the model parameters are discussed.

Inaccuracies in design calculations can often be traced to inaccurate prediction of properties, which in turn, can be traced (in some cases) to the wrong selection of thermodynamic models. For example, the UNIFAC-LLE model should not be used for VLE calculations even though this model has the available group interaction parameters while the WILSON model cannot be used for liquid-liquid phase split computations. Also, the range of validity of the UNIFAC-LLE model parameters is from 282 K to 313 K. Cox (1993) has pointed out some of the dangers of using models without considering the implications (for example, some of the methods for predicting the behavior of a surfactant molecule will yield incorrect predictions if water is present). This indicates that the mixture (or species) for which the property is to be estimated, needs to be carefully analysed. From the models presented in the above sections, it is obvious that aqueous mixtures of alcohols, polymers and salts need different models for the estimation of the same properties. Also, the sensitivity of the model parameters may vary for the same property and mixture (or species). The solution of Eq. 3 may become very sensitive to numerical errors near the critical point.

It is also necessary to be aware of the range of validity of the model parameters and how the model parameters were determined. The extrapolation capability of the model needs to be carefully evaluated for untried systems and conditions outside the range of the experimental data (used for parameter estimation). If the confidence region of the parameters is very small, even interpolation of the model parameters can be risky.

Finally, the problem of unavailable parameters needs to be discussed. In many cases, the experimental phase equilibrium data are available at conditions which do not coincide with the conditions maintained in plant operation. Also, in many cases, a subset of the model parameters may not be available. Can the existing model parameters be used or should new parameters be estimated? If experimental data are not available, how can the new parameters be estimated? What happens if the mixture-related parameters are available but some pure component parameters are not available? Some questions users may need to ask are: How should the pure component property parameters be estimated? Using the mixture properties? General answers to these question unfortunately do not exist because different models behave differently even for the same property prediction problem (but for different mixtures). Thus, it is important to analyze the model parameter information and study the sensitivity of the model parameters before attempting to find a method of solution. Thermodynamic consistency should also be considered during the validation of computed property values.

Three application examples using some of the models described in the section on mixture properties are presented in the form of computed phase diagrams, since they play an important role in the design and synthesis of chemical and biochemical processes. The first example (Fig.1) concerns the prediction of UCST, LCST and hourglass phase diagrams with the Entropic-FV model for the system polystyrene/acetone. It can be seen that Entropic-FV predicts qualitatively well UCST and hourglass curves and quantitatively well LCST phase envelopes.

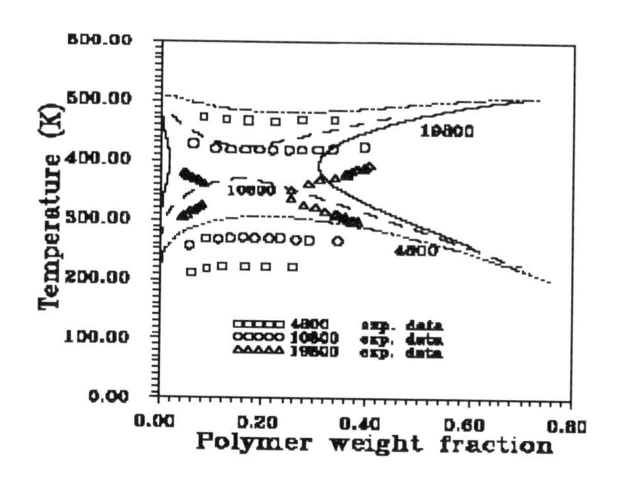

Figure 1. Experimental and predicted LLE phase diagram for polystyrene/acetone (at three polymer molecular weights) with Entropic-FV. The points are the experimental data and the lines are the model's predictions.

The second example (Fig.2) concerns the application of the aqueous electrolyte model by Nicolaisen *et al.* for the determination of the phase diagram (salt precipitation curves) for the system SO_4^{2-}, Cl^-, Na^+, K^+/water. The model yields satisfactory predictions of multicomponent equilibria.

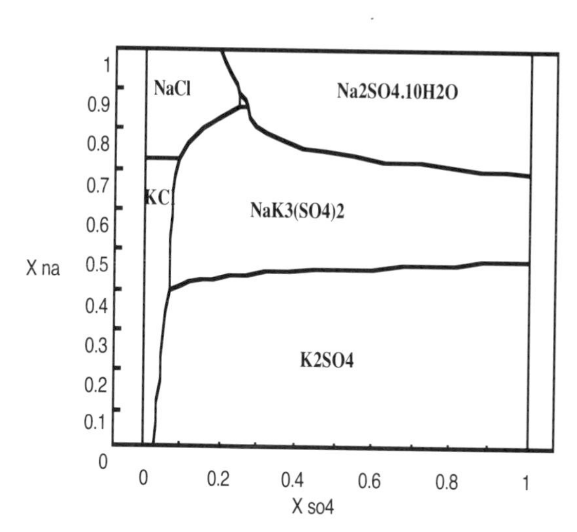

Figure 2. Quaternary SLE diagram on a water-free basis for the system SO_4^{2-}, Cl^-, Na^+, K^+ /water at 289.09 K calculated with the model of Nicolaisen et al. (The experimental data, not shown in the figure, almost coincide with model's predictions).

The third example (Fig. 3) concerns the use of the MHV2 model for the prediction of VLLE for the system propane-ethanol-water at P = 35 atm. It can be seen that the ethanol-water azeotrope has disappeared at this pressure. Also, that the vapor phase is very close to the liquid-liquid phase boundary.

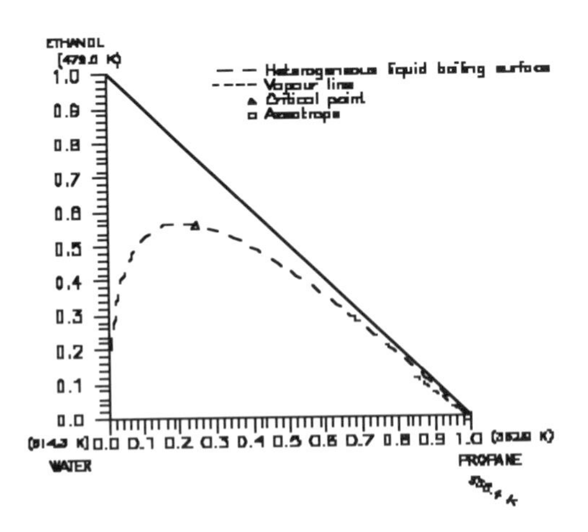

Figure 3. Prediction of VLLE for the ternary system propane-ethanol-water at P=35 atm. using the MHV2 model.

Conclusions and Future Directions

The need for reliable data and predictive methods for the estimation of pure component and mixture properties for phase equilibrium calculations is illustrated. Several models for solving the phase equilibrium equations have been briefly presented. Special attention is given to the GC approach and related models. In particular, the various versions of UNIFAC and the MHV2 mixing rule are discussed in detail and both their achievements and limitations are clearly stated.

It can be concluded that:

1. It seems difficult to use a single model for solving many different phase equilibrium problems and for a variety of mixtures.
2. The GC approach is a powerful technique for preliminary design calculations, but accurate or reliable predictions cannot be always obtained, while there are several additional problems (e.g., LLE calculations at high temperatures, mixtures with isomers).
3. Difficult systems (e.g., mixtures with polymers, electrolyte solutions, high pressure complex equilibria) can be described reasonably well with relatively simple models (similar to UNIQUAC/UNIFAC), often based on the principles of the GC approach.
4. The recent EoS/G^E mixing rules deserve special attention. These mixing rules allow cubic EoS to be used for describing highly non-ideal sys-

tems, or equivalently extend the applicability of classical low pressure activity coefficient models like UNIFAC to high pressure applications including gas solubilities. Future work on these mixing rules is necessary and should focus on improved understanding of their capabilities as well as to extend their applicability to difficult systems (e.g., mixtures with associating fluids, electrolytes and polymers, including polymer blends, high pressure LLE, solid/supercritical fluid equilibria).

Future work should further focus on solving problems related to the biotechnology and food industries, environmental applications, new materials (e.g., co-polymer blends), as well as quantitative representation of condensed phase equilibria (e.g., LLE, solid solutions). In those cases where experimental data (needed in the design of processes and models) are not available, Molecular Simulation (MS) techniques might be particularly useful in the future. MS may provide guidelines for the development of better models or help in the estimation of parameters for equilibrium models using only pure component data (e.g., see Jonsdottir *et al.*, 1994). MS will also help in the investigation of the extrapolation performance of existing models in areas where experimental measurements are scare, very difficult or expensive (e.g., critical properties of heavy compounds) or even it may help in explaining discrepancies between different sets of experimental data (e.g., see the work of Siepmann *et al.*, 1993 on the critical density of heavy alkanes).

Finally, a note regarding the estimation of pure component properties, which are often of great importance for phase equilibrium calculations and product design, needs to be made. For many compounds related to industrially important applications these properties are not available experimentally and must be estimated, usually by GC methods. Novel estimation methods should focus not only on the description of classical critical properties, but also on temperature dependent properties (e.g., vapor pressures, second virial coefficient) and, if possible, their applicability should be extended in the estimation of a number of important non-equilibrium properties (e.g., surface tension, viscosity, diffusion coefficient, glass transition temperature). Finally, it will be important to see if these novel GC methods will cover all the molecular weight range for which experimental data are available, i.e., from low molecular weight solvents up to oligomers and polymers.

References

Aspen Plus (1982). Technical reference manual (Aspen Technology Inc, Cambridge, Massachusetts, USA).

Bogdanic, G. and Aa. Fredenslund (1994). Revision of the Group-Contribution Flory Equation of state for phase equilibria calculations in mixtures with polymers. 1. Prediction of Vapor-Liquid Equilibria for polymer solutions. *Ind. Eng. Chem. Res.*, **33**(5), 1331-1340.

Boukouvalas, C., N. Spiliotis, P. Coutsikos, N. Tzouvaras, and D.Tassios (1994). Prediction of vapor-liquid equilibrium with the LCVM model: a linear combination of the Vidal and Michelsen mixing rules coupled with the original UNIFAC and the t-mPR equation of state. *Fluid Phase Equilibria*, **92**, 75-106.

Buck, E., and T. E. Daubert (1990). Project 801: The DIPPR data compilation project In T. B. Selover (Ed.), *Design institute for physical property Data: Ten years of accomplishment, AIChE Sympo. series,* 275, **86**, 1-4.

Cox, K. R. (1993). Physical property needs in industry. *Fluid Phase Equilibria,* **82**, 15-26.

Cao, W., K. Knudsen, Aa. Fredenslund, and P. Rasmussen (1993). Group-Contribution viscosity predictions of liquid mixtures using UNIFAC-VLE parameters. *Ind. Eng. Chem. Res.*, **32**(9), 2088-2092.

Chen, C.C., H.I.Britt, J.F.Boston, and L.B. Evans (1982). Local composition model for excess Gibbs energy of electrolyte systems. *AIChE J.*, **28**(4), 588-596.

Coniglio, L., K. Knudsen and R. Gani (1994). Model prediction of supercritical fluid-liquid equilibria for carbon dioxide and fish oil related compounds. Submitted to *Ind. Eng. Chem. Res.*

Constantinou, L, and R. Gani (1994). A new group-contribution method for the estimation of properties of pure compounds. Accepted for publication in *AIChE J.*

Constantinou, L, R. Gani, and J. P. O'Connell (1994). Estimation of the acentric factor and the liquid molar volume at 298K through a new group contribution method. Submitted to *Fluid Phase Equilibria*.

Coutsikos, P., N.S. Kalospiros, and D. Tassios (1994). Capabilities and limitations of the Wong-Sandler mixing rules. Submitted for publication.

Dahl, S., Aa. Fredenslund, and P. Rasmussen (1991). The MHV2 model: a UNIFAC-based equation of state model for prediction of gas solubility and vapor-liquid equilibria at low and high pressures. *Ind. Eng. Chem. Res.*, **30**, 1936-1945.

Dahl, S., and M.L. Michelsen (1990). High-pressure vapor-liquid equilibrium with a UNIFAC-based equation of state. *AIChE J.*, **36**(12), 1829-1836.

Danner, R. (1994). Paper presented at the 14th European Seminar on Applied Thermodynamics, Athens, Greece (June 9-12).

Elbro, H.S., Fredenslund, Aa., and Rasmussen, P. (1990). A new simple equation for the prediction of solvent activities in polymer solutions. *Macromolecules*, **23**(1), 4707-4714.

Elbro, H.S., Aa. Fredenslund, and P. Rasmussen (1991). Group contribution method for the prediction of liquid densities as a function of temperature for solvents, oligomers and polymers. *Ind. Eng. Chem. Res.*, **30**(12), 2576-2582.

Fredenslund, Aa., J. Gmehling, and P. Rasmussen (1977). Elsevier (Ed.), *Vapor-Liquid Equilibria using UNIFAC*, Amsterdam pp. 1-380.

Gmehling, J. (1985). Dortmund Data Bank-Basis for the development of prediction methods, CODATA Bulletin 58.

Gmehling, J. (1986). Group contribution methods for the estimation of activity coefficients. *Fluid Phase Equilibria*, **30**, 119-134.

Gmehling, J., T. Anderson, and J.M. Prausnitz (1978). Solid-Liquid Equilibria using UNIFAC. *Ind. Eng. Chem. Fundam.*, **17**(4), 269-273.

Hansen, H.K., P. Rasmussen, Aa. Fredenslund, M. Schiller, and J. Gmehling (1991). Vapor-Liquid Equilibria by UNIFAC Group-Contribution. 5. Revision and Extension. *Ind. Eng. Chem. Res.*, **30** (10), 2352-2355.

Hansen, H.K., B. Coto, and B. Kuhlmann (1992). UNIFAC with linearly temperature dependent group-interaction parameters. *Internal publication of IVC-SEP center No. 9212*, Institut for Kemiteknik, The Technical University of Denmark.

Heidemann, R.A., and Aa. Fredenslund (1989). Vapor-liquid equilibria in complex mixtures. *Chem. Eng. Res. Des.*, **67**, 145-158.

High, M.S., and R.P. Danner (1990). Application of the group contribution lattice-fluid EoS to polymer solutions. *AIChE J.*, **36**(11), 1625-1632.

Hysim Manual (1992), Hyprotech Ltd., Canada.

Iwai, Y. and Arai, Y. (1989). Measurement and prediction of Solubilities of hydrocarbon vapors in molten polymers, *J. Chem. Eng. Japan*, **22**(2), 155-161.

Jonsdottir, SO., K. Reassumes, and AA. Fredenslund (1994). UNIQUAC-parameters determined by Molecular Mechanics. Accepted for publication by *Fluid Phase Equilibria*.

Kontogeorgis, G.M., Aa. Fredenslund, and D. Tassios (1993). Simple activity coefficient model for the prediction of solvent activities in polymer solutions. *Ind. Eng. Chem. Res.*, **32**, 362-372.

Kordas, A., K. Tsoutsouras, S. Stamataki, and D. Tassios (1994). A generalized correlation for the interaction coefficients of CO_2-hydrocarbon binary mixtures. *Fluid Phase Equilibria*, **93**, 141-166.

Kikic, I., Alessi, P., Fredenslund, Aa., and P. Rasmussen (1980). On the combinatorial part of the UNIFAC and UNIQUACK models, *Can. J. Chem. Eng.*, **58**, 253-258.

Larsen, B.L., P. Rasmussen, and Aa. Fredenslund (1987). A modified UNIFAC group-contribution model for prediction of phase equilibria and heats of mixing. *Ind. Eng. Chem. Re.*, **26**, 2274-2286.

Magnussen, T., P. Rasumssen, and Aa. Fredenslund (1981). UNIFAC Parameter table for prediction of liquid-liquid equilibria. *Ind. Eng. Chem. Proc. Des. Dev.*, **20**, 331-339.

Michelsen, M.L. (1990). A modified Huron-Vidal mixing rule for cubic equations of state. *Fluid Phase Equilibria*, **60**, 213-219.

Nicolaisen, H., P. Rasmussen, and J. M. Sørensen (1993). Correlation and prediction of mineral solubilities in the reprocal salt system (Na^+,K^+)(Cl^-,SO_{42-})-H_2O at 0-100 °C. *Chem. Eng. Sci.*, **48**(18), 3149-3158.

O'Connell, J.P. (1983). Structure of thermodynamics in process calculations. Proc. 2nd Int. Conf. FOCAPD, CACHE, Snowmass, CO, June 19-24.

Oishi, T. and J.M. Prausnitz (1978). Estimation of solvent activities in polymer solutions using a group-contribution method. *Ind. Eng. Chem. Process Des. Dev.*, **17**, 333-339.

Pro/II Manual (1991), Simulation Sciences Inc., USA, version 3.02.

Sander, B., Aa. Fredenslund, and P. Rasmussen (1986). Calculation of vapor-liquid equilibria in mixed solvent/salt systems using an extended UNIQUAC equation. *Chem. Eng. Sci.*, **41**(5), 1171-1183.

Scott, A. C. (1987). PPDS, the U.K. Physical Properties Data Service for the chemical and process industries. In R. Jowitt (Ed.), *Physical properties of foods*, Vol. 2, Elsevier, London, pp. 503-521.

Siepmann, J.I., S. Karaborni, and B. Smit (1993). Simulating the critical behavior of complex fluids. *Nature*, **365**, 330-332.

Suarez, J.T., C.Torres-Marchal, and P. Rasmussen (1989). Prediction of surface tensions of nonelectrolyte solutions. *Chem. Eng. Sci.*, **44**(3), 782-786.

Vidal, J. (1978). Mixing rules and excess properties in cubic equations of state. *Chem. Eng. Sci.*, **33**, 787-791.

Wu, D.T. (1986). Prediction of viscosities of liquid mixtures by a group contribution method. *Fluid Phase Equilibria*, **30**, 149-156.

Zeck S., and D. Wolf (1993). Requirements of thermodynamic data in the chemical industry. *Fluid Phase Equilibria,* **82**, 27-38.

CACHE'S ROLE IN COMPUTING IN CHEMICAL REACTION ENGINEERING

H. Scott Fogler
University Of Michigan
Ann Arbor, Michigan 48109

Abstract

Throughout the last 25 years CACHE has either initiated, nurtured, or financially supported projects to enhance courses in chemical reaction engineering and, as a result, has been instrumental in the evolution of chemical reaction engineering education. Five major CACHE-assisted projects involving computer modules, and/or interactive simulations are discussed. Specifically, we focus on early computer programs, the University of Michigan Projects, the Purdue University Modules, POLYMATH, and the University of Washington's Chemical Reactor Design Tool.

Introduction

CACHE has played many roles in the incorporation of computing in Chemical Reaction Engineering (CRE). It has been an initiator of projects, such as the computer programs in chemical reaction engineering (1972), and the set of six CHEMI books on Kinetics (1976-1979), it has been a distributor of programs such as POLYMATH and the Michigan Interactive Kinetics modules. CACHE has also been a mentor and facilitator in the development of these materials, providing both financial, and intellectual resources through collaboration, not only with CACHE trustees, but also with many colleagues in the United States, Canada, and overseas. This paper highlights five areas in which CACHE has played a major role in the advancement of chemical reaction engineering education: the computer programs, the University of Michigan's Interactive Kinetics modules, the Purdue University's Simulations, the University of Washington's Chemical Reactor Design Tool, and POLYMATH.

Early Computer Programs

One of the very first CACHE projects involving the development of computer programs in the areas of chemical reactor design, control design, transport, stoichiometry, and thermodynamics. The programs were written by college faculty and sent on punched cards for evaluation to the appropriate area editor. Professor Matt Reilly was the area editor for kinetics and reactor design, and set the standards for accepting programs by running them in every way possible to try to get them to fail. If a program passed Matt's rigorous testing, a source program listing and sample output were included in the published volume (Reilly, 1972). A list of the programs and authors is shown in Table 1.

Table 1. CACHE Computer Programs for Chemical Engineering Education: Kinetics (1972)

1. BATCH REACTORS
   - 1.1 Complex Reaction System in a Batch Reactor by Jaime P. Ampaya and Robert G. Rinker
   - 1.2 Batch Decomposition of Acetylated Castor Oil by Joseph J. Perona
   - 1.3 Series of Reactions in a Batch Reactor or in a Sequence of Stirred Tank Reactors by R. Rajagopalan
2. CONTINUOUS STIRRED-TANK REACTORS
   - 2.1 Molecular Weight Distribution in a Polymerization Reactor Sequence by James L. Kuester
   - 2.2 Dynamic Mass and Energy Balances in a CSTR by Ephraim Kehat
3. HOMOGENEOUS TUBULAR REACTORS
   - 3.1 Tubular Reactor Design with Two Consecutive Reactions by E. H. Crum
   - 3.2 Chemical Equilibrium and Reaction Rate Conflict by Billy C. Crynes and Barney L. Ghiglieri
   - 3.2 Plug Flow Reactor with Dispersion by George M. Homsy
4. HETEROGENEOUS CATALYSIS
   - 4.1 Effectiveness Factor for a Spherical Catalyst Pellet by Joseph S. Naworski
   - 4.2 Non-Isothermal Catalyst Pellet by Ran Abed and Robert G. Rinker
   - 4.3 Computer-Administered Programmed Instruction in Effectiveness Factors by James T. Cobb, Jr., and Thomas Gestrich
5. CATALYTIC REACTORS
   - 5.1 Analysis of Catalytic Reactions in a Packed Bed Reactor by Thomas Z. Fahidy
   - 5.2 Design of a Fixed-Bed Reactor by Charles N. Satterfield and Russell L. Jones
   - 5.3 Design of Non-Isothermal Flow Reactor by William J. Hatcher, Jr.
   - 5.4 Adiabatic Catalytic Reactor Design for Methanol Synthesis: High Pressure by Kermit L. Holman
6. REACTOR DESIGN AND OPTIMIZATION
   - 6.1 Air Sparged Reactor Design by Louis L. Edwards
   - 6.2 SO_2 Oxidation Reactor Simulation Program by Donald B. Wilson and Robert L. Hair
   - 6.3 Optimal Selection of a Blend of Two Catalysts in a Tubular Reactor by Michael S. K. Chen
7. EXPERIMENTAL DATA ANALYSIS
   - 7.1 Data Correlation via Non-Linear Regression by George W. Roberts
   - 7.2 Residence-Time Distribution Computation for Flow Processes by William C. Clements, Jr.

The University of Michigan Projects

In 1976, approximately three years after the computer programs listed in Table 1 were published and distributed to chemical engineering departments across the nation, a second project was completed. With the aide of a grant from the Olin Corporation, five interactive programs, listed below, were developed at the University of Michigan by Professor H. Scott Fogler and a junior in chemical engineering, T. Michael Duncan (now a professor at Cornell).

1. Rate Laws
2. Stoichiometry
3. Heterogeneous Catalysis
4. Columbo (Murder in a CSTR)
5. Death by Dying (Analysis of Rate Law Data)

These programs were made available for testing by students in the chemical reaction engineering course during the 1976-77 academic year. The response and enthusiasm for the modules was overwhelming. These modules were demonstrated at the Summer School for Chemical Engineering Faculty held in Snowmass, Colorado during the summer of 1977. The response of the faculty viewing the modules was equally enthusiastic; following the summer school, a number of faculty at different universities requested copies of the modules for use in their classes. At this point, we encountered our first major hurdle. The uniqueness of the University of Michigan computer graphics systems was such that it could not be run easily, even on similar computers at other universities. A few faculty at other universities (in particular, David Himmelblau at the University of Texas and Dick Mah at Northwestern) even hired or encouraged students to modify parts of the code to get the programs to work at their universities, with little or no success. Consequently, the interactive computing modules went into hibernation until the portability problem could be solved.

The introduction of the IBM PC breathed new life into the modules as the portability problem was solved almost overnight. Between 1982 and 1985, four of the modules developed for the mainframe computer were rewritten for the PC, and a heterogeneous catalysis module and three modules from other areas of chemical engineering were distributed by the CACHE Corporation to all departments in the U.S. The four kinetics modules, along with a Kepner-Tregoe plant design module developed by the author (Reilly, 1972). served as the basis for a proposal to the National Science Foundation in the summer of 1987; the proposal was funded a year later (Fogler, 1992). The funds from this grant were used to develop nine modules in chemical reaction engineering and 15 in other areas of chemical engineering. During the five-year development process, we incorporated the pedagogical expertise gained through our interaction with the Center for Research on Learning and Teaching at the University of Michigan, review of the literature, and our previous experiences with interactive computing modules. Extensive testing at the University of Michigan and many other universities has allowed us to distribute modules which, we believe, address important issues that ensure success in interactive computer learning:

- Ease of use
- Maintaining focus on the concepts
- Minimal tediousness
- Promoting learning
- Individual guidance

These modules run on IBM-PCs and compatibles with EGA or better graphics and minimal system requirements of DOS 5.0 or later, a 20Mh 80386 processor, an 80387 math co-processor and at least 512 KB of RAM. (HETCAT - the Heterogeneous catalysis module -requires approximately 570 KB of memory to perform properly). The interactive modules in reaction engineering currently available are described below.

KINCHAL1: Kinetics Challenge 1 - Introduction to Kinetics

This module allows the students to test their knowledge of the general mole balance equation and reaction rate laws, as well as types of reactions and reactors. The interaction occurs in the form of an interactive game (see Fig. 1) with timed responses and computer-generated competitors. Twenty questions are selected from a pool of approximately 100 multiple choice questions (see Fig. 2). Students can choose questions from any of four categories (mole balance, reactions, rate laws, and reactor types) and five difficulty levels (100 - 500 points):

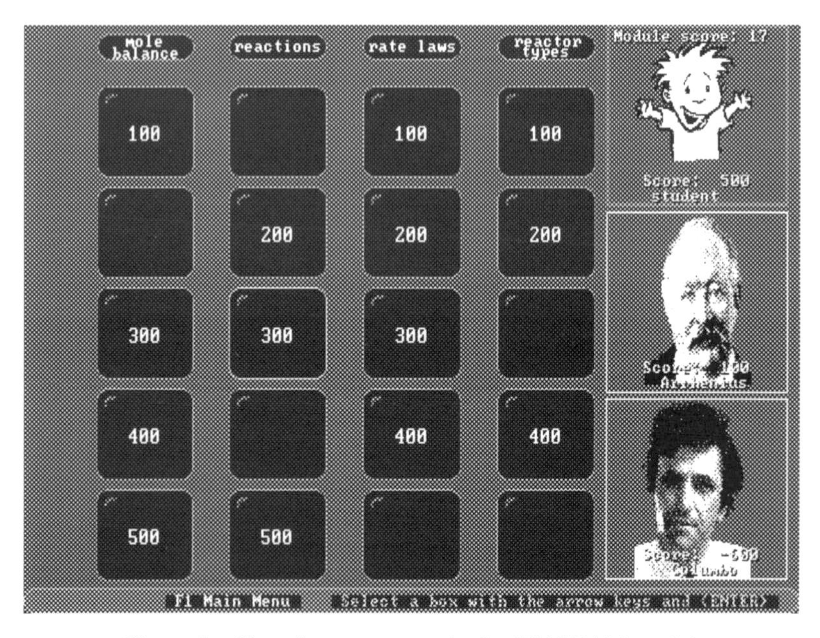

Figure 1. Choosing a category in the KINCHAL1 module.

The student has one minute to choose the correct answer. The module responds to the student's choice, either reinforcing the reasoning for a correct answer, or immediately clarifying a misunderstanding if an incorrect answer is entered. If no response is entered within the time limit, or if an incorrect response is entered, the points are lost and one of the computer competitors tries to answer the question.

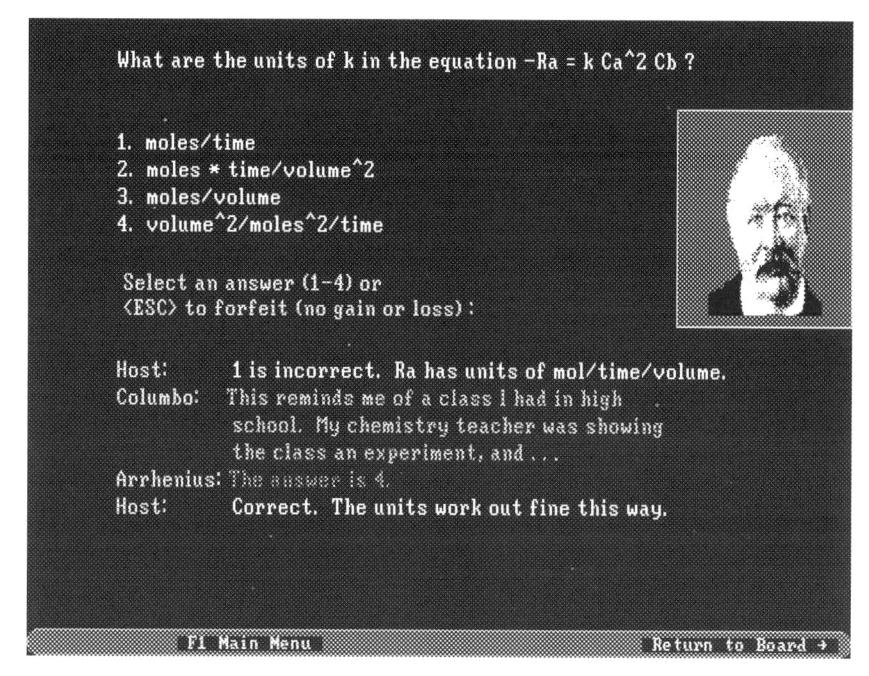

Figure 2. A question and response from the KINCHAL1 module.

The competitor who last answered a question correctly gets to pick the following category and degree of difficulty (note that this will not necessarily always be the student). In addition to regular questions, one question is randomly assigned as the "Double Challenge," in which the student has the option of betting points. After all twenty questions have been answered, the contestants with positive scores go on to the "Final Challenge" question, in which they are also allowed to bet points.

STAGING: Reactor Staging and Optimization

In the interaction portion of this module the student must arrange a group of five reactors – CSTRs and PFRs – in the order that will result in at least 75% conversion, while maximizing the product flow rate, for the reaction A → B. Both the $-F_{A0}/r_A$ vs. X (conversion) graph and the reactor volumes are specified, and many arrangements of reactor order and inlet flow rate can be tested using an interactive simulation.

The student may at any time access a reference section that reviews the derivation of the design equations for PFRs and CSTRs, clarifying the change in conversion down a PFR, and the well-mixedness of the CSTR.

The reactor optimization simulator can also be run independently of the scenario. This allows the professor to present the student with a variety of open-ended problems to be investigated using the simulator.

KINCHAL2: Kinetics Challenge 2 - Stoichiometry and Rate Laws

This module focuses on rate laws and stoichiometry, allowing the student to master the elements of the stoichiometric table. The interactive portion of the module is similar to that in Kinetic Challenge 1. Students can choose from four categories (reactants, products, rate law, potpourri) and four levels of difficulty (200-1,000 points).

The key focus in this module is to provide students with practice so that they will avoid the more prevalent mistakes (expressing the reaction rate law for an irreversible reaction as if it were reversible, and using the ideal gas law for liquid-phase reactions).

COLUMBO: CSTR-Volume Algorithm - A Murder Mystery

The principal purpose of this module is to allow students to practice the algorithm for CSTR design.

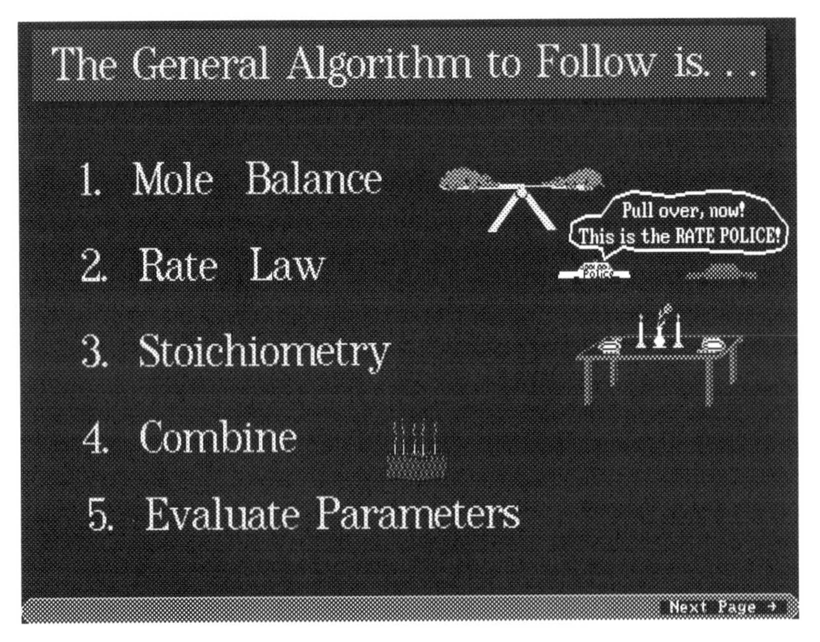

Figure 3. Frame from the COLUMBO module.

In the interactive portion of the module the student must solve a murder mystery, with the aid of Lieutenant Columbo. It seems that overnight there was a slight irregularity in the conversion in the reactor at the Nutmega company (see Fig. 4).

It is feared that one of the employees may have been murdered by a fellow employee, and the body left in the reactor. By analyzing the conversion data and using personnel information and knowledge of CSTR reactor design, the student must determine the identity of both the murderer and the victim. Help may be obtained by questioning the suspects.

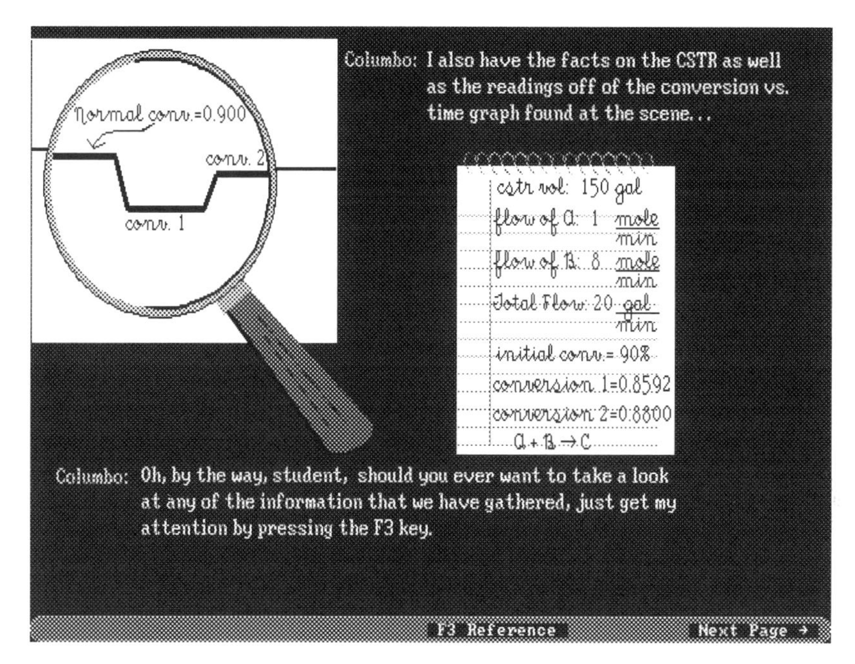

Figure 4. Frame from the COLUMBO module.

TICTAC: Ergun, Arrhenius, and Van't Hoff Equations in Isothermal Reactor Design

This module allows the student to examine nine reactor design problems, and investigate the effect of varying reactor parameters on process performance. The problems are organized as in a tic-tac-toe board (see Fig. 5). The reactors covered by these problems include PFRs, CSTRs, packed bed reactors and semi-batch reactors.

The student must master the concepts in enough squares (three adjacent squares horizontally, vertically, or diagonally) to successfully win the tic-tac-toe game. Each problem allows the student the opportunity to examine the effect of a specified operational parameter on reactor performance, using simulators:

After performing the "experiments," the student proceeds to answer three questions that examine the effects observed. These effects can be explained through the Ergun, Arrhenius, and Van't Hoff equations. In many cases, competing effects are highlighted. The square is "won" by answering two out of the three questions correctly.

ECOLOGY: Collection and Analysis of Rate Data - Ecological Engineering

The student, as an employee of a company trying to meet environmental regulatory agency standards, must sample concentration data for a toxic material found in a wetlands channel between a chemical plant upstream and a protected waterway downstream, and analyze the rate of decay of the toxic material.

The wetlands are modeled as a PFR. The student must first develop the necessary reactor

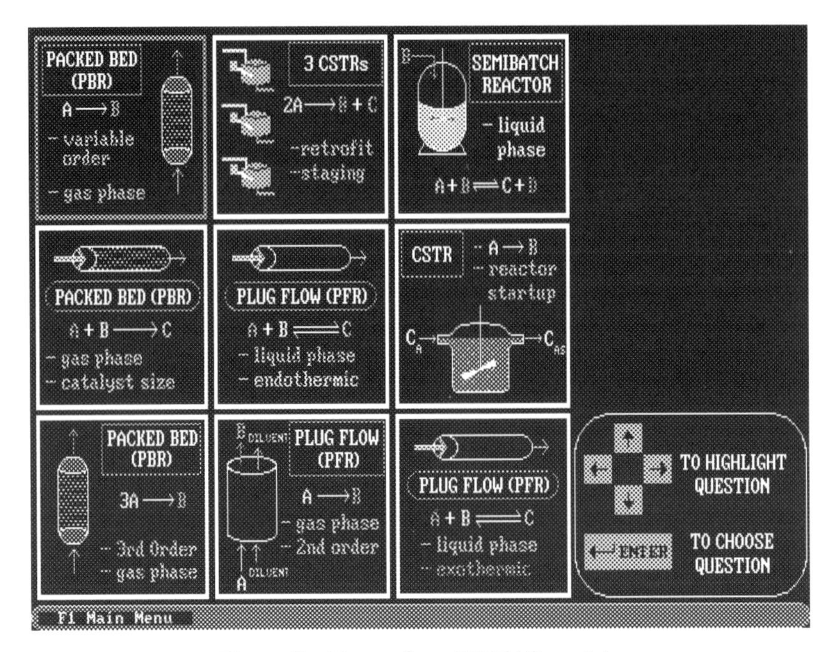

Figure 5. Frame from TICTAC module.

design equation for a PFR, then start to collect data. These concentration data, which include experimental error, are then analyzed in various ways (polynomial fit of the data followed by differentiation of the resulting equation, difference equations, etc.) to determine the rate law, the rate constants and the reaction order. Students must determine which points (if any) are to be excluded from the analysis and which points may be resampled.

The student then analyzes this information and submits a memo with the requested parameters. This information is reviewed by the boss, who evaluates the parameter values and makes recommendations.

HETCAT: Heterogeneous Catalysis

The review section of this module first covers the essential elements of heterogeneous catalysis. The student must derive the rate equation for a given reactive system by analyzing the rate data obtained in a differential reactor. The student then chooses which experiments to run, that is, the entering pressures of each species and total flow rate. In order to obtain the dependence of the rate equation on the pressure of a given species, the student selects which of the points are to be included in a plot of reaction rate vs. species partial pressure. Given the requested plot, the student must determine the form of the dependence of the rate law on the pressure of the given species.

Once all dependencies have been established, the student must decide which rate law parameters can be determined, through judicious plotting of the experimental data. The review section also outlines the derivation of the governing equations of heterogeneous catalysis.

HEATFX-1: Simulation - Mole and Energy Balances in a CSTR

This module allows students to investigate the effect of parameter variation on the operation of a non-isothermal CSTR. An extensive review section derives the energy balance for the CSTR, and also describes the terms in the mole balance that are temperature dependent.

A simulator is included in the review section. This allows the student to vary parameters and observe the effects on the conversion-temperature relationships (see Fig. 6) as described by both the mole balance and the energy balance. The parameters that may be varied include feed flow rate and temperature, the reversibility/irreversibility of the reaction, heat of reaction, heat exchanger area, and heat transfer fluid temperature. The operating conditions can be determined from the intersections of the mole balance and energy balance curves.

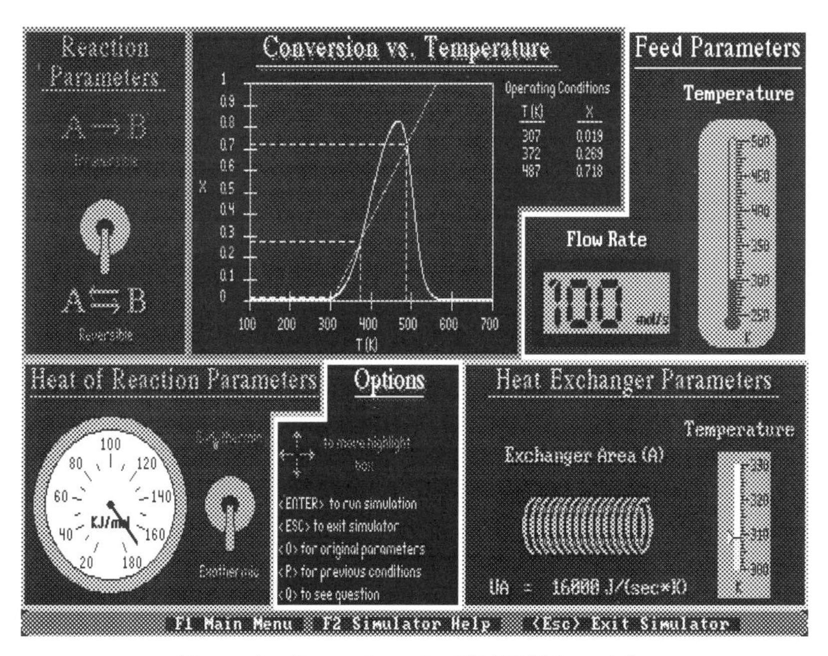

Figure 6. Frame from the HEATFX-1 module.

The module can also be run in the interactive mode, in which the scenario takes the student to a basketball tournament. The student has the choice of two-point and three-point questions. The simulator is available to help in answering the three-point questions.

HEATFX-2: Simulation - Mole and Energy Balances in a PFR

This simulation allows the student to explore the effects of various parameters on the performance of a non-isothermal plug flow reactor. The student may choose from eight simulations, that span all combinations of exothermic/endothermic, reversible/irreversible conditions, as well as one simulation that includes the effect of pressure drop. The parameters that may be varied include heat transfer coefficient, inlet reactant and diluent flow rate, inlet temperature, and ambient temperature.

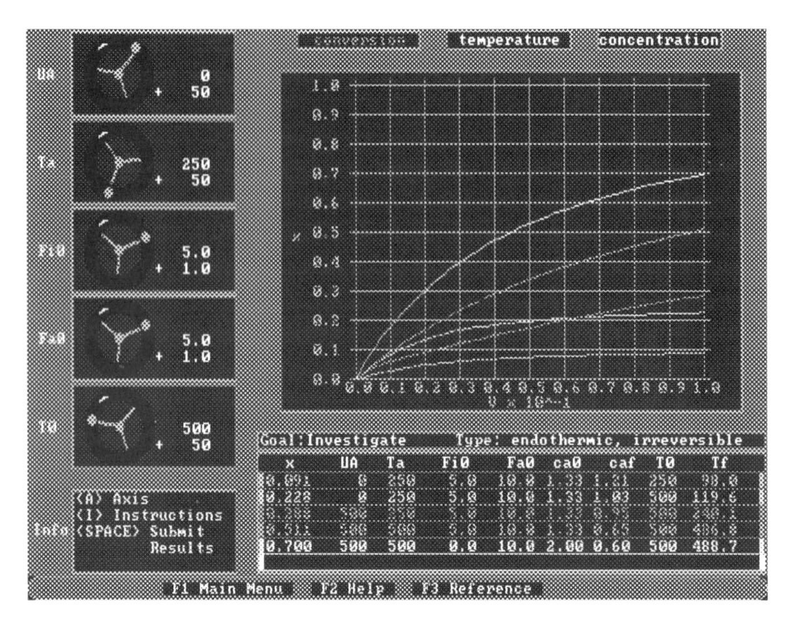

Figure 7. Frame from the HEATFX-2 module.

The results from the simulator may be analyzed in the form of plots of concentration, conversion or temperature (see Fig. 7) as functions of reactor volume. The module may also be run in the interactive mode, in which the student must achieve specific goals (e.g., achieve a given conversion without exceeding a given temperature within the reactor), in order to get to the center of the reactor complex.

Each interactive module has the following format:

- Menu
- Review of engineering principles
- Demonstration
- Interactive simulation
- Evaluation

In August of 1993, these interactive modules were sent by the CACHE Corporation to every chemical engineering department in the U.S. and to departments in foreign countries that are CACHE subscribers.

A Paradigm Shift

With the emergence of extremely user-friendly software packages, students can explore problem solutions much more effectively, to develop an intuitive *feeling* for the reactor/reaction behavior, and obtain more practice in creative problem solving.

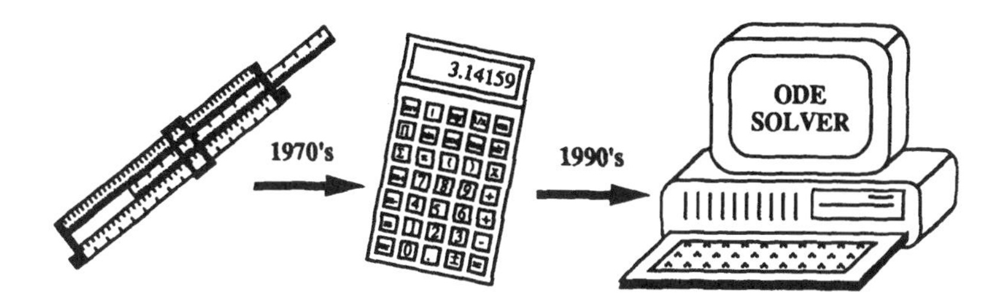

Figure 8. Paradigm shifts in chemical engineering education.

One of the most user friendly ODE solvers is POLYMATH, developed by Professors Michael Cutlip of the University of Connecticut and Mordechai Shacham of Ben Gurion University (Shacham and Cutlip, 1981a, 1981b, 1982, 1983). CACHE provided partial development support for the program and currently licenses POLYMATH to the industrial and university communities.

The numerical package developed by Cutlip and Shacham had its origins in a computer-based course on chemical reaction engineering using the PLATO (Programmed Logic for Automated Teaching Operations) System. This course, first offered in the late 1970s, was self-paced and had the following components:

- Self-Paced textbook (Fogler) 40%
- Homework assignments (on PLATO) 15%
- PLATO lessons 15%
- Videotape lectures 15%
- Reaction modeling and simulation (on PLATO) 7%
- Exams (on PLATO) 8%

This course was very successful when offered at both the University of Connecticut and the University of Michigan. As material on reaction engineering for the individual lessons became sufficiently developed, Professors Cutlip and Shacham focused their efforts on developing a numerical package that could be used with the course. Specifically, they began developing software to solve the non-linear ordinary differential equations typical of those found in chemical reaction engineering when heat effects are important. In addition to these ODE solvers, data analysis and polynomial fitting routines were developed for use in the course. These numerical packages were the origins of the current version of POLYMATH.

Unfortunately, the course that was developed for the PLATO system ceased to exist in the mid-1980s when support for it was withdrawn by its commercial supported. Fortunately, Professors Cutlip and Shacham were able to transport POLYMATH software from PLATO to IBM PCs. As a result, POLYMATH is now widely used throughout the US, Canada, Asia and

Europe. With POLYMATH, tedious and time consuming computer programming is eliminated, and the reaction engineering student can explore complex problems by varying the parameters and the operating conditions. Consequently, virtually every problem or homework assignment in chemical reaction engineering can be turned into an open-ended problem that will allow the students to practice their creative modeling and synthesis skills.

To illustrate this point, consider a gas phase exothermic reaction

$$A \rightarrow B + C$$

carried out in a plug flow reactor with heat exchange.

For non-isothermal reaction in CRE we must choose which form of the energy balance to use (e.g., PFR, CSTR) and which terms to eliminate (e.g., Q=0 for adiabatic operation). The structure introduced to study these reactors builds on the isothermal algorithm by introducing the Arrhenius equation, and $k = A\,e^{-E/RT}$ in the *rate law* step, which results in *one* equation with *two* unknowns, X (conversion) and T (temperature), when we finish with the *combine* step (Reilly, 1972). For example, using the PFR mole balance and conditions, we have, for constant pressure:

$$\frac{dX}{dV} = \frac{A\,e^{-E/RT}(1-X)}{u_o(1+eX)}\left(\frac{T_o}{T}\right) \tag{1}$$

We now recognize the necessity of performing an energy balance on the reactor to obtain a second equation relating X (conversion) and T (temperature). An energy balance on a PFR with heat exchange yields the second equation we need, relating the independent variables X and T:

$$\frac{dT}{dV} = \frac{\left[UA_c(T_a - T) + (r_A)(\Delta H_R)\right]}{F_{A0}\,C_{P_A}} \tag{2}$$

These simultaneous differential equations can be solved readily with an ODE solver, as discussed below.

Obtaining the temperature and concentration profiles requires the solution of two coupled non-linear differential equations such as those given by Equations (1) and (2). In the past it would have been necessary to spend a significant amount of time choosing an integration scheme and then writing and developing a computer program before any results could be obtained. With the available software programs, especially POLYMATH, it rarely takes more than 10 minutes to type in the equations and obtain a solution (Sacham and Cutlip, 1982). As a result, the majority of the time for the exercise can be spent exploring the problem through parameter variation and analysis of the corresponding observations. For example, in the above exothermic reaction in a PFR with heat exchange the students can vary such parameters as the ambient and entering temperatures, the flow rates, and the heat transfer coefficient and look for conditions where the reaction will "ignite" and conditions for which it will "run away." By trying their own different combinations and schemes, the students are able to carry out open-ended exercises that allow them to practice their creativity and better understand the physical characteristics of the system.

The following example, which was given as 50% of a 2 hour final exam at the University of Michigan, illustrates how significantly more complex problems can be rapidly solved with POLYMATH.

The elementary irreversible gas *exothermic* phase reaction

$$A \rightarrow 2B$$

is carried out in a packed bed reactor. There is pressure drop in the reactor and the pressure drop coefficient is 0.007 kg^{-1}. Pure A enters the reactor at a flow rate of 5 mol/s, at a concentration of 0.25 mol/dm^3, a temperature of 450 K and a pressure of 9.22 atm. Heat is removed by a heat exchanger jacketing the reactor. The coolant flow rate in the jacket is sufficient to maintain the ambient temperature of the heat exchanger at 27°C. The maximum weight of catalyst that can be packed in this reactor is 50 kg. The term giving the product of the heat transfer coefficient and area per unit volume divided by the bulk catalyst density is given by:

$$\left(\frac{Ua}{\rho_B}\right) = \frac{5\ \text{Joule}}{\text{kg cat}\cdot\text{s K}} \tag{3}$$

a) Plot the temperature, conversion X, and the pressure ratio ($y = P/P_0$) as a function of catalyst weight. b) At what catalyst weight down the reactor does the rate of reaction ($-r_A$) reach its maximum value? c) At what catalyst weight down the reactor does the temperature reach its maximum value? d) What happens when the heat transfer coefficient is doubled? e) What happens if the heat coefficient is halved? f) Discuss your observations of the effects on reactor performance (i.e., conversion, temperature and pressure drop).

Additional Information:

ΔH_R = -20,000 J/mol A at 273 K, C_{P_A} = 40 J/mol K, C_{P_B} = 20 J/mol K, E = 31.4 kJ/mol, and

$$k = 0.35\ \exp\frac{E}{R}\left(\frac{1}{450} - \frac{1}{T}\right)\ \frac{dm^3}{\text{kg.cat. sec}} \tag{4}$$

The relevant equations are:

$$\frac{dX}{dW} = -r'_A / F_{A0}$$

$$\frac{dP}{dW} = \frac{-\alpha}{2}\frac{T}{T_0}\frac{P_0}{(P/P_0)}(1+\varepsilon X) \quad \underline{\text{or}} \quad \frac{dy}{dW} = \frac{-\alpha}{2}\frac{T}{T_0}\frac{(1+\varepsilon X)}{y}$$

$$\frac{dT}{dW} = \frac{\frac{U_a}{P_B}(T_a - T) + (-r'_A)(-\Delta H_r)}{F_{A0}\left(C_{P_A} + X\left(2C_{P_B} - C_{P_A}\right)\right)}$$

$$-r_A = k\,C_A$$

$$C_A = C_{A0}\frac{(1-X)}{(1+X)}\frac{T_0}{T}\frac{P}{P_0}$$

$$k = 0.35\ \exp\left[3776.76\left(\frac{1}{450} - \frac{1}{T}\right)\right]$$

The POLYMATH solutions for this problem are shown in Figs. 9-12.

```
                    Pb 8-7 & 8-8 Final Exam Winter 1992

The equations:
    d(x)/d(w)=-ra/fa0
    d(y)/d(w)=-alpha*(1+x)*(t1/450)/2/y
    d(t1)/d(w)=((2.5*(300-t1)+ra*dhr))/fa0/cpa
    k=.35*exp(3776.6*((1/450)-(1/t1)))
    ca0=.25
    ca=ca0*(1-x)*(450/t1)/(1+x)*y
    ra=-k*ca
    fa0=5.
    dhr=-20000
    cpa=40
    rate=-ra
    alpha=.007
Initial values: w0= 0.0, x0= 0.0, y0= 1.0000, t10= 450.00
Final value:  wf= 50.000

                           CHANGE OPTIONS
 e. Enter/Change/Delete equations.     t. E/C/D the title.
 i. Change initial values.             f. Change final value.
             r. Restart from the current conditions.
   F8 when done.                       F6 for helpful information.
```

Figure 9. POLYMATH equations.

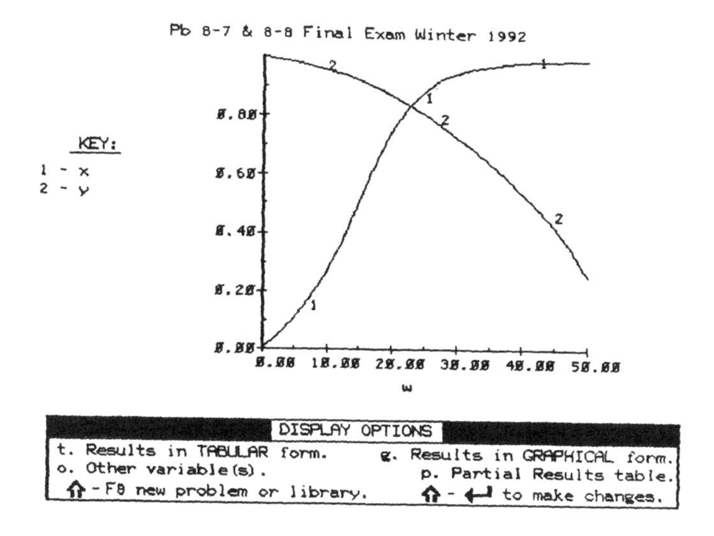

Figure 10. Conversion (x) and pressure ratio (y) as a function of catalyst weight.

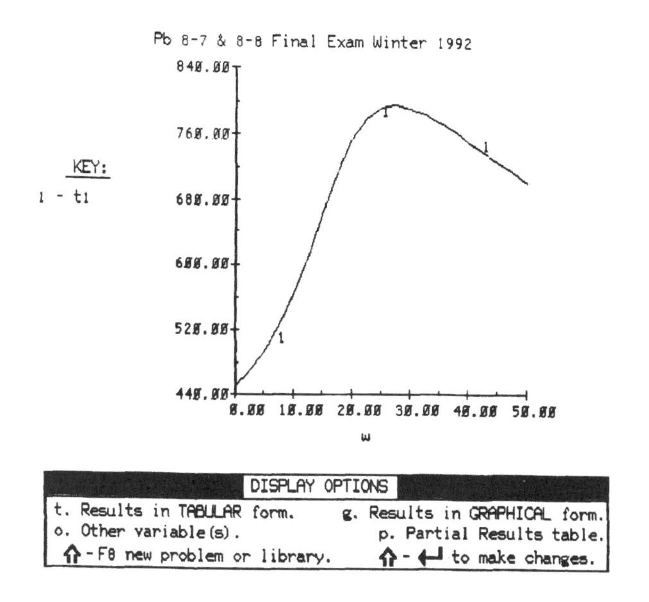

Figure 11. Temperature as a function of catalyst weight.

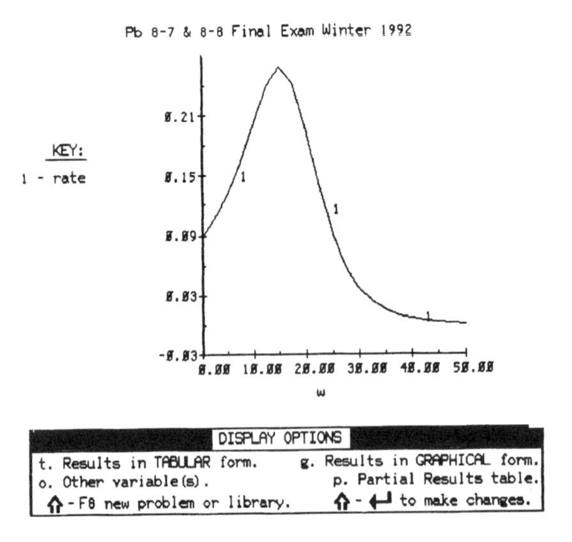

Figure 12. Reaction rate as a function of catalyst weight.

The Purdue-CACHE Modules

The Purdue-Industry Computer Simulation Modules are another educational innovation that CACHE fostered through Task Force support and distribution to virtually all chemical engineering departments in the US and Canada. Since 1986, a series of computer modules of state-of-the-art chemical engineering industrial processes have been and are being developed at Purdue University (Squire et al., 1992, Anderson et al., 1992). Each module has an industrial sponsor who furnishes data on a process on which the simulation is based and also produces a 20 minute videotaped "tour" of the real process. A number of these modules focus on chemical reaction engineering.

The modules are meant to supplement, not to replace, traditional laboratory experiments. Computer-simulated experiments have a number of advantages over traditional experiments:

- Processes that are too large, complex, or hazardous for the university laboratory can be simulated with ease on the computer.
- Realistic time and budget constraints can be built into the simulation, giving students a taste of "real world" engineering problems.
- The emphasis of the laboratory can be shifted from the details of operating a particular piece of laboratory equipment to more general considerations of proper experimental design and data analysis.
- Computer simulation is relatively inexpensive compared to the cost of building and maintaining complex experimental equipment.
- Simulated experiments take up no laboratory space and are able to serve large classes because the same computer can run many different simulations.

Each module is written as an industrial problem caused by a change of conditions in an existing process, requiring an experimental study to re-evaluate the characteristic constants of the process. These might include, for example, reaction rate constants, equilibrium constants, heat transfer and mass transfer coefficients, and phase equilibrium constants. The student teams are expected to design experiments that will enable them to evaluate the needed constants. This is referred to as the *measurements* section of the problem.

After the constants have been determined, the students must validate them by using them in an existing computer model of the process, and comparing the simulated and experimental results. When they are convinced that their constants are reliable, the students must use these constants to predict some other specific process performance characteristics. This is called the *applications* section.

Each process is made to seem realistic not just by the videotaped "tour" but also by the assigned financial budget and time constraints. The problems are open-ended in that the experimental conditions such as temperature, pressure, flow rates, and compositions are under the student's control. The cost and the associated duration of running experiments vary with the type of experiment and are also functions of operating conditions. Instructor-controlled statistical fluctuations are built into the simulations so that the results of duplicate experiments are not identical. The students must plan their experiments to obtain data from which, with proper analysis, the required constants may be determined without exceeding their budgetary and time constraints.

In the Purdue course, students work in teams of three and have eight 3-hour classes to complete the assignment. After an introductory 2-hour lecture, they are on their own. If they have queries, they are free to ask the consultant (the instructor, of course), but are charged a fee which is deducted from their budget. Written reports and a 20-minute oral presentation (video recorded for later analysis by the instructor and the student) are required.

Chemical kinetics has played a large role in most of the modules. The following section gives a brief outline of the chemical kinetics application of each of the Purdue modules.

Amoco Resid Hydrotreater (Squires et al., 1991)

This module requires the students to design a series of experiments (in a pilot plant, using one, two, or three CSTR's in series) to determine the rate constants of a series-parallel network of seven pseudo-first-order non-catalytic irreversible reactions. In addition, the rate constants must be determined for the catalytic desulfurization reaction, which follows Langmuir-Hinschelwood kinetics. This study is complicated by the fact that the catalytic activity deactivates with time.

Once the constants have been determined, the students are asked to start up the plant. This is complicated, since the reaction network is unstable with multiple steady-state solutions. Non-optimal control strategies may easily lead to temperature runaways.

Dow Styrene-butadiene Copolymerization (Jayakumar et al., 1995)

The students must determine the rate constants of the four propagation reactions and two chain transfer reactions, by designing experiments to be run in a two-gallon laboratory reactor.

Once the constants are determined, students are asked to predict the performance of a 10,000 gallon plant reactor. This is complicated, since the reactor has insufficient surface area for heat transfer. Students must determine:

a. how much additional heat transfer surface area is needed, or
b. the amount of feed pre-cooling required to achieve controllability.

Mobil Catalytic Reforming (Jayakumar et al., 1994)

In this study, the process is simplified by considering the catalytic reforming of only the C_6 range of hydrocarbons. The students must determine the catalyst deactivation parameter of a series of four coupled reforming reactions.

Once the parameters are known, students are asked to predict the hydrogen-to-hydrocarbon ratio in the feed that will optimize the annual profit of the process.

Eastman Chemical Reactive Distillation (Jayakunmar et al., 1993)

One step in Eastman Chemicals acetic anhydride from coal process, involves the sulfuric acid catalyzed reaction:

$$\text{acetic acid} + \text{methanol} \rightarrow \text{methylacetate} + \text{water}$$

This reaction is normally equilibrium limited. The key concept of the reactive distillation process relies on methylacetate being more volatile than the other reactants and products. The reaction occurs on a distillation tray, and a significant amount of methylacetate will vaporize, forcing the reaction to the right and increasing the yield beyond the normal equilibrium limitation.

Air Products Hydrogen Reactive Cooling Process

Hydrogen at room temperature is an equilibrium mixture of 25% para and 75% ortho. When cooled below 30°R the equilibrium mixture is almost 100% para-hydrogen. If the hydrogen is cooled in the absence of a catalyst, the ortho-para reaction will not take place and the resulting liquid hydrogen will still be 75% ortho. In the liquid hydrogen storage vessel the ortho-para reaction will occur, and the exothermic heat of this reaction will cause the vaporization of much of the liquid hydrogen.

In order to avoid the boil-off losses, it is necessary to catalyze the ortho-para hydrogen reaction during the cooling process rather than later in the storage vessel. The students must determine the rate constants for the ortho-para reaction.

Once the constants are determined, the students are asked to determine the optimum number of side reactors required for the reactive cooler design.

Concluding Comments

In July 1990, and again in July 1991, three-day workshops were held at Purdue University at which the chemical engineering faculty participants were given hands-on experience with all the modules. Faculty representatives from 56 chemical engineering departments in the United

States and Canada participated in these workshops. Many of these schools are now using some of the modules. It is particularly interesting to note that several of the other schools (notably Georgia Tech., Carnegie Mellon, West Virginia) are using the materials as problems in reactor design courses. The authors also presented a workshop on the project at the ASEE Chemical Engineering Faculty Summer School in August 1992. The modules are currently being used by 25 schools, including five schools in foreign countries.

The modules were originally created for execution on Sun (UNIX) workstations. NSF has recently funded a proposal to port the modules to other workstations (such as DEC, IBM, HP, and Silicon Graphics). Nine other schools have agreed to participate as beta-test sites in this work, bringing the total number of user schools to 34. Once the modules are available on these other computers, there should be a significant increase in the number of participating schools

The University Of Washington – CACHE Project

The Chemical Reactor Design Tool is a set of computer programs that permit a student to design chemical reactors, including the realistic transport effects that are frequently present. The interface is written in X-windows so that the student can include complications easily; the program automatically uses the correct, robust tools to solve the problem. Results are displayed graphically, which makes comparison studies especially easy. The programs were developed by Professor Bruce Finlayson under sponsorship of the National Science Foundation and the University of Washington, and are made available to universities through CACHE.

CRDT Educational Goals

Introductory textbooks concentrate on problems that can be solved analytically. Recent textbooks include material for problems that can be solved with an ordinary differential integrator. These include batch reactors and simple plug flow reactors; extensions from one reaction to several reactions are possible, but time-consuming. When attempting to solve real problems, students are faced with several difficulties, which are mostly difficulties in manipulation and book-keeping rather than conceptual. Phenomena that might be important include:

- Multiple reactions (lots of bookkeeping)
- Temperature of catalyst and fluid may be different
- There may be internal mass transfer (requires solving an effectiveness factor problem)
- There may be cooling at the wall (leads to radial dispersion)

Students and design engineers may not be able to make realistic estimates of which phenomena must be included. In some cases it is necessary to calculate with a suspected phenomenon included to see if it is important. That has been difficult to do, because each phenomenon creates problems that require special techniques to solve. Used in the CRDT are ordinary differential equation integrators, the orthogonal collocation method, the finite difference method, techniques to convert partial differential equations to sets of ordinary differential equations, iterative techniques to solve large sets on nonlinear equations, and linear programming methods to guide initial guesses for integrative techniques. All these methods are transparent to the user.

Key Aspects

The user can examine effects very easily and make their own deductions about the importance of physical phenomena. Phenomena that can be easily included are:

- Different reactors: CSTR, batch, plug flow
- Axial dispersion, radial dispersion
- Intraparticle heat and mass transfer
- Significant mole changes
- Significant pressure changes

The cases when it makes a difference are:

- Selectivity, especially in non-isothermal cases
- Non-isothermal problems

None of these complications is too complicated for the student to do, but there is not enough time to do so. In design problems, though, some of these complications are necessary.

Textbook Supplements

An important feature is the Textbook Supplement, which gives any new equations in the notation of that textbook. Since the programs are much more general than can be treated in most undergraduate books, it is necessary to explain the problems continuing the notation of the textbook being used by the student. In addition, a two-page handout provides hints on the best way to approach problems, summaries of the equations for easy reference, and standard correlations for some of the transport properties. This handout is designed for quick reference while using the program.

Design Decisions

Design decisions may revolve around conflicting constraints, none of which can be easily handled if one has to write the program:

- Use a small catalyst diameter to avoid diffusion resistance; but this increases the pressure drop.
- A recycle compressor may be an expensive component in a gas-phase reaction system.
- An adiabatic reactor avoids radial dispersion, but the temperatures may be too big; cooling at the wall usually makes radial dispersion important.

By using the Chemical Reactor Design Tool these realistic complications can be treated by the student-designer.

Output

Output is presented graphically in addition to printed form. The user can call for the following plots:

- line plots
- 3D perspective views
- 2D contour plots
- solution variables: concentration, molar flow rate, temperature wall flux
- each term in the equations: diffusion, reaction, convection terms

These plots are created automatically, but the user has some control over them either before they are created (contour and 3D views) or after they are created (line plots).

Conclusions

What will the next 25 years bring? We are already beginning to see the development of multimedia modules with the newest CACHE/NSF/University of Michigan initiative under the direction of Professor Susan Montgomery. Part of this project will involve development of kinetic and bioreactor modules which will incorporate video-clips showing growing bacteria as well as the transport of bacteria in porous media.

Another exciting area on the horizon is virtual reality (VR). CACHE has recently formed a Virtual Reality Task Force; within the next five years we will see a significant number of VR modules developed. One module currently under simultaneous development and testing at the University of Michigan is the prototype of a chemical plant that uses a straight-through transport reactor with a coking catalyst. Here the student uses VR to enter the plant lobby where he or she is given an overview of the process and is free to explore various parts of the room and video tapes at will, simply by moving a joy stick. After this introduction, the student enters the reactor room where he or she can change the operating parameters and see their effect on the reaction variables such as degree of coking, conversion, etc. The student can enter the reactor to observe the coking and catalyst transport, and, in addition, can enter the catalyst pellet to view the internal pore space and reactions occurring on the surface. This visualization of the process and reaction mechanisms will greatly enhance the students' understanding and appreciation of this reaction engineering process.

In addition to POLYMATH, the use of other software packages is on the rise in CRE courses. MATLAB, MAPLE, and Mathematica are becoming increasing user friendly and also are now being introduced in the freshman calculus or required computing courses at many universities. These packages will provide greater flexibility and a higher level of sophistication in the type and degree of complexity of problems the students can solve. In addition to these ODE solvers, we can expect partial differential equation (PDE) solvers to be available in the not too distant future. The PDE solvers will allow the students to explore radial as well as longitudinal gradients in packed bed reactors.

Finally, we can expect our future textbooks to be on CD-ROMs, so that the student can interact with the book while reading it. Interactions will include video-clips, audio, animation of mechanisms and equations, and much more. The CACHE CD-ROM task force has already prepared and distributed two CD-ROMS containing many instructional modules, video-clips, POLYMATH, and early versions of some of the papers in this monograph. There are great excitements ahead in computing in chemical reaction engineering.

References

Andersen, P.K., S. Jayakumar, R. G. Squires, and G.V. Reklaitis (1992). Computer Simulations in Chemical Engineering Education. Proceedings *on Frontiers in Education Conference*, Nashville, TN.

Fogler, H.S, S.M. Montgomery, and R.P. Zipp (1992). Interactive Computer Modules for Undergraduate Chemical Engineering Instruction. *Computer Applications in Engineering Education*, **1**, 11-24.

Jayakumar, S., R.G. Squires, G.V. Reklaitis, P.K. Andersen, and L.R. Partin (1993). Purdue-Industry Chemical Engineering Laboratory Computer Module-II. Eastman Chemical Reactive Distillation Process. *Chemical Engineering Education*. **27**(2), 136-139.

Jayakumar, S., R.G. Squires, G.V. Reklaitis, P.K. Andersen, B.C. Choi, and K.R. Graziani (1994). The Use of Computer Simulations in Engineering Capstone Courses: A Chemical Engineering Example–The Mobil Catalytic Reforming Process Simulation. *International Journal Engineering Education*, **9**(3), 243-250.

Jayakumar, S., R.G. Squires, G.V. Reklaitis, P.K. Andersen, and B.K. Dietrich (1995). The Purdue-Dow Styrene-Butadiene Polymerization Simulation *Journal of Engineering Education*, **84**(3), 271-278.

Reilly, M.J. (1972). *Computer Programs for Chemical Engineering Education: Volume II - KINETICS*, CACHE, Houston, TX.

Shacham, M. and M.B. Cutlip (1981). Computer-Based Instruction: Is There a Future in ChE Education? *Chemical Engineering Education*, 78.

Shacham, M. and M.B. Cutlip (1981). Educational Utilization of PLATO in Chemical Reaction Engineering. *Computers & Chemical Engineering*. **5**(4), 215-224.

Shacham, M. and M.B. Cutlip (1982). A Simulation Package for the PLATO Educational Computer System, *Computers & Chemical Engineering*, **6**(3), 209-218.

Shacham, M. and M.B. Cutlip (1983). Chemical Reactor Simulation and Analysis at an Interactive Graphical Terminal. *Modeling and Simulation in Engineering*, 27.

Squires, R.G., G.V. Reklaitis, N.C. Yeh, J.F. Mosby, I.A. Karimi, and P.K. Andersen (1991). Purdue-Industry Computer Simulation Modules–The Amoco Resid Hydrotreater Process. *Chemical Engineering Edition*. **32**, 98-101.

Squires, R.G., P.K. Andersen, G.V. Reklaitis, S. Jayakumar, and D.S. Carmichael (1992). Multi-Media Based Education Applications of Computer Simulations of Chemical Engineering Processes. *CAEE*, **1**(1).

TRANSPORT PHENOMENA

Bruce A. Finlayson
University of Washington
Seattle, Washington 98195

Andrew N. Hrymak
McMaster University
Hamilton, Ontario L8S 4L7 Canada

Abstract

This chapter describes computer software that is available for solving transport problems, including those with fluid flow. Included are programs that run on PCs, programs that need workstations, and commercial codes that run on large workstations. All programs described here were generated in the 1980s and 1990s.

Introduction

Chemical engineering education was revolutionized in the 1960s by the introduction of transport phenomena, as advanced originally by the seminal book *Transport Phenomena*, by Bird, Stewart and Lightfoot (1960). The subject requires more mathematics than other parts of the curriculum, and typical transport courses invoke heavy use of mathematics. Because mathematics in the 1960s was mostly done analytically, problems treated in transport courses have focused on problems that are simple enough to be solved analytically. Usually that means the problems are one-dimensional and linear, e.g., fully developed flow in a pipe. When flow occurs it is usually laminar. Situations involving two dimensional flows or turbulent flows are not handled except in extremely simple cases, e.g., flow past a sphere at zero Reynolds number, or with correlations, e.g., Nu = f(Re, Pr).

With modern numerical tools it is possible to solve more realistic models. Thus the student must be able to formulate the problem in a reasonable way. In fact, industrially the problem may be solved by a packaged program, or a program written by a computer scientist. Thus the formulation of the problem, and the need to verify and understand assumptions, is especially important as we look into the future. Of all the areas in the curriculum, transport phenomena is probably the one that has been influenced least by the growth of computer power; it thus stands to gain the most by the introduction of computer tools.

In the 1980s CACHE established a task force to develop IBM PC lessons for chemical engineering courses other than design and control. This task force was under the chairmanship of Professor Warren Seider of the University of Pennsylvania and developed a number of modules which were distributed to all Universities supporting CACHE. Only one of the lessons involved transport - design of a slurry pipeline, and it is described below. Throughout the 1980s the CACHE News ran a column announcing programs written by professors, edited by Professor Bruce Finlayson of the University of Washington. Only one of those programs falls into the class of a transport program - solving the convective diffusion equation in one-dimension and time, and it, too, is described below. A CACHE task force was formed to develop specific lessons for the IBM PC that could be used in transport courses. Most of these were eventually abandoned, it is very time-consuming to generate a decent computer lesson! However, Professor Scott Fogler at the University of Michigan persisted and developed several modules that are described below. Finally the use of spreadsheets can be advantageously applied to transport problems, and Professor Finlayson described those in a chapter in the second edition of a transport book by Professor Ray Fahien (1995) of the University of Florida. A brief summary of what can be done is provided here.

When looking to the future, an on-going project is described that makes available finite element tools to seniors to solve transport problems that are extensions of those in their texts. This is an advanced topic that may be most relevant to senior students and beginning graduate students. Described here is the philosophy of the Transport Module being prepared by Professor Finlayson, and then a review of commercial computational fluid dynamics codes is given for those that have access to them.

PC Programs

Design of a Slurry Pipeline

The slurry pipeline computer program was written by William Provine, Benny Feeman, Gregory Dow, and Professor Morton Denn at the University of California at Berkeley. The emphasis was to provide a design problem that students could solve using the theory and understanding they had achieved in their fluid mechanics course. The problem is to design a slurry pipeline for transporting material under specified conditions. The students can choose to dilute it, to reduce its viscosity, operate it in laminar or turbulent flow, and must avoid settling of the suspended solids. Simplifications are made - a maximum pressure is specified so that the cost factors involved in thicker pipe walls need not be included, the suspension exhibits no yield stress and no shear rate dependence - but the essential part of the problem remains. Economic data are supplied, along with a program to do some of the technical calculations.

The basic equations are

$$\eta = \frac{9}{8}\eta_{SF}\frac{(\phi/\phi_s)^{1/3}}{1-(\phi/\phi_s)^{1/3}}$$

for the slurry viscosity,

$$\text{laminar: } f = \frac{16}{Re}$$

$$\text{turbulent: } f = \frac{0.046}{Re^{0.2}}$$

for the friction factor, and

$$v_R = 1.3 \left[2gD \left(\frac{\rho_S}{\rho_{SF}} - 1 \right) \right]^{1/2}$$

for the minimum velocity for reentrainment of a sedimenting slurry. The problem statement then leads the student through exercises establishing the minimum power for turbulent flow, for laminar flow, and for any flow. The computer program runs on an IBM compatible computer under the DOS operating system.

CDEQN

This program was written by Professor Bruce Finlayson to solve the transient convective diffusion equation in one space dimension.

$$\frac{\partial c}{\partial t} + Pe \frac{\partial c}{\partial x} = \frac{\partial^2 c}{\partial x^2}$$

The methods used include the finite difference method and the Galerkin finite element method and the solution is plotted automatically at various times. A variety of numerical choices are available to the student, so that they can explore how those choices affect their solution. Such matters are very important when they turn to simulations of the flow of contaminants underground, for example; improper use of the packaged programs can lead to misleading results. Later this program was expanded into a commercial program, CONVECT, which has many more numerical methods. The program runs on a Macintosh computer with any operating system (although System 7 users must turn off the cache memory).

PC Lessons

A series of PC lessons has been developed at the University of Michigan under the direction of Professor Scott Fogler. The program "Shell: Shell Momentum and Energy Balances" leads the student through the exercise of making shell balances for three problems: water flowing down a vertical flat surface, water flowing through a vertical circular tube, and heat conduction in an electric wire. If the correct shell balance is achieved, the solution to the problem is displayed. The program "Simp: Simplification of the equations of motion and energy" works on the problem in the reverse order: the student chooses which terms to leave out of the equations to simplify them to model one-dimensional transport. This module uses the language of video games. It is called the "Equation Avenger," and the student 'shoots' the unnecessary terms. Since there seems to be a definite bias between boys and girls playing video games, and the type of game they prefer, this module may run afoul of the Politically-Correct Police and be inappropriate for the men and women in our classes! The program "Visc: Rheology - Identification of Liquids" helps the student review viscometer principles for identifying the type of non-Newtonian fluid. "Patch: Diffusion - Drug Patch Design" requires the student to design a drug patch that supplies a drug to astronauts in a space shuttle mission. The drug flow rate must fall within a specified range, and this flow rate must be maintained for a required amount of time. The student can use the simulator to experiment with different materials (diffusivities), patch thicknesses, and different reservoir drug concentrations. Thus the student is exposed to open-ended design problems in the transport course, which is a trend of growing importance. "Thermowell: Conduction, Convection and Radiation" allows students to investigate the effects of various thermowell parameters on the temperatures measured by a thermocouple. Again the problem is open-ended and the computer allows students to try many choices easily and quickly.

Transport Using Spreadsheets

Professor Bruce Finlayson at the University of Washington has written a chapter entitled "Numerical Methods for Solving Transport Problems" for the second edition of the book *Fundamentals of Transport Phenomena*, by Ray W. Fahien, McGraw-Hill, New York (first edition: 1983). This chapter describes methods for solving typical and extended transport problems using the numerical tools that exist today.

The first section treats a heat transfer problem. First the problem is taken as a linear problem, so that the finite difference method can be described. Then complications are added to the problem: a thermal conductivity that depends on temperature and a heat generation rate that is not constant. The equations solved are

$$\frac{1}{r}\frac{d}{dr}\left(k\,r\frac{dT}{dr}\right) = -\frac{2\,\Phi_0}{k}\left(1-\frac{r^2}{R^2}\right)$$

$$k = k_0[\,1 + a\,(T - T_R)\,]$$

Finally, the solution of these problems is described using a spreadsheet program with iteration capabilities, and detailed information is given about how to organize the calculations and check them. The next section considers transient heat transfer.

$$\frac{\partial T^*}{\partial t} = \alpha\frac{\partial^2 T^*}{\partial x^2}$$

The finite difference method is applied to reduce the problem to a set of ordinary differential equations (using the method of lines), and packages such as MATLAB are used to solve them. The final section considers the more complicated situation of heat conduction in both directions.

$$\frac{\partial^2 T}{\partial x^2} + \frac{\partial^2 T}{\partial y^2} = 0 \text{ in } 0 \le x \le 1,\ 0 \le y \le 1$$

Such problems can be solved using iterative methods as long as the domain has straight edges (either constant x or constant y, but there can be many of them). A problem is worked showing the effect of having a hole in the middle of the domain. The finite element method is also applied to the same problem so that students can see the limitations of using a spreadsheet program for such problems.

When solving differential equations numerically, though, there are still approximations made, and the possible error must be assessed. Throughout the chapter methods are given for organization of the calculations, often with solution of a simpler problem which is easily checked, and with complications introduced one by one with checking at each step. Of even more importance, though, is the use of the truncation error to determine the error of the numerical solution. Since the error in the solution is proportional to some power of Δx, for example, the solution values should follow that behavior when the problem is resolved with a different Δx. By using this information, it is possible to assess the accuracy of the numerical solution, and improve it if necessary (it seldom is).

Workstation Programs

Brief Description of Transport Module

The Transport Module is being developed by Professor Bruce Finlayson at the University of Washington to allow students to solve fluid mechanics and transport problems in laminar two-dimensional flow situations. The program uses the finite element method to discretize the mesh. The user sketches the domain on the screen, identifies the fixed boundaries, flow boundaries, etc., sets the boundary conditions, and then instructs the program to solve the problem. The finite element mesh is constructed automatically, and the results are displayed graphically as contour plots. The user interface is constructed in X-windows for use on Unix machines.

Goals

With the Transport Module students will be able to investigate fluid flow phenomena, such as how vorticity is generated, convected and diffused, and how this influences the flow field, heat and mass transfer. They will be able to make quantitative estimates - such as: "Is it isothermal?", and "If not, how much error is introduced?" It is even possible that students can tackle design problems in their senior year that involve small-scale processes and transport limitations, including the interaction of flow, heat, and mass transport, such as might occur in chemical vapor deposition. As they solve problems under different assumptions, students will gain intuition rather than just make assumptions because the instructor says so, or because of mathematical convenience. Examples can be more exciting and can involve the newer technologies. A textbook supplement will provide problems that are tied to existing textbooks, making it easy to incorporate the new ideas in existing courses.

Detailed Description of the Transport Module

The three pillars on which the Transport Module is built are the user interface (to specify the problem), the finite element programs, and the graphics display of the results.

The user interface allows the user to sketch the domain as a series of lines or quadratic curves. Boundary conditions are specified for each of these lines in a natural way. Whereas most finite element codes allow you to specify each variable at each boundary node, this generality can also lead to gross modeling errors. Thus students are restricted to choosing boundary conditions that make sense: solid boundary, flow boundary, centerline (symmetric) boundary, etc. This restriction prevents gross errors, and it also makes the data entry considerably simpler. The fluid flow and heat transfer properties are easily entered in an X-windows interface.

The finite element codes allow a variety of problems to be solved, but are restricted to laminar flow in two dimensions, either Cartesian or cylindrical geometry. The Navier-Stokes equations can be solved, including the possibility of the viscosity depending on shear rate, so that simple non-Newtonian fluids like power law and Bird-Carreau fluids can be modeled. The energy and transport equation can also be added, in which case the viscosity can depend on temperature as well. The user specifies a viscosity subroutine and a heat generation and rate of reaction term as well. The primary aim is to allow flow and heat transfer, rather than a complete inclusion of chemical reaction phenomena. Only one chemical species can be included, allowing for dilute systems or systems in which the number of moles does not change. Chemical reactor models are best handled with the Chemical Reactor Design Tool. The energy equation can be solved by itself for heat transfer problems, and the energy and mass transfer equation can be solved with a specified velocity for chemical reactor situations.

Following solution of the finite element problem, various post-processing programs are invoked to make the results more meaningful. First, vorticity and streamline are determined, if desired. Then various plotting options are chosen. The flow situation can be examined by plotting the steamlines and vorticity, but more detail is also available. The different terms in the equation can be plotted individually; thus the convective and the viscous terms can be compared for flow around a cylinder, and the student can explore the Stokes paradox. The stress components can be plotted, as can the viscosity throughout the domain. Both of these plots are useful for gaining insight to non-Newtonian flows. When the energy equation is solved, the viscous dissipation term can be plotted and compared with the diffusion or convection term; this helps the student see the validity of the constant temperature approximation. These pictures are especially meaningful when done in a 3D perspective view. In addition, certain properties can be viewed along boundaries: the drag forces and heat flux.

Because of the processing power needed to run the finite element codes, the actual calculations can best be done on a large computer, and the Unix environment makes this especially easy. With the X-windows display on a local computer screen, the graphics can be viewed locally (and printed locally), but the speed of the central processor is especially welcome. Of course, calculations can be done at the workstation, too, but with more delay. When computer power increases it will be possible to do the processing locally, too.

Of course it is necessary for educational purposes that the student be able to solve many problems they have seen before. Most of the problems they solved in transport are one-dimensional, but these can be solved in the Transport Module as well (with some loss of efficiency). Couette flow, Poiseuille flow, and combined Couette-Poiseuille flows are all solvable. The Graetz problem can be solved with fully developed flow, with developing flow, with Newtonain and non-Newtonian fluids, and with temperature-dependent viscosity. Laminar boundary layer problems can be solved, as can flow past spheres and cylinders. The effect of walls around the spheres and cylinders can be explored, provided the geometry still retains a two-dimensional character.

In the time since the project was defined the power and availability of commercial codes has increased (see below) to the point that this program may never be made available. However, the method of using them in class and interacting with the existing curriculum have value that is not provided by commercial codes. Such interactions are described below.

Sample Questions which can be Explored

Consider any flow problem, e.g., flow past a sphere. What is the effect of Reynolds number? How small is small, so the inertial effects can be ignored? How large is large, so the inertial effects must definitely be included? How big are the inertial terms (which one would like to ignore) compared with the diffusion terms? These are questions which can be answered by students using the Transport Module. Once their intuition has been developed, the Transport Module can be used in more design-type situations. They can compare the effect of different shapes (at least cylindrical shapes with different cross sections).

In the heat transfer problems the students can see the effect of Prandtl number, in that it effects the boundary layer thickness, which they will see in the plots. They can allow the thermal conductivity to depend on temperature if they think that is important, and compare results to simulations in which the thermal conductivity is constant. The heat transfer module can also be used to solve Laplaces's equation, making it suitable for some problems in electronic materials processing. There the effect of different geometries can be easily explored.

Sample Extension Problems

Listed below are several problems of the kind that can extend those in current textbooks. The examples are some of those related to Denn's book (1980), but textbook supplements can easily be prepared for all major textbooks. For the Chemical Reactor Design Tool only two textbooks supplements were prepared, because there are only two major books used by universities, but the added work to prepare a second supplement is small compared to the work necessary to prepare the first one.

> Flow of a sphere in a tube, p. 61. How far away does the outer boundary have to be for less than 5% effect? Check the correlation in Figure 4-6. Extend that correlation as a function of Reynolds number.
>
> Flow past objects that are not regular, p. 66. Compare the drag of a sphere with that of a disk oriented with the flat edge forward. Consider other shapes. Prepare a universal graph (and test it) by using the surface to volume ratio to obtain an effective diameter.
>
> Calculate pressure drops in a reaction injection molding device, p. 136.
>
> Flow in a manifold, p. 125. With one inlet and several outlets, how much flow goes out each?
>
> Determine the errors incurred when measuring the pressure at the bottom of a hole or slit.
>
> Entrance pressure loss in contracting flows, p. 336. Correlate it, design a shape that will minimize it, do for power law fluids, too.
>
> Flow Distribution in a single screw extruder, p. 197. Do a two dimensional analysis and determine the errors in the one dimensional analysis.
>
> Boundary layer flow, p. 286. Do an complete analysis for flow past a flat plate and examine the terms that have been neglected in the boundary layer analysis. Are they really small? How small? How large a region is affected?

Examples

Examples of the problems that can be solved are given here.

The problem of heat transfer to flow past a sphere is commonly treated in textbooks. For slow flow (Re < 0) the asymptotic formulas have been derived for large and small Peclet number.

$$Nu = 2 + \frac{Pe}{2} + \frac{1}{4} Pe^2 \ln Pe + 0.03404\ Pe^2 + \frac{1}{16} Pe^3 \ln Pe, \ Pe<1, \ Re\ small$$

$$Nu = 0.991\ Pe^{1/3}, \ Pe\ large, \ Re\ small$$

An undergraduate and Professor Finlayson used the finite element code to generate solutions for Peclet numbers between 1 and 1000, as well as Reynolds numbers between 1 and 50. That data was then correlated in the following form

$$\frac{1}{Nu - 2} = \frac{1}{Pe/2} + \frac{1}{0.9\ Pe^{1/3}\ Re^{0.11}}$$

This correlation agrees with the asymptotic formulas in their region of validity. Interestingly, it is an even better correlation of the heat transfer experimental data than the usual formulas derived from the data in the form

$$Nu = 2 + [\ 0.4\ Re^{1/2} + 0.06\ Re^{2/3}\ ]\ Pr^{0.4}$$

See Finlayson and Olson (1987). Westerberg and Finlayson (1990) showed that for very small Reynolds number the term $0.9Pe^{1/3}Re^{0.11}$ should be replaced by $0.89Pe^{1/3}$. The temperature profiles in different regimes clearly show the meaning of the Peclet number as a ratio of convection to conduction of energy. Figure 1 and 2 show temperature contours for cases with small and large Pe.

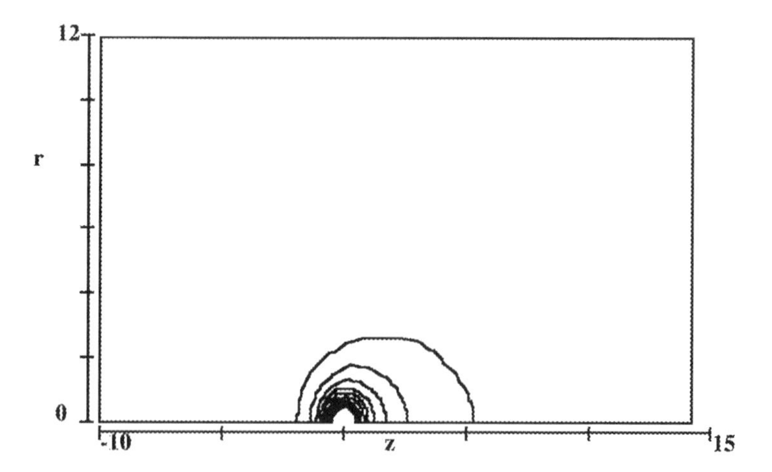

Figure 1. Temperature contours for Re = 1, Pe = 1.

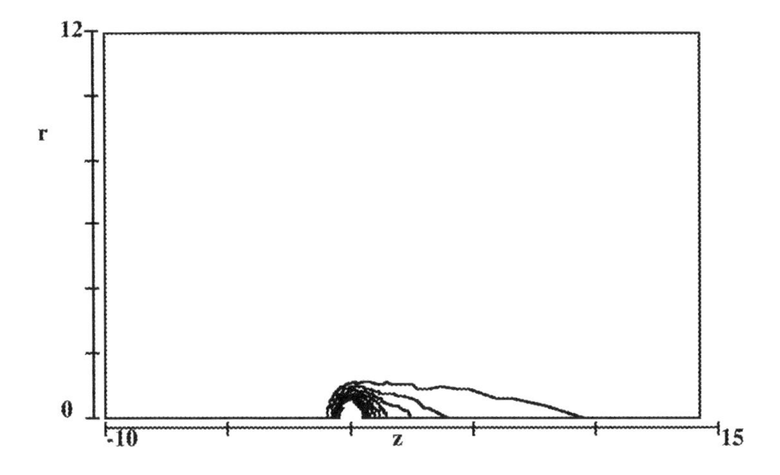

Figure 2. Temperature contours for Re=1, Pe=20.

Another example is the study of concentration distribution of adenosine diphosphate which is released from a platelet membrane during thrombus growth (Folie and McIntire 1989). The goal was to model thrombi of various shapes and dimensions. Because the finite element method can easily model changes in shape it is possible to study the effect of geometry.

Another example of a problem involving complicated geometry is the design of a thermal conductivity cell, as shown in Fig. 3. Calculations can determine the errors caused by conduction through the plexiglass and the heat losses to the sourrounds. One dimensional calculations should be done first to give an estimate, and the estimate can be checked with the more detailed calculations.

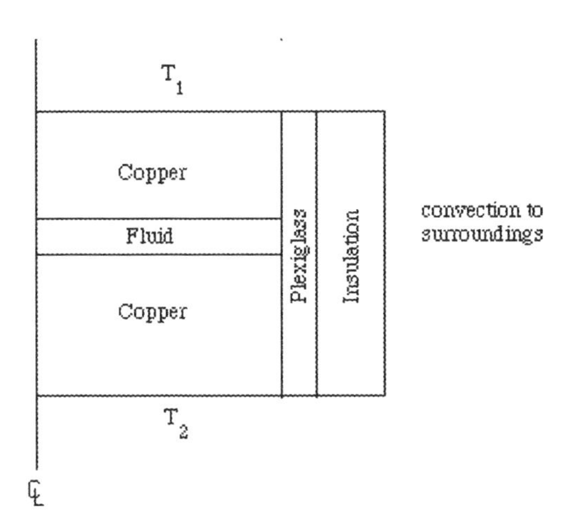

Figure 3. Thermal conductivity cell.

The last example is a problem to design a slotted-electrode electrochemical cell, as described by Orazem and Newman (1984). The problem can be reduced to that shown in Fig. 4, which is easily solved using the code for heat conduction. The student can then consider other designs (geometries) to insure that the current is uniform along the face.

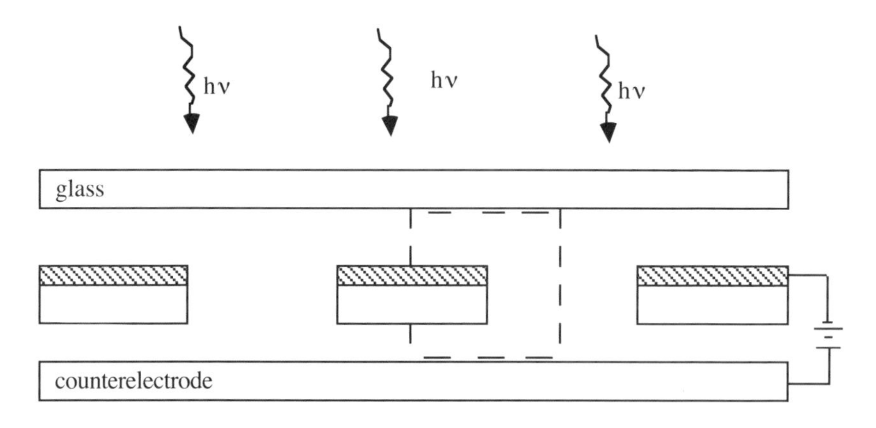

Slotted-electrode electrochemical cell

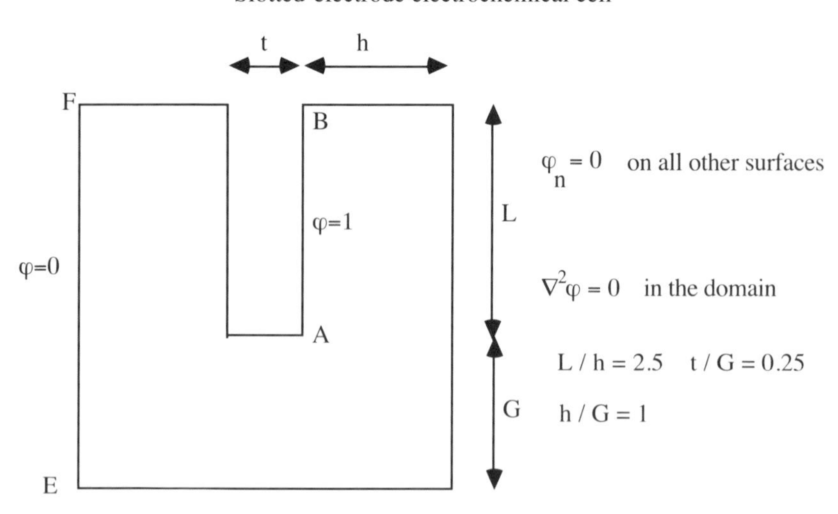

Figure 4. Problem description for slotted-electrode electrochemical cell.

Computational Fluid Dynamics

At the advanced undergraduate and graduate levels, the equations needed for the study of transport phenomena become more difficult. Two-dimensional transient, three spatial dimensions, nonlinear physical properties and convective model components are a few of the complications that arise with more realistic problem specifications. The first method of attack is to simplify the problem so that perturbation methods, Green's functions or transforms can be used (Finlayson, 1980; Denn, 1980; Leal, 1992).

In general, numerical methods will be needed to solve problems with complicated geometry or nonlinear effects. Computational fluid dynamics (CFD) is the name commonly applied to the study of fluid flows using computer simulation methods. Other transport processes, such as heat and mass transfer, are also included in CFD since many problems involve the solution of coupled transport processes. The use of numerical methods to solve transport process problems, with either coupled or uncoupled phenomena, has a long history that precedes the use of computers. The current availability of powerful computers and the development of sophisticated discretization algorithms for the solution of the partial differential equation (PDE) sets found in transport problems has led to the wide use of CFD in industry and academe.

The conservation equations for mass, momentum and energy require constitutive relations to form a closed set of equations. In general, the exact solution cannot be found and we seek an approximate solution to the state variables which will be defined at discrete points, or nodes, within the domain. There may be local approximating polynomials which interpolate the state variables between sets of neighboring nodes. The discretization method replaces the PDE set by an algebraic or differential equation set. For steady state problems, the PDE would be converted to a set of algebraic equations, the discrete equation set, with an equation for each state variable at each node. In transient problems, algorithms exist to generate algebraic equations for each state variable at each node for a given point in time. The method of lines, as applied to transient PDE sets, generates a set of differential equations which define the value of each state variable at each node as a function of time.

We can write a general steady state differential equation for the state variables as follows, which applies to the conserved species:

$$\nabla \bullet (\rho u \phi + (-\Gamma \nabla \phi)) - S = 0 \tag{1}$$

Given an approximation to ϕ called θ, which is inserted into the governing PDE,

$$\nabla \bullet (\rho u \theta + (-\Gamma \nabla \theta)) - S = R \tag{2}$$

The magnitude of R, the residual error, varies locally and measures how well the conservation equation is being satisfied at a particular point in space. The Method of Weighted Residuals (MWR) defines a set of algebraic relationships between the values θ_i (which are the approximated values of the state variables at each node) and its neighboring nodal values. The weighted integral of R is forced to zero through

$$\int_V WRdV = 0 \tag{3}$$

The choice of W, the weighting function, determines the type of method. The numerical schemes that are most often used include: finite differences, finite volumes, finite elements and collocation methods. The *finite difference* method has a value of W=1 at node i, zero elsewhere, and thus the residual is forced to zero at each node. The derivatives of the state variables are approximated using Taylor Series expansions. If the approximation to θ is given by a polynomial and the weighting function is W=1 at specified sampling points, the method is called *collocation* (Finlayson, 1980). *Finite element* methods use a definition of a polynomial for θ over a small region within the domain with the values of the state variables at defined nodes such that

$$\theta(x,y,z) = \sum_i^{(e)} N_i(x,y,z)\,\theta_i^{(e)} \tag{4}$$

The terms N_i are known as shape functions and are defined within each element such that their value is 1 at the node i, a value between 0 and 1 within the element and zero outside the element. The choice of W defines the particular type of finite element method. For example, W= N_i defines the widely used Galerkin finite element method.

$$\int_{\forall(e)} N_i \left[\nabla \bullet \left(\rho u \left(\sum_i^{(e)} N_i^{(e)} \theta_i \right) + \left(-\Gamma \nabla \left(\sum_i^{(e)} N_i^{(e)} \theta_i \right) \right) \right) - S \right] dV = 0 \tag{5}$$

Finally, *finite or control volume* methods can be derived by starting with W=1 within a control volume around the node i and zero outside of this subdomain (Patankar, 1980). The integral can then be broken into a volume integral and a surface integral which satisfies the integral form of the conservation equations

$$\int_{A^i} (\rho u \theta + (-\Gamma \Delta \theta)) \bullet n dA - \int_{V^i} S dV = 0 \tag{6}$$

Variations on the finite volume methods depend on how the control volume is chosen around the nodal points and the manner in which gradient terms are approximated using neighboring nodes and their respective control volumes. Thus, from a basic definition, one can derive the basic versions of all the popular methods used in solving advanced CFD problems. The methods share the common feature of gridding where the network of nodes, which will define the points at which the state variables will be determined, is chosen and some methods require the definitions for the elements or subdomains. The basic steps for all the methods are the same: choose the conservation equations and terms within the equations appropriate to the problem, divide the domain into an appropriate set of nodes and subdomains/elements, assemble the discretized forms of the PDEs as (non)linear algebraic equations and solve the equation set.

Graduate courses which use the concepts described above vary widely in content depending on the objectives for the student. One approach is to do an overview of the different discretization methods and then provide a series of problems which require a numerical analysis. Commercial CFD simulators are widely used both in industry and academe, for example: FLUENT/BFC which uses the finite volume method (Fluent Inc., 1990), FIDAP which uses the finite element method (FDI, 1993) and NEKTON which uses the spectral element method (Fluent Inc., 1992). Readers are referred to the annual Software Directory published by *Chemical Engineering Progress* for current listings of available software applicable to chemical engineering problems and the article by Wolfe (1991). Commercial simulators have a variety of tutorial problems based on practical problems taken from the literature. It is common to ask students to modify existing meshes or modify boundary conditions to simulate another problem. It is important to recognize the amount of time it takes to generate a good mesh for the solution of realistic problems. The general rule-of-thumb is that 80% of the user's time will be in the development of the mesh. In posing problems for students it is important to keep the time requirements in mind when dealing with complicated geometry when an existing mesh is not available to the student. The problems that are currently available in commercial CFD package libraries span a wide spectrum of applications and include: crystal growth, polymer flows, porous media flows, electronic packaging cooling and particle-laden fluid flows. All the packages have an interface to develop the mesh and post-processing capabilities to visualize the resulting flow velocities and other state variables in the domain. Though implementations exist for the personal computer platform (PC-DOS and Macintosh), most realistic problems require a workstation.

It is very important to build up the expertise of the student by encouraging the solution of problems for which experimental data are available. The amount of data produced by a CFD code can be overwhelming, particularly in 3-D problems, and so a thorough grounding in the expected flows and fluxes of well-characterized problems provides a good basis for analysis of more complicated problems. The effects on the calculated state variable values of mesh density, coupling of transport processes and sensitivity of the solution to physical property parameters requires skills gained through experience. It is too easy to determine a steady-state solution to a problem for which no steady-state exists!

Another approach for teaching CFD courses is to develop a particular method in the classroom and have the students implement subroutines as they learn the important concepts of the method. For example, students may implement a basic meshing algorithm, the formulation of the discretized equations and the linear equation solver. Visualization software should be used to simplify analysis of the results. This type of course gives the student a better perspective on the development of the equations and the problems in solving the final set of equations, but limits their exposure to more complicated problem formulations. Sample reference textbooks for each of the methods as applied to CFD include: Anderson et al. (1984) for finite differences, Baker (1983) and Dhatt and Touzot (1984) for finite elements and Patankar (1980) for control volumes. Minkowycz et al. (1988) and Fletcher (1988) provide chapters dedicated to the use of different numerical schemes common in CFD applications.

In summary, CFD and the numerical solution of the transport equations are important fields which are widely taught at the graduate level and are beginning to be exposed in the undergraduate curriculum. The wider use of CFD has been accelerated by the combined availability of faster computers, more sophisticated numerical solution techniques and graphic tools to visualize the results.

References

Anderson, D.A., J.C. Tannehill and R.H. Pletcher (1984). *Computational Fluid Mechanics and Heat Transfer*, Hemisphere, New York.

Baker, A.J. (1983). *Finite Element Computational Fluid Mechanics,* Hemisphere, New York.

Bird, R.B., W.E. Stewart, E.N. Lightfoot (1960). *Transport Phenomena*, Wiley, New York.

Denn, M.M. (1980). *Process Fluid Mechanics*, Prentice-Hall, New York.

Dhatt, G., and G. Touzot (1984). *The Finite Element Method Displayed*, John Wiley & Sons, New York.

Fahien, R. (1983). *Fundamentals of Transport Phenomena*, McGraw-Hill, New York.

Finlayson, B.A. (1980). *Nonlinear Analysis in Chemical Engineering*, McGraw-Hill, New York.

Finlayson, B.A. and J.W. Olson (1987). Heat Transfer to Spheres at Low to Intermediate Reynolds Numbers. *Chemical Engineering Communications*, **58**, 431-447.

Fletcher, C.A.J. (1988). *Computational Techniques for Fluid Dynamics, Volumes 1 and 2*, Springer-Verlag, New York.

Fluent Inc. (1980). *FLUENT and FLUENT/BFC 4.0 Manual*, Lebanon, New Hampshire.

Fluent Inc. (1992). *NEKTON 2.85 Manual*, Lebanon, New Hampshire.

Fluid Dynamics International, Inc. (1993). *FIDAP 7.0 Manuals*, Evanston, Illinois.

Folie, B.J. and L.V. McIntire (1989). Mathematical Analysis of Mural Thrombogenesis. *Biophys. J.*, **56**, 1121-1141.

Leal, L.G. (1992). *Laminar Flow and Convective Transport Processes: Scaling Principles and Asymptotic Analysis*, Butterworth-Heinemann, Boston.

Minkowycz, W.J. et al. (1988). *Handbook of Numerical Heat Transfer*, John Wiley & Sons, Inc., New York.

Orazem, M. and J. Newman (1984). Primary Current Distribution and Resistance of a Slotted-Electrode Cell. *J. Electrochem. Soc.*, **131**, 2857-2861.

Patankar, S.V. (1980). *Numerical Heat Transfer and Fluid Flow*, Hemisphere, New York.

Westerberg, K.W. and B.A. Finlayson (1990). Heat Transfer to Spheres from a Polymer Melt, *Numerical Heat Transfer, Part A*, **17**, 329-348.

Wolfe, A. (1991). CFD Software: Pushing Analysis to the Limit, *Mechanical Engineering*, January, 48-54.

SEPARATIONS PROCESSES

Ross Taylor
Clarkson University
Potsdam, NY 13699

Abstract

The equilibrium stage has been used for modeling separations process problems for nearly 100 years. The model equations are particularly amenable to computer solution, and teachers of separations courses were among the first to introduce computers into undergraduate chemical engineering curricula. CACHE has supported development and use of interactive instructional modules for the separations area, and has promoted distribution of the powerful *ChemSep* multicomponent distillation design tool for use in separations courses.

In recent years, it has become possible to model separations processes using mass-transfer, rate-based *nonequilibrium* models. These nonequilibrium models are likely to see increased use in the future, particularly for simulating nonideal systems, systems with reaction, and/or processes with multiple feeds and products. This trend toward introduction of nonequilibrium (and also dynamic) models should act as a spur for closer integration of separations and mass-transport courses.

Introduction

A recent advertisement for a journal claims that there are 40,000 distillation columns and that operating these columns requires 7% of all of the energy consumed in the United States alone. When we consider all of the distillation columns elsewhere in the world, as well as all the closely related operations of absorption, stripping and extraction, it is clear that, even if these figures are significantly in error (other published figures put the energy consumption at about 3%), the classical separation processes are unusually important unit operations.

Chemical engineers (whether they are professionals in industry or university students) have been solving separation process problems using the *equilibrium stage model* for about 100 years (since Sorel first used the model for the distillation of alcohol). The key assumption in the equilibrium stage model is that the vapor and liquid streams leaving a stage are in equilibrium with each other. The equations that model equilibrium stages are known as the *MESH* equations. The *M* equations are the Material balance equations, the *E* equations are the equilibrium relations, the *S* equations are the mole fraction summation equations, and the *H* equations are the enthalpy balance equations. The unknown variables determined by solving these equations are the mole fractions of both phases, the stage temperatures, and the flow rate of each phase.

There can be few other mathematical models in any branch of engineering which are so well suited to computer solution and that have prompted the development of so many truly different algorithms as have the *MESH* equations of the equilibrium stage model. Indeed, it would not be too far from the truth to claim that it is equilibrium stage calculations that brought computing into chemical engineering (or should that be chemical engineers to computers?). Since computers became available in the late 1950s hardly a year has passed without the publication of at least one new algorithm for solving the equilibrium stage model equations and, in many years, several new algorithms have appeared. Most of the better numerical methods (and some of the not so good ones) have been used in the solution of equilibrium stage problems. Indeed, a good test of a numerical method might therefore be to see if it can be used to solve distillation problems.

It is not our intention to provide a complete review of computer-based methods for separation process calculations; readers can consult one of several textbooks that includes some discussion on computer-based methods (Smith, 1964; King, 1980; Henley and Seader, 1981). Other books have focused almost entirely on the computational aspects (Holland, 1963, 1975, 1981) and Seader (1985) gives an interesting history of equilibrium stage separations calculation methods with more detail than is appropriate this brief article.

CACHE Contributions

Nowadays, even chemical engineering students solve their distillation problems using computer software. In this section we review the part that CACHE has played in making this possible.

While not a CACHE product, it would be remiss of us not to mention the collection of programs for separation process simulation by Hanson, Duffin, and Somerville (1962). Their book included several programs in Fortran for solving a wide variety of multistage separation process problems including distillation, absorption, stripping and liquid-liquid extraction. A chapter at the end of the book discusses the peculiar difficulties associated with interlinked columns. Several introductory chapters serve to introduce the reader to "a sufficient number of techniques... to make possible convergent solutions to *any* problem" (emphasis added). The optimism expressed in the above quotation is interesting for, despite many developments that have taken place over the past 30 years, we still have not reached that happy (for some) situation where all of our equilibrium stage separation process problems can be solved the first time they are attempted. We are, however, a great deal closer to that goal.

Included in the CACHE series *Computer Programs for Chemical Engineering Education* published in 1972 (and mentioned elsewhere in this volume as well) was a volume entitled *Stagewise Computations* (Christensen, 1972). The paperback edition included descriptions, examples of usage, and listings of 17 Fortran programs for modeling such operations as counter-current leaching, liquid-liquid extraction, batch distillation, and multicomponent distillation, stripping and absorption. The programs in this book represented a cross-section of computerized implementations of old-fashioned calculation methods, originally devised for solving separations process problems by hand, and programs that implemented methods that better represented the state of the art as it was at the time. These programs were used at many universities around the world.

Among the programs in the CACHE collection are two by N.S. Berman and O.C. Sandall that implement the stage-to-stage calculation method developed by Lewis and Matheson (1932) (and refined by others). D.M. Watt provided a program that used another algorithm from the same era, the method of Thiele and Geddes (1933). L.L. Hovey provided an implementation of the Amundsen - Pontinen (1957) method. This was one of the first methods to exploit matrix algebra in the solution of counter-current separations problems.

`WHENDI` by P.J. Johansen and J.D. Seader implements the Wang and Henke (1966) bubble point method for distillation, which is a modification of the Amundsen - Pontinen (1957) method. `STAB`, by the same authors plus T. Shinohara, solves absorption and stripping problems using the sum-rates method (Sujata, 1961; Burningham and Otto, 1967). Bubble point and sum-rates methods belong to a class of algorithm known as equation-tearing methods and have been used in industry for several decades. These methods pair model equations and variables in two loops and alternately solve the equations in each loop until convergence is reached. Bubble point methods use temperatures as inner loop variables and the total vapor flow profile is adjusted in an outer loop by solving the energy balances. Sum-rates methods fix the temperature and pressure profile for the column and the mass balance and phase equilibrium equations are solved simultaneously for the component flow rates in an inner loop. The energy balance equations are solved in an outer loop in order to adjust the temperature profile.

J.W. Tierney contributed a program that could handle interlinked columns. The program employed an algorithm that falls into the class of 2N Newton methods that Tierney and his co-workers helped to develop in the late 1960s and early 1970s (Tierney and Bruno, 1967; Tierney and Yanosik, 1969). In these methods the temperatures and flows are adjusted simultaneously using Newton's method.

Newton's method now is widely used in commercial simulation programs (although it has taken many years longer than it should to have gained the measure of acceptance it now enjoys). To the best of our knowledge, a method to solve *all* the MESH equations for all stages at once using Newton's method was first described by Whitehouse (1964). Among other things, Whitehouse's code could solve problems involving purity specifications or of T, V, L or Q on any stage. Interlinked systems of columns and nonideal solutions also could be handled. Unfortunately, Whitehouse's work was published in a rather obscure proceedings volume (Stainthorp and Whitehouse, 1967) and had essentially no influence on the development of computer-based simulation methods. The credit for showing us how to use Newton's method for separation process calculations, therefore, goes to Naphtali and Sandholm, (1971) and to Goldstein and Stanfield (1970) for distillation and to E.C. Roche (1971), who applied the method to liquid-liquid extraction and who contributed the progran`LIQLIQ` that is included in the CACHE collection.

In 1987 CACHE distributed to supporting departments a collection of computer-based lessons containing "open-ended" design type problems for use in courses other than the capstone design course (Seider, 1987). Among the six lessons were three for separations courses: Supercritical fluid extraction (by J. Kellow, M.L. Cygnarowicz, and W.D. Seider, Pennsylvania), Gas Absorption with Chemical Reaction (by K. Nordstrom and J.H. Seinfeld, Cal Tech), and Design of Flash Vessels and Distillation Towers (by B.A. Finlayson, E.W. Kaler, and W.J. Heideger, Washington) based on short-cut methods of column design. The major change in computing practices brought about by the personal computer in the decade of the 1980s is evident in this release; the material was provided on floppy disks for use with IBM PC compatible computers.

In 1992 CACHE made available *ChemSep*, a computer software system developed by R. Taylor and H.A. Kooijman (1992). As a piece of software *ChemSep* is much more closely related to flowsheet simulation programs than to the computer-based lessons and modules discussed above. The program was created for use in university separations courses (but may also be useful in thermodynamics and design) but also can be (and is) used by professionals in industry. *ChemSep* features a menu-driven user interface and simulation programs that can handle flash, and single distillation, absorption and liquid-liquid extraction columns with multiple feeds and sidestreams. Many of the most widely used thermodynamic models are available in *ChemSep* and Newton's method is used to solve the large, sparse system of nonlinear equations. The CACHE version of the package includes a database for 189 different chemicals and can handle up to 10 components and 100 stages.

H.S. Fogler and S. M. Montgomery (1993) have created a collection of interactive (personal) computer modules for chemical engineering instruction. There are five of these modules for separations courses including: BASIS - An introduction to Separation Processes, CASCADES - Liquid-liquid Extraction, MCCABE - Binary Distillation (via the McCabe-Thiele method), ABSORP - Packed absorber design, and MEMBRANES - Spiral membrane process optimization. These modules feature animations within a graphical user interface that allow students to interactively review material and to carry out simulated experiments (see Figs. 1 and 2).

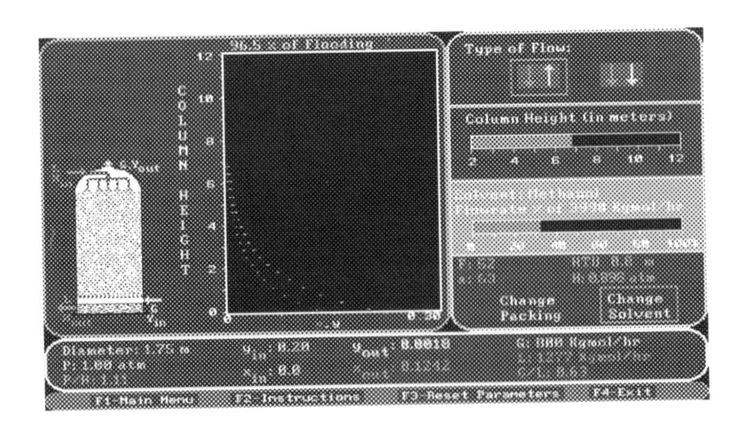

Figure 1. Screen image from CACHE module ABSORP for absorption column design.

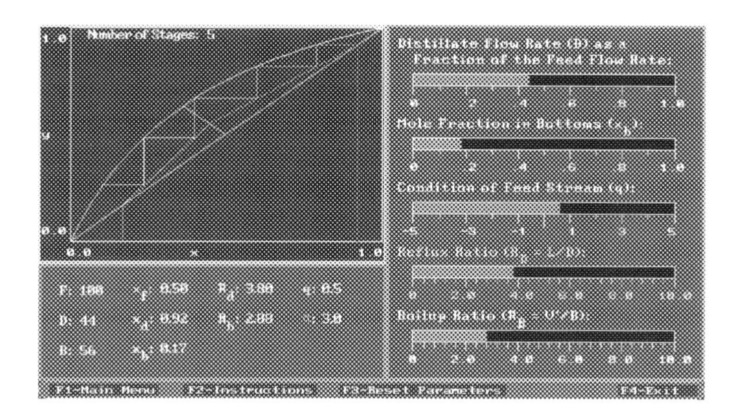

Figure 2. Screen image from CACHE module MCCABE for binary distillation.

It is interesting to contrast the first CACHE programs with their more recent offerings. `WHENDI`, for example, is just 635 lines of Fortran. Contrast this with *ChemSep* which consists of about 100,000 lines of Fortran (of which about 10,000 is just the part that computes physical properties) and Turbo Pascal. The comparison is, of course, not entirely fair since `WHENDI` was never intended to be able to handle the range of problems that can be tackled with *ChemSep*. Nevertheless, it serves to illustrate one of the negative aspects of modern software, a trend to ever larger software systems; a trend that, fortunately, is countered by the ever increasing speed and capacity of computer hardware.

The Impact of Computer-Based Tools on Separations Courses

Perhaps nowhere else in the chemical engineering curriculum has the impact of computers been greater than in the teaching of separations. The availability of computer-based tools such as *ChemSep* make it possible to include realistic multicomponent distillation design problems in the first undergraduate course on separations. In spite of these developments, the graphical McCabe-Thiele method, devised in the 1920s, combines simplicity and elegance and remains a useful method for the analysis of distillation type operations, even for multicomponent systems, as shown in Fig. 3. Indeed, graphical techniques (McCabe-Thiele diagrams and column profiles such as those in Figs. 4 and 5) for the visualization of numerical simulation results are extremely valuable instructional tools. (With the exception of Figs. 1 and 2, all of the illustrations in this article were created by *ChemSep*, printed to a file using one of several graphic image formats supported by the program, and imported into the word processor used to create this article.)

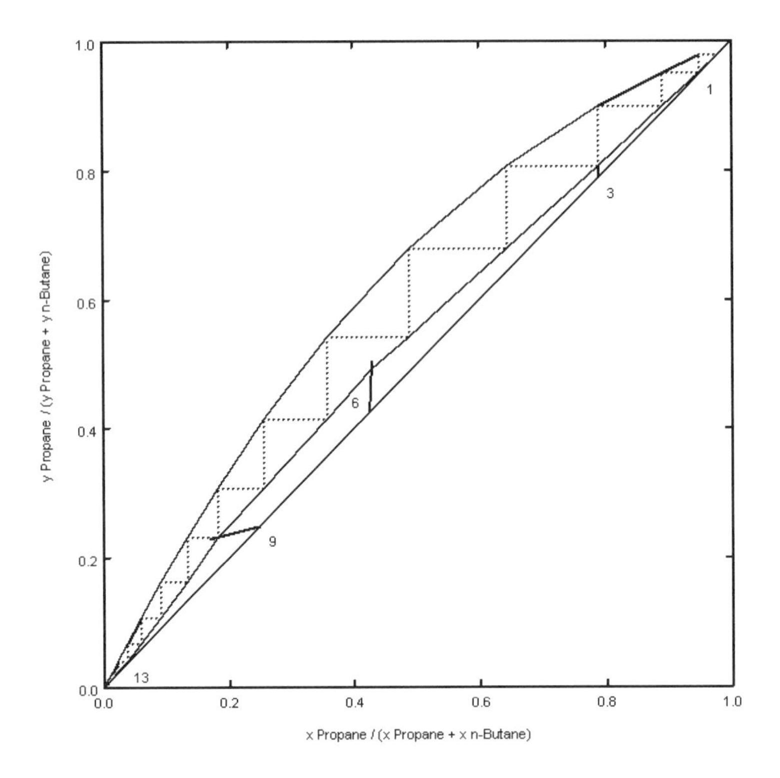

Figure 3. Modified McCabe-Thiele diagram for an example involving a five component system. This particular example accompanies the CACHE program`WHENDI` *(see Christensen, 1972, p. 384; Henley and Seader, 1981, p 568) although, in this case, the results were obtained with the CACHE program* ChemSep.

Several topics that are covered at length in some separations texts no longer need to be covered in the first course on separations. Once-popular graphical methods (Ponchon diagrams, for example) are being dropped from undergraduate courses as computer-based methods are used more frequently. It is no longer necessary to teach undergraduate students all (or even any) of the gory details behind the algorithms used to solve multicomponent distillation problems; computer software has become sufficiently reliable that our students no longer need to know *how* the model equations are solved. It is, however, useful to provide some sort of instruction on how to use computer tools to solve open-ended separation process engineering problems. For example, students need to know what specifications are most likely to allow computer-based algorithms to find converged solutions (and what specifications may lead them into trouble). What can we do in the event that convergence is not obtained? What process variables can be changed in order to improve the operation and/or design? Short-cut methods, based on several more or less limiting assumptions, remain useful for preliminary design.

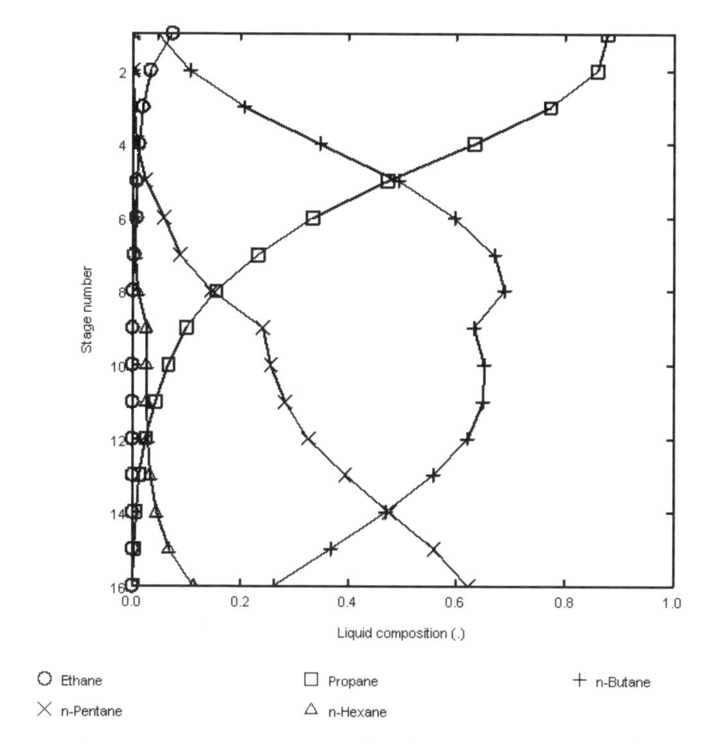

Figure 4. Composition profiles for the column in Fig. 3.

Material that can be relegated to a second (elective) course on separation process modeling and simulation includes details of the algorithms that are used to solve the *MESH* equations, methods for the solution of large sparse linear systems of equations, Newton's method for solving systems of nonlinear equations (although, in view of the versatility and power of this approach, a case could be made for covering the method elsewhere in the curriculum), problems associated with Newton's method such as its sensitivity to the initial estimates, and homotopy-continuation methods for solving difficult problems.

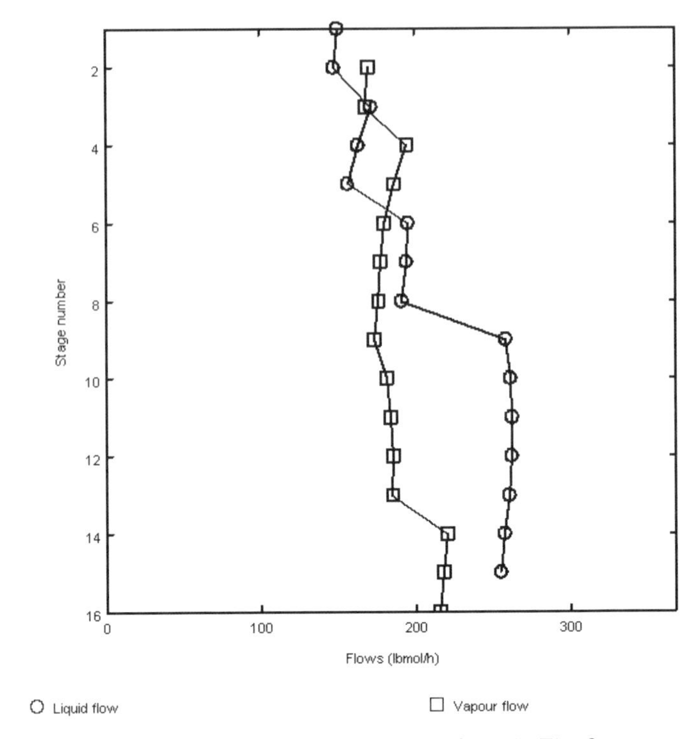

Figure 5. Flow profiles for the column in Fig. 3.

Future Trends

As noted above, the equilibrium stage model has been part of chemical engineering - and of a chemical engineer's education - for over a century. It is, of course, well known that the trays in multicomponent distillation columns do not operate at equilibrium. The usual way around this problem is to use an efficiency factor of some kind and separations texts discuss the evaluation of efficiencies at some length. Using efficiencies introduces new problems and the drawbacks of these quantities also are well known (see, for example, Seader, 1989).

In recent years it has become possible to model separation processes as the mass transfer rate-based operations they really are (Krishnamurthy and Taylor, 1985; Taylor and Krishna, 1993). Efficiencies are not used in these *nonequilibrium* (or *mass transfer rate-based*) *models* (although they may be calculated *after* a simulation has been completed, Fig. 6). The building blocks of these models include the mass and energy balances, equilibrium and summation equations that are familiar to us from the equilibrium stage model. There is, however, a fundamental difference in the way these equations are used. In a nonequilibrium model the balance equations are written for each phase rather than for the stage as a whole. The K-values are evaluated at the temperature, pressure, and composition of the interface, which is assumed to be an equilibrium surface, offering no resistance to mass transfer. Equations that model the mass and energy transfer across the phase boundary are included in the model and solved simultaneously with the other equations. Mass and energy are transferred across the interface at rates that depend on the extent to which the phases are *not* in equilibrium with each other. These rates are calculated from models of mass transfer in multicomponent systems (see Taylor and Krishna (1993) for an extended discussion of such models). Mass and heat transfer coefficients and interfacial areas must be computed from empirical correlations or theoretical models. These coefficients depend on the column design as well as on its method of operation. Nonequilibrium models can be used to simulate packed columns just as easily as they can tray columns, thereby avoiding the use of HETPs (Height Equivalent to a Theoretical Plate). A nonequilibrium column model was included in Version 3 of the CACHE product ChemSep.

It is very likely that nonequilibrium models will see increased use in the future; they will be particularly useful for simulating nonideal systems, systems with chemical reaction, and low-efficiency systems (this includes some nonideal separations and many gas absorption processes), and columns with multiple feeds and products. A thorough understanding of mass transfer (with or without simultaneous chemical reaction) in multicomponent systems will be essential for the engineer (and student) working with nonequilibrium models. It is pertinent to point out that mass transfer in binary systems constitutes a very special case in that none of the interesting phenomena that can take place in systems with more than two components can occur. Multicomponent mass transfer effects can manifest themselves in interesting ways, including causing distillation point efficiencies of different components to *not* be equal from component to component or from tray to tray (as they are so often assumed to be) as is clearly evident in Fig. 6 and, to a lesser degree, Fig. 7. Any emphasis on mass transfer rate based models in the future should act as a spur to educators to more tightly integrate courses on separations and mass transport; the latter will need to be modified in order to cover the elements of multicomponent mass transfer (see Taylor and Krishna, 1993).

The equations that describe multicomponent mass transport (the Maxwell-Stefan equations) have, in fact, been with us even longer than has the equilibrium stage model although their application to modeling stagewise separations is a relatively recent development. However, not only do the Maxwell-Stefan equations allow us to model mass transfer in conventional operations such as distillation, absorption, and liquid-liquid extraction, they also describe mass transfer in all of the less common separation operations such as membrane processes, the ultracentrifuge, thermal diffusion columns, and many more. Indeed, the Maxwell-Stefan formulation of mass transfer can provide a basis for unifying the treatment of separation processes (Krishna, 1987).

Models of the dynamic behavior of column performance are being used more frequently in industry. Improvements in hardware and software will mean that ten years from now there will be no excuse for not using a dynamic model to explore the operation of a real distillation column. Dynamic models will become available that are based on fundamental mass and energy transfer processes. These models will also be used in undergraduate courses and will make it possible to provide more convincing software demonstrations of how real columns behave.

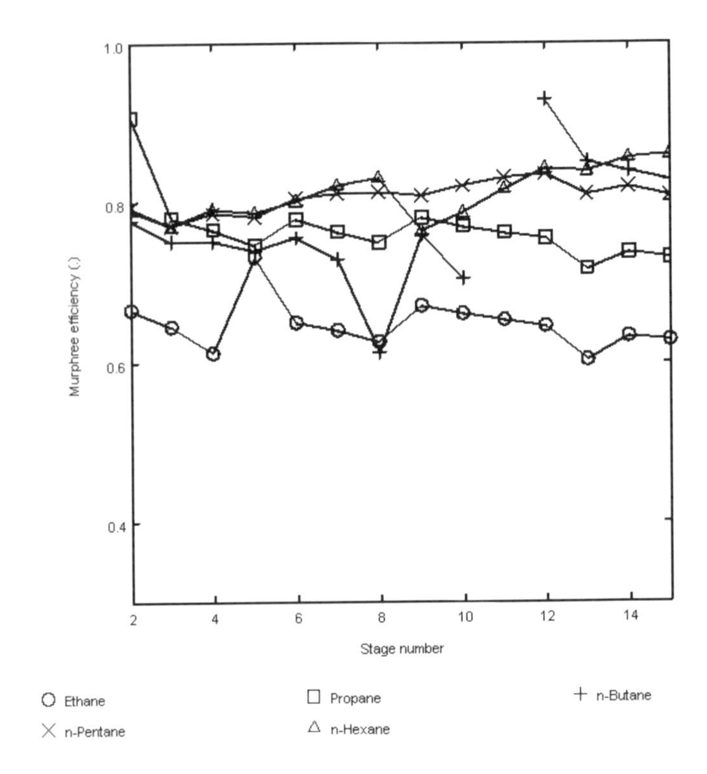

Figure 6. Efficiency profiles calculated following a nonequilibrium simulation of the column in Figs. 3 to 5. The usual assumption of equal component efficiencies is clearly inappropriate.

Of course, models based on equilibrium stage concepts will not be abandoned, nor is there any need to do so. New approaches to process visualization (see Fig. 8), and design and analysis of separation processes using equilibrium stage models have been pioneered by M.F. Doherty, M.F. Malone and their coworkers at the University of Massachusetts (see, for example, Fidkowski *et al*., 1991; Doherty and Buzad, 1992; Julka and Doherty, 1993). At the time of writing, however, there is no software product that implements their methods that is available from CACHE.

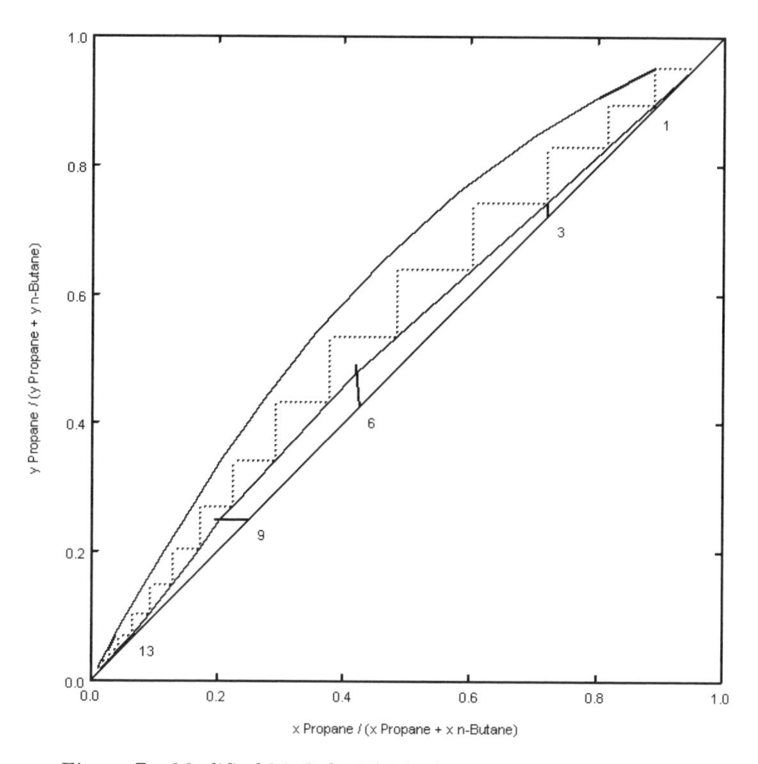

Figure 7. Modified McCabe-Thiele diagram provides an interesting view of the effects of mass transfer in stagewise separations. Note how the triangles representing the trays do not reach the equilibrium line, demonstrating that they are not equilibrium stages.

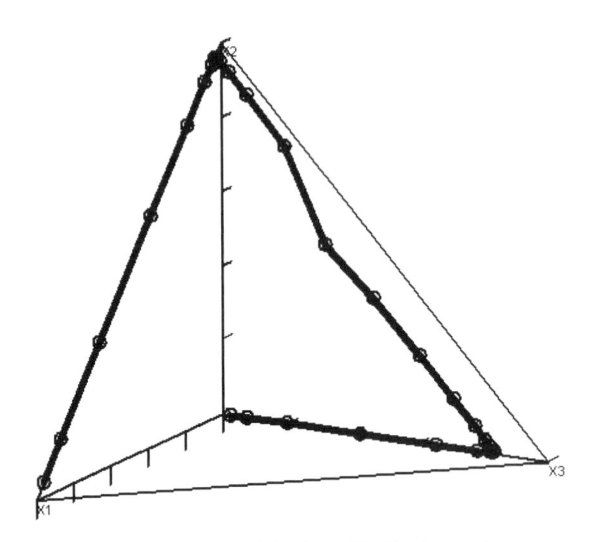

Figure 8. Composition profiles in a distillation column processing a four component system. This illustration is based on an exercise in Henley and Seader (1981). Graphics inspired by, for example, Julka and Doherty (1993).

References

Amundsen, N.R., and A.J. Pontinen, Multicomponent Distillation Calculations on a Large Digital Computer, *Ind. Eng. Chem.*, **50**, 730 (1958).

Burningham, D.W., and F.D. Otto, Which Computer Design for Absorbers?, *Hyd. Proc.*, **40** (10), 163 (1967).

Christensen, J. (Ed.) *Computer Programs for Chemical Engineering Education: Stagewise Computations*, CACHE (1972).

Doherty, M.F. and Buzad, G., Reactive Distillation by Design, *Trans. I. Chem. E.*, **70**, 448-458 (1992).

Fidkowski, Z.T., Malone, M.F., and Doherty, M.F., Nonideal Multicomponent Distillation: Use of Bifurcation Theory for Design, *AIChEJ*, **37**, 1761-1779 (1991).

Fogler, H.S. and Montgomery, S.M. *Interactive Computer Modules for the IBM-PC: Separations*, CACHE (1993).

Fogler, H.S. and Montgomery, S.M. Interactive Computer Modules for the IBM-PC, CACHE News, **37**, 1-5 (1993).

Goldstein, R.P. and R.B. Stanfield, Flexible Method for the Solution of Distillation Design Problems using the Newton-Raphson Technique, *Ind. Eng. Chem. Process Des. Dev.*, **9**, 78 (1970)

Hanson, D.N., Duffin, J.H., and Somerville, G.F. (1962). *Computation of Multistage Separation Processes*, Rheinhold, New York.

Henley, E.J., and J.D. Seader (1981). *Equilibrium-Stage Separation Operations in Chemical Engineering*, Wiley.

Holland, C.D. (1963). *Multicomponent Distillation*, Prentice Hall, Inc., NJ.

Holland, C.D. (1975). *Fundamentals and Modeling of Separation Processes*, Prentice Hall, Inc., NJ.

Holland, C.D. (1981). *Fundamentals of Multicomponent Distillation*, McGraw-Hill, Inc.; New York.

Julka, V. and Doherty, M.F. (1993). Geometric Nonlinear Analysis of Multicomponent Nonideal Distillation: A Simple Computer-Aided Design Procedure, *Chem. Eng. Sci.*, **48**, 1367-1391.

Kooijman, H.A. and Taylor, R. (1992). *ChemSep* - Another Software System for the Simulation of Separation Processes, *CACHE News*, **35**, 1-9.

Krishna, R. (1987). A Unified Theory of Separation Processes Based on Irreversible Thermodynamics, *Chem. Eng. Commun.*, **59**, 33-64.

Krishnamurthy, R. and Taylor, R. (1985). A Nonequilibrium Stage Model of Multicomponent Separation Processes. I - Model Development and Method of Solution, *A.I.Ch.E.J.*, **31** (3), 449-456.

Lewis, W.K. and G.L. Matheson (1932). Studies in Distillation: Design of Rectifying Columns for Natural And Refinery Gasoline, *Ind. Eng. Chem.*, **24** (5), 496.

Naphtali, L.M. and D.P. Sandholm (1971). Multicomponent Separation Calculations by Linearization, *AIChE* J., **17** (1), 148.

Roche, E.C. (1971). General Design Algorithm for Multistage Counter Current Equilibrium Processes, *Brit. Chem. Eng.* **16**, 821.

Seader, J.D. (1985). The BC (Before Computers) and AD of Equilibrium Stage Operations, *Chem. Eng. Ed.*, Spring, 88.

Seader, J.D. (1989). The Rate-Based Approach to Modeling Staged Separations, *Chem. Eng. Progress*, 41-49, October.

Seider, W.D. (1987). *CACHE IBM PC Lessons for Chemical Engineering Courses Other Than Design and Control*, CACHE.

Stainthorp, F.P., and Whitehouse, P.A. (1967). General Computer Programs for Multistage Counter Current Separation Problems - I: Formulation of the Problem and Method of Solution, *I. Chem. E. Symposium Series*, **23**, 181.

Smith. B.D. (1964). *Design of Equilibrium Stage Processes*, McGraw-Hill, New York.

Sujata, A.D. (1961). Absorber-Stripper Calculations Made Easier. *Hyd. Proc.*, 40 (12), 137.

Taylor, R. and Krishna, R. (1993). *Multicomponent Mass Transfer*, Wiley, New York.

Thiele, E.W., and R.L. Geddes (1933). Computation of Distillation Apparatus for Hydrocarbon Mixtures, *Ind. Eng. Chem.*, **25**, 289.

Tierney, J.W. and J.A. Bruno (1967). Equilibrium Stage Calculations, AIChE J., **13** (3), 556.

Tierney, J.W., and J.L. Yanosik (1969). Simultaneous Flow and Temperature Correction in the Equilibrium Stage Problem, *AIChE J.*, **15** (6), 897.

Wankat, P.C. (1988). *Equilibrium Staged Separations*, Elsevier, New York.

Wang, J.C. and G.E. Henke (1966). Tridiagonal Matrix for Distillation, *Hyd. Proc.*, **45** (8), 155.

Whitehouse, P.A. (1964). PhD Thesis in Chemical Engineering, University of Manchester Institute of Science and Technology, April.

CONCEPTUAL DESIGN AND PROCESS SYNTHESIS

James M. Douglas
University of Massachusetts
Amherst, MA 01003

Jeffrey J. Siirola
Eastman Chemical Company
Kingsport, TN 37662

Abstract

Systematic approaches for the invention of conceptual chemical process designs have been proposed and discussed for more than twenty-five years, and have now been developed to the point of industrial application. Process synthesis has also now become an integral part of many chemical engineering design curricula. CACHE has contributed to the promotion and advancement of conceptual design and process synthesis methods through case studies, educational software products, symposia, and the sponsorship of process synthesis sessions within the *Foundations of Computer-Aided Process Design* conference series. Although much progress has been made, continuing research advances suggest that significant improvements in industrial and educational conceptual design and process synthesis methods and tools can be expected in the near future.

Introduction

When CACHE was established in 1969, design was the one aspect of chemical engineering that had most readily incorporated digital computing into the curriculum. Algorithms and software were becoming widely used for both the design of individual unit operations, and for the simulation of entire process flowsheets.

At that time, design was taught basically as a course in analysis. A flowsheet was given and students were asked to solve for the process material balances, energy balances, equipment sizes, utility flows, capital and operating costs, and the profitability of the process. The AIChE Student Contest problems and the Washington University Design Case Study Series followed this general format. Simulators were ideally suited to problems of this type. At this level of detail, more or less rigorous models were developed for the most expensive process units in order to optimize the design. The generation of process flowsheet alternatives themselves, however, received very little attention in design textbooks except to note that the activity requires creativity.

About 1970 academics became interested in the possibility of developing systematic procedures for inventing process flowsheets. This effort was called process synthesis because the inputs to and outputs from the process were fixed, and the goal was to find flowsheets that could transform the inputs into the outputs. Preliminary analysis steps (often based on short-cut calculations) accompanied the synthesis steps to help make choices among alternative flowsheet structures. CACHE recognized the potential of this research and sponsored the first *Symposium* in the emerging field of computer-aided chemical process synthesis in 1972 (Seader, 1972).

Early Process Synthesis Explorations

Rudd (1968) suggested perhaps the first organized framework for design based on the notion of *systems decomposition*. He suggested that for any original design problem, a number of simpler specific subproblems should be identified the solutions to which when put together would plausibly solve the original problem. Typical chemical process design subproblems might include raw material receiving and storage, feed conditioning, reaction, product isolation, purification, storage, and shipping, and so on. Some sort of systems decomposition is a common conceptual design practice.

King and coworkers (1972) proposed an alternative *evolutionary modification* synthesis approach patterned on another industrial conceptual design practice. This approach starts with an existing flowsheet for the desired product or a similar product and then identifies aspects of the design that could be changed to greatest advantage to better meet the objectives of the specific case at hand as well as alternatives to implement that change. Various heuristic and analytical methods were proposed to identify evolutionary opportunities, but the success of the scheme depended strongly on the initial flowsheet.

As a third approach, Ichikawa and coworkers (1972) viewed process synthesis as an optimization over process structure and approached conceptual design through the application of operations research and mathematical programming techniques. This analysis-dominated approach starts with a larger superflowsheet which contains embedded within it many redundant alternatives and interconnections among them and then systematically strips the less desirable parts of the superstructure away. This *superstructure optimization* offers the promise of simultaneous optimization of structural as well as other design parameters. However, it requires a starting superstructure from somewhere (which for some simple problems, however, may be implicit in the formulation), as well as very extensive computational capability since the superstructure optimization problem is in general nonlinear, nonconvex, and involves both continuous and discrete variables.

Siirola (1971) suggested an alternative decomposition scheme based on an hierarchical ordering of physical properties. In the resulting *systematic generation* scheme, property differences between raw materials and desired products were detected and resolved using the *means-ends analysis* paradigm. The result was the application of chemical technologies in a sequence such that the raw materials become transformed into the desired products. The *property hierarchy* was molecular identity first, followed by species amount, concentration, phase, temperature, pressure, and possibly other form properties. Species identity differences were attacked first and resolved by the application of reaction technologies. Species amount differences were identified next and resolved by splitting or mixing or purchase. Concentration differences were identified next and generally resolved by separation technologies. Finally phase, temperature, and pressure differences were detected and resolved by a variety of enthalpy-changing technologies. If a difference-resolving technology were identified but could not be directly applied to a given stream (for example, reaction that could not be applied because feeds were not pure enough, or a separation technology could not be applied because the feed stream was the wrong phase), recursive design subproblems were identified with the subgoal of producing stream conditions suitable for the application of the suspended technology, generally, but not always, by altering properties lower in the hierarchy. After all identified property differences were resolved, the property-changing tasks were integrated into actual processing equipment. The procedure may be repeated if desired at increasing levels of detail making use of screening results obtained at lower levels of detail.

Alternative solutions are generated when more than one technology is identified that can be applied to reduce or eliminate a property difference. The selection of which technology to choose might be made on the basis of rules or some evaluation at the time the technologies are being examined. Alternatively, each may be chosen separately and the consequences followed separately (leading to alternative design solutions) and then each final solution is evaluated. One additional possibility is that all feasible alternative technologies are selected and applied in parallel leading to a redundant design or superstructure. At the end of the design process, the superstructure is reevaluated in its entirety, and the less economical redundant portions eliminated.

The hierarchical decomposition paradigm became the basis of a series of flowsheet synthesis programs including AIDES (Siirola, 1970), BALTAZAR (Mahalec and Motard, 1977), and PIP (Kirkwood, Locke, and Douglas, 1988), as well as the Pinch Technology Onion (Linnhoff and Ahmad, 1983). It has also formed the basis for the present classification of the major process synthesis subproblems: Reaction Path Synthesis, Species Allocation (or Input-Output and Recycle Structure), Separation Scheme Synthesis, and Heat and Power Integration.

In 1990, the *PIP* (Process Invention Procedure) flowsheet synthesis software became available through CACHE. This software follows closely the hierarchical design procedure described by Douglas (1988) which involves alternating rule-based synthesis and short-cut analysis and economic evaluation which enables rough estimates of the optimum design. The method generates for each reaction step in a process, an input-output and recycle structure, a vapor recovery system for the reactor effluent, a liquid separation system for the reactor effluent, and a heat recovery network. Currently an expert system implementation of PIP is under development.

Heat Exchanger Network Synthesis

The pairing of complementary enthalpy-increasing tasks with enthalpy-decreasing tasks so as to optimally recover and reuse energy within a process while decreasing dependence on external utilities is known as *heat integration*. The synthesis of heat exchanger networks is the process synthesis subproblem that has received perhaps the most attention. This synthesis problem generally occurs late in the synthesis hierarchy after the various hot and cold streams have been fairly well defined with respect to amount, composition, available temperature, desired temperature, and other physical properties. Also, the tradeoffs between capital and operating cost are well understood.

It turns out that the maximum amount of heat integration and the minimum heating and cooling utility requirements can be determined for a process without actually designing a heat integration network through a technique called *Targeting*. Targeting procedures also exist for estimating the optimal number of exchangers and the total heat transfer area. These estimates can also be made in the light of constraints requiring or forbidding energy matches between specific pairs of streams. CACHE has made available for educational use one such program for calculating these estimates, *TARGET II*, developed by Linnhoff-March Inc.

Linnhoff and Hindmarsh (1983) popularized important representations including Composite Curves and Grand Composite Curves for visualizing the thermodynamics governing the heat exchanger network synthesis problem. They also introduced the concept of *Pinch Technology*, based on the point in the design where heat is transferred at the minimum approach temperature, which is especially useful in constraining the selection of energy matches which preserve overall minimum utility usage. Pinch technology also provides guidance on splitting streams for parallel heat exchangers, minimizing the total number of exchangers, proper incorporation of heat pumps, and integrating heat and power tasks. Within these constraints, procedures for the actual synthesis of heat exchanger networks are somewhat less well developed, but all of the standard process synthesis approaches including systematic generation, evolutionary modification, and superstructure optimization using all the standard tools of heuristics, expert systems, deterministic algorithms, and mathematical programming have been suggested. *THEN*, a heat exchanger matching algorithm developed by Carl Knopf, is available through CACHE. The straightforward application of heat integration technology to a typical process design generally reduces the net present cost on the order of 10%, due largely to sensible heat recovery.

As mentioned, heat exchanger networks have also been a testbed for superstructure optimization approaches to process synthesis. In this case, techniques from operations research have been used to essentially convert the synthesis problem into one of analysis that can be solved by mathematical programming. CACHE's sixth Process Design Case Study, *Chemical Engineering Optimization Models with GAMS*, contains three heat exchanger network synthesis examples including a linear programming transshipment model for predicting minimum utility consumption with stream matching constraints, a nonlinear programming model for global optimum search in the synthesis of heat exchanger networks, and a mixed-integer nonlinear programming model for simultaneous synthesis and optimization of heat exchanger networks.

Heat-Integrated Distillation System Synthesis

Synthesis methods for separation tasks depend in part on the nature of the system to be separated. In the case of distillative separations of relatively ideal systems, simple volatility ordering may be sufficient to represent the solution thermodynamics. Alternative separation trains can be synthesized using list processing and either heuristic rules (Seader and Westerberg, 1977) or dynamic programming techniques (Hendry and Hughes, 1972). If the number of components is not too large, all possible separation orderings may be generated and evaluated, as is implemented in the PIP software.

Distillation schemes involve a fair number of auxiliary enthalpy-changing tasks including reboilers, condensers, feed quality adjustment, and product coolers. Such distillation schemes can be heat integrated with any of the standard techniques. However, since the temperatures of these enthalpy-changing tasks depend upon the pressures, purity specifications, and other operating conditions of each column in a distillation sequence, it is advantageous to simultaneously perform heat integration within the column pressure optimization (Siirola, 1981). Utility requirement reductions in the neighborhood of 50% and net present cost savings of 35% are typical for such simultaneous distillation sequence synthesis, operating parameter optimization, and heat integration. CACHE's sixth Process Design Case Study also includes an example of a GAMS-based MINLP approach for optimal sequencing and heat integration in the synthesis of sharp separation sequences.

For binary separations or processes that contain a single or a particularly dominant distillation, significant energy recovery by the previous approach may not be possible. However, a single distillative separation may be implemented as two columns in such a way that the latent heat rejected from one might be recycled to the other, approximately halving the net utility requirement. In order for this to be accomplished, the condenser of one column must be at a higher temperature than the reboiler of the other. This may be done by operating the columns at different pressures, possibly assisted by the degree of separation designed for each column. Blakely (1984) compared a number of such multiple effect designs against a single column as a function of feed flowrate, feed composition, and relative volatility. The specific results depend on the costs and temperatures of available utilities as well as on the capital cost correlations. In fact, at low flowrates the economics of the single column design are generally superior to all of the multiple effect designs largely because at those scales the capital costs of the columns dominate the energy costs. However at higher flowrates, for all conditions of feed composition and relative volatility, there exists one or more multiple effect designs with lower net present costs by as much as 35%, similar in magnitude to the savings achievable in the heat integration of other multiple distillation separations.

Nonideal Distillation Scheme Synthesis

For systems which exhibit more nonideal behavior involving azeotropes and regions of immiscibility, a more detailed representation of the thermodynamics may be necessary even for the synthesis of separation schemes at the targeting level of detail. One such representation is the *Residue Curve Map* developed by Doherty and coworkers (1985). Residue curve maps are derived from an analysis of a single-stage batch still, but may be applied to the understanding of the thermodynamic behavior of solutions and the behavior of continuous distillative separation operations on those systems. Residue curve maps define regions of feed compositions from which it is possible to reach the same lowest-boiling composition and the same highest-boiling composition by alternative operation of a single simple distillative separation task. Some systems involving azeotropes exhibit more than one such region separated by boundaries which normally can not be crossed by simple distillation. However techniques have been devised for overcoming these distillation boundaries including pressure shifting, exploiting boundary curvature, and exploiting separation technologies not involving volatility such as liquid-liquid decantation, solid-liquid crystallization, and kinetic phenomena such as membrane diffusion. The preparation of residue curve maps as well as the design of distillative separations consistent with these maps is being implemented in the MAYFLOWER software under development at the University of Massachusetts.

An industrial separation synthesis procedure for systems exhibiting nonideal thermodynamics has been described in the recent CACHE-sponsored conference, *Foundations of Computer-Aided Process Design* (Siirola, 1994). Key features of the method include the *Residue Curve Map* representation of solution thermodynamics overlaid with pinched regions and liquid-liquid (or solid-liquid) equilibria data; identification of thermodynamic *Critical Features* to be avoided (pinched regions), overcome (distillation boundaries), or exploited (for example, liquid-liquid tie lines); a *Strategic Hierarchy* to address issues raised by the critical features first; the identification of *Interesting Compositions* (useful as mass separation agents which must be regenerated and recycled within the process); *Opportunistic Resolution* of remaining concentration property differences by mixing, decant, or distillative separation as permitted by distillation boundary constraints; and pursuit of *Multiple Alternatives* (heaviest underflow as well as lightest overhead) for every multicomponent distillative separation. The method also includes methods for pressure shifting distillation boundaries, exploiting boundary curvature, producing compositions which are neither the highest-boiling nor the lowest-boiling in a region, extractive distillation, handling recycle, removing redundancy, and heat integration. A similar rule-based system called SPLIT (Wahnschafft, Jurain, and Westerberg, 1991) is currently being incorporated into the ADVENT software by Aspen Technology, Inc. CACHE is working to make methods and tools such as these and other aspects of Residue Curve Map technology available for educational use.

Outlook for the Future

Many of the methods and tools being developed in the process synthesis research community are still more concerned with analysis (targeting, representation, operability, etc.) than with alternative generation. More effort will be directed specifically to alternative generation for conceptual design at all levels of detail.

Many of the applications discussed here were based on algorithmic systematic generation approaches or heuristic rule based or expert systems based approaches to process synthesis. Because of combinatorial difficulties, neither the algorithmic nor the heuristic methods carry any guarantees of structural optimality.

Superstructure approaches, on the other hand, while not especially practical at the present time, do offer with the promise of structural optimality and have tremendous potential for the future. As mentioned, several glimpses of how such approaches might be formulated are included in CACHE's sixth Process Design Case Study. There are significant challenges remaining to improve computational efficiency to overcome combinatorial difficulties and to develop global optimization strategies for the generally nonconvex mixed-integer nonlinear problems involved. Steady progress is being made on these fronts, and other process synthesis problems such as reaction path and reactor scheme synthesis, design for waste minimization, and consideration of other important design criteria such as safety, operability, and controllability. CACHE will continue its efforts to make advances in all of these process synthesis methods and tools available to the educational and industrial communities.

Conclusions

Systematic approaches to conceptual design and process synthesis have begun to have measurable industrial success. Although there are many restrictions, these techniques have led to higher value, lower energy, lower environmental impact, and sometimes even novel design alternatives. Process synthesis techniques have been successfully applied to the conceptual design of total process flowsheets, as well as to specific design subproblems including heat integration, heat-integrated distillation trains, multiple-effect distillation, and the separation of azeotropic systems. Typical energy savings of 50% and net present cost reductions of 35% have been achieved using systematic process synthesis methodologies.

Certain features of systematic approaches to process synthesis appear to have special merit. These include architectures which are hierarchical in scope and iterative in level of detail, targeting as a useful bounding exercise, means-ends analysis as a recursive problem-solving paradigm with hierarchical as well as opportunistic goals implemented by an iterative formulation-synthesis-analysis-evaluation design strategy, thinking specifically in terms of tasks to be accomplished before equipment to be employed, the importance of representations which encapsulate analysis, and the advantages of solving related synthesis problems simultaneously.

Advances in problem formulation and in computational hardware and software capability offer the promise of a new generation of practical process synthesis techniques based directly on structural optimization. Even greater benefits are expected to be realized as this next generation of approaches are transferred to industry and incorporated into design education. Soon the goal of synthesizing provably superior conceptual process flowsheets may be at hand.

References

Blakely, D.M. (1984). *Cost Savings in Binary Distillation through Two-Column Designs*. M.S. dissertation, Clemson University, Clemson.

Doherty, M.F. and G.A. Caldarola (1985), *Ind. Eng. Chem. Fundam.*, **24**, 474.

Douglas, J.M. (1988). *Conceptual Design of Chemical Processes*. McGraw-Hill, New York.

Foucher, E.R., M.F. Doherty, and M.F. Malone (1991). Automatic Screening of Entrainers for Homogeneous Azeotropic Distillation. *Ind. Eng. Chem. Res.*, **30**, 760.

Hendry, J.E. and R.R. Hughes (1972). Generating Separation Process Flowsheets. *Chem Eng. Progr.*, **68**(6), 69.

King, C.J., D.W. Gantz, and F.J. Barnes (1972). Systematic Evolutionary Process Synthesis. *Ind. Eng. Chem. Process Design Develop.*, **11**, 271.

Kirkwood, R.L., M.H. Locke, and J.M. Douglas (1988). A Prototype Expert System for Synthesizing Chemical Process Flowsheets. *Comput. Chem. Engng.*, **12**, 329.

Linnhoff, B. and S. Ahmad (1983). *Towards Total Process Synthesis*. AIChE Meeting paper 26d, Washington.

Linnhoff, B. and E. Hindmarsh (1983). The Pinch Design Method of Heat Exchanger Networks. *Chem. Eng. Sci.*, **38**, 745.

Mahalec, V. and R.L. Motard (1977). Procedures for the Initial Design of Chemical Processing Systems. *Comput. Chem. Engng.*, **1**, 57.

Morari, M. and I. E. Grossmann, Eds (1991). *Chemical Engineering Optimization Models with GAMS*. CACHE Design Case Studies Volume 6, CACHE, Austin.

Rudd, D.F. (1968). The Synthesis of Process Designs, I. Elementary Decomposition Principle. *AIChE J*, **14**, 342.

Seader, J.D. (1972). *Computer-Aided Chemical Process Synthesis*. CACHE-Sponsored Symposium, CACHE, Cambridge.

Seader, J.D. and A.W. Westerberg (1977). A Combined Heuristic and Evolutionary Strategy for the Synthesis of Simple Separation Sequences. *AIChE J.*, **23**, 951.

Siirola, J.J. (1970). *The Computer-Aided Synthesis of Chemical Process Designs*. Ph.D. dissertation, University of Wisconsin, Madison.

Siirola, J.J. and D.F. Rudd (1971). Computer-Aided Synthesis of Chemical Process Designs. *Ind. Eng. Chem. Fundam.*, **10**, 353.

Siirola, J.J. (1981). Energy Integration in Separation Processes. In R.S.H. Mah and W.D. Seider (Eds.), *Foundations of Computer-Aided Process Design*, Vol. II. Engineering Foundation, New York. 573.

Siirola, J.J. (1994). An Industrial Perspective of Process Synthesis. In L.T. Biegler and M.F. Doherty (Eds.), *Foundations of Computer-Aided Design. AIChE Symposium Series*, Vol. 91, No. 304.

Umeda, T.A., A. Hirai, and A. Ichikawa (1972). Synthesis of Optimal Processing System by an Integrated Approach. *Chem. Eng. Sci.*, **27**, 795.

Wahnschafft, O.M., T.P. Jurain, and A.W. Westerberg (1991). SPLIT: A Separation Process Designer. *Comput. Chem. Engng.*, **15**, 656.

PROCESS SIMULATION

Lorenz T. Biegler
Carnegie Mellon University
Pittsburgh, PA 15213

J. D. Seader
University of Utah
Salt Lake City, UT 84124

Warren D. Seider
University of Pennsylvania
Philadelphia, PA 19104

Abstract

This brief discussion of process simulation is written from a standpoint of the impact of the contributions of CACHE on the educational community. Emphasis is on steady-state process simulation and optimization with FLOWTRAN. Some suggestions are made for needed future developments.

Introduction and Historical Development

One of the essential early steps in process design or analysis, following the synthesis of a process flowsheet and the establishment of a preliminary set of operating conditions suitable for a base-case study, is the computation of material and energy balances so as to determine: (1) for every major stream in the process, the flow rate, composition, temperature, pressure, and phase condition; and (2) for every major piece of equipment, the overall size and the shaft work or heat transfer requirements. Before the availability of digital computers, these calculations were done manually with a slide rule or mechanical desk calculator. If the process involved recycle streams, multistaged, multicomponent separation operations, and/or reactors with multiple reactions, the manual computations could be extremely tedious and time-consuming. Often, the computations would be done by approximate methods and incomplete closure of material balances was sometimes acceptable. If optimization studies were made, they were usually restricted to individual operations using one or two design variables, because a more global approach was not practical.

Two significant events occurred in the mid-1950s that were to drastically alter the approach to process simulation and design. The first was the introduction of the IBM 704 digital computer with its built-in floating-point arithmetic. The second was the development of the easy-to-use, high-level, procedure-based programming language, of J. W. Backus and others from IBM, called FORTRAN. The IBM 704 came with a FORTRAN language compiler and subroutines written in that language were automatically handled. Almost overnight, chemical engineers in several large companies and in universities began writing large-scale FORTRAN computer programs to replace the manual methods. At first, these programs were called computer-aided, steady-state chemical process design programs. Later, they would be referred to as simply simulation or flowsheeting programs. The ability to replace tedious hand calculations by computerized calculations also spurred tremendous development in the areas of thermodynamic properties of liquids and gases, rigorous calculation of equilibrium-stage operations, and design of nonisothermal, multiple-reaction reactors, including effects of transport phenomena. At the same time, it was quickly realized that the digital computer was not limited to linear systems. Consequently, there was an explosion in the development of numerical methods for solving all sorts of nonlinear, as well as linear, problems.

By 1967, six commercial and two academic steady-state simulation programs, all written in FORTRAN, were in publicized use. Undoubtedly, other unpublicized commercial simulators were also in use. One of the widely publicized academic programs, PACER (Process Assembly Case Evaluator Routine), developed by Professor Paul T. Shannon at Purdue University, as reported in 1963, was made available to chemical engineering departments in 1967, following an instructional PACER workshop at Dartmouth College, April 23-28, 1967, which was attended by 20 chemical engineering professors from the United States and Canada. In 1971, the author of PACER together with Professors Crowe, Hamielec, Hoffman, Johnson, and Woods from McMaster University published the first book on simulation, entitled *Chemical Plant Simulation - An Introduction to Computer-Aided Steady- State Process Analysis*, which provided the necessary background to use PACER (Crowe, et al., 1971).

The academic version of PACER was mainly a modular executive program for directing material and energy calculations for flowsheets, which could include recycle streams. PACER had a limited library of unit-operation models and lacked a physical property estimation package. However, it was open-ended so that users could add operation and property models. In 1968, Professors R. L. Motard and Ernest J. Henley, at a workshop at the University of Houston that was attended by 42 professors, introduced a more advanced simulation program called CHESS (Chemical Engineering Simulation System). Like PACER, CHESS was based on a sequential modular architecture, in which models of the operating units were solved one at a time. For each unit, the inlet stream conditions were known or assumed and the outlet stream conditions and heat and work transfer rates were computed by the operation model. If the process did not involve recycle loops, the unit calculations began at the feed end of the process and progressed to the product end, so that the calculated outlet stream(s) from one unit became the inlet stream(s) to the next unit(s). If the process included one or more recycle loops, the user could provide a guess for the conditions of one stream in each loop and the program would automatically iterate each loop to convergence to some specified tolerance. The CHESS program included an expanded library of unit-operation models and a physical property package, making it easy to use. However, it lacked a general, rigorous, multistage, multicomponent separation unit for modeling distillation-type operations.

By the early 1970s, a number of new and mostly commercial steady-state simulation programs that contained rigorous absorption/stripping, distillation, and liquid-liquid extraction models began to emerge. The CACHE (Computer Aids for Chemical Engineering Education)

Committee, which had been formed in 1969, directed its Large-Scale Systems Task Force, chaired by Professor J. D. Seader, and the Program Distribution Task Force, chaired by Professor Warren D. Seider, to conduct a survey of academic and commercial simulation programs that might be made available at a reasonable cost for use by chemical engineering faculty and students. That survey identified a number of such programs and these two Task Forces then evaluated the programs with a comprehensive test problem that determined their ease of use and robustness. The FLOWTRAN program of Monsanto Company emerged as the preferred program.

FLOWTRAN was conceived in 1961 by the Applied Mathematics Department of Monsanto Company when they, like many large chemical processing companies, began contemplating the development of a generalized modeling computer program for process design and simulation. The project gained considerable impetus late in 1964 when Robert (Bob) H. Cavett, a brother of the entertainer Dick Cavett, left Pure Oil Company and joined the Applied Mathematics Department of Monsanto Company. While at Pure Oil, Bob Cavett had developed a physical property package for vapor-liquid systems and had investigated numerical analysis procedures for converging nonlinear systems of equations and recycle loops. Monsanto purchased from Pure Oil all rights to Cavett's work. By 1965, the name FLOWTRAN had been selected for the name of Monsanto's program, which was to be based on the sequential modular architecture. In 1966, the initial version of the program was available for use by chemical engineers at Monsanto. From 1969 to 1973, more than 70 outside companies used FLOWTRAN through Monsanto Enviro-Chem Systems, Inc., via commercial computer networks. However, this type of outside use was terminated by Monsanto in 1974 in favor of licensing the program.

On June 13, 1973, in a letter to Mr. John W. Hanley, President of Monsanto Company, CACHE requested that Monsanto consider allowing chemical engineering faculty and students to use FLOWTRAN in course work and research. With the assistance of Professor John J. McKetta of the Advisory Committee of CACHE and F. E. Reese, M. C. Throdahl, J. R. Fair, and S. I. Proctor of Monsanto, approval of the use of FLOWTRAN, via a national computer network, was granted by Monsanto Company on December 10, 1973. This approval included assistance in implementation of the system by providing a grant and loaning specialists from the Monsanto Corporate Engineering Department. Starting on May 10, 1974, Dr. Allen C. Pauls of Monsanto directed a three-day training course on FLOWTRAN, which was attended by Richard R. Hughes, H. Peter Hutchison, J. D. Seader, Warren D. Seider, and Arthur W. Westerberg. On June 5, 1974, at a meeting of the CACHE trustees in Seven Springs, Pennsylvania, Dr. Proctor of Monsanto formally presented to CACHE the Monsanto grant and conditions for making FLOWTRAN available to universities.

On July 1, 1974, with significant contributions from Pauls and Hughes, Seader and Seider completed the draft of a student textbook entitled *FLOWTRAN Simulation - An Introduction*. That book included material on the principles of simulation, detailed instructions on the application of FLOWTRAN to a wide variety of problems including a comprehensive design study by Hughes, numerous examples and homework problems, and descriptions of all of the equipment-unit and physical-property models. By August, 1974, this book was published by CACHE in a soft-bound edition and stocked for sale by Ulrich's Bookstore in Ann Arbor, Michigan.

The academic version of FLOWTRAN was almost identical to the commercial version that was used by Monsanto and licensed to other companies. The program consisted of 9 executive routines tailored to the particular computer on which the program was to run, 7 main FORTRAN programs, 2 data lists, 86 FORTRAN subroutines, and input data and output listings for 27 test problems that illustrated most of the features and capabilities of FLOWTRAN. The program was the result of 60 man-years of effort at a cost to Monsanto of more than two million dollars. At the time of the approval for university use, Monsanto engineers were running about 700 FLOWTRAN jobs per month. The application of FLOWTRAN to design and simulation was restricted to vapor-liquid systems. The library of operations models, called blocks, included 5 flash blocks, 5 distillation blocks, 2 absorber-stripper- extraction blocks, 8 heat exchanger blocks, 3 reactor blocks, 2 pump and compressor blocks, 5 stream mixing-dividing-manipulation blocks, 4 control blocks, 1 recycle convergence block, 16 equipment cost blocks, 4 financial analysis blocks, and 6 report blocks. Included in two of the flash blocks was the first algorithm for calculating a vapor-liquid-liquid (three-phase) flash, which was described in the 1969 textbook, "Material and Energy Balance Computations" by Professor E. J. Henley and Dr. E. M. Rosen of Monsanto Company. The control blocks allowed the user to manipulate certain equipment unit parameters to achieve desired values of stream variables associated with that unit or at another upstream unit.

With respect to thermodynamic properties, the academic version of FLOWTRAN included a property-constant data bank for 180 chemical species. Mixture property models were provided to estimate density, enthalpy, and K-values using the gamma-phi formulation. The Redlich-Kwong equation of state was used for non-ideality in the vapor phase and the equations of Scatchard-Hildebrand (regular solutions), van Laar, Wilson, and Renon-Prausnitz (NRTL) were available for non-ideality in the liquid phase. The latter three equations required binary interaction parameters that had to be supplied by the user. A special program, called VLE, included in the FLOWTRAN system, could be used to compute these parameters from experimental phase equilibria data. For species not in the data bank, the user could add their property constants to a private file. If the constants were not available, they could be estimated with a special program, called PROPTY, which was also included in the FLOWTRAN system.

When the FLOWTRAN program was being conceived, considerable effort was directed to the choice of architecture. With respect to the handling of recycle streams, both the sequential modular and simultaneous modular (two-tier) methods were debated. Although the latter was preferable for processes with a large number of recycle loops, the former was selected because it could be more readily understood and appeared to be more efficient on a number of test problems. A very flexible recycle convergence block was written, which could converge up to three recycle loops simultaneously and provided a number of options to the user, including the methods of successive substitution, bounded Wegstein, delayed Wegstein, and damped Wegstein. The user was required to specify the location of the convergence block in the recycle loop(s). A novel feature of the FLOWTRAN architecture was its variable internal structure. User input consisted of both data and commands in the form of ordered keywords pertaining to selection of chemical species and thermodynamic property models, and equipment unit blocks. Also, the user could insert FORTRAN statements among the block commands. The input was parsed into data and commands. The data were stored and the commands were translated into a main FORTRAN program, consisting mainly of block calls, which was then compiled and linked to just those subroutines that were needed for that particular simulation. This type of architecture made FLOWTRAN efficient in mass storage requirements and open-ended so that user-supplied equipment unit subroutines could be readily added.

On August 13, 1974, Hughes, Seader, and Seider, with Dr. Pauls and Professor Richard S. H. Mah, held, at Northwestern University, the first of three three-day FLOWTRAN workshops, sponsored by the AIChE. The workshop was attended by 38 faculty members representing 35 different universities in the United States and Canada. One attendee remarked that FLOWTRAN was destined to alter the way process design was being taught to senior students in chemical engineering. Today, 20 years later, that prediction has come true; most, if not all, chemical engineering departments now teach computer-aided process design with any of a number of available steady-state process simulators.

The 1974 edition of *FLOWTRAN Simulation - An Introduction* was followed by a second edition in February, 1977, which added a few subroutines and a new chapter on writing cost blocks. A major revision was published in May, 1987, which included a new SQP optimization capability, a new Broyden-based convergence block, and a new chapter on optimization, all by Professor L. T. Biegler, as discussed in the section on optimization below. In January, 1975, CACHE published the booklet *CACHE Use of FLOWTRAN on UCS* by Hughes. This booklet, which was revised in January, 1982 with the assistance of Biegler, provided all the instructions necessary to access the FLOWTRAN program over the United Computing Systems (UCS) network, with computers located in Kansas City, Missouri. In December, 1977, CACHE published, *Exercises in Process Simulation using FLOWTRAN*, edited by Dr. J. Peter Clark. This book, which was revised in December, 1980, with the assistance of Dr. T. P. Koehler and Professor Jude T. Sommerfeld, contained problem statements and solutions for 8 short exercises, 14 workshop exercises, and 5 more comprehensive problems.

By mid-1975, one year after the publication of the FLOWTRAN textbook and the completion of the faculty workshop at Northwestern University, 25 universities were using FLOWTRAN. During the period from 1974 to 1983, an average of 21 universities/year were using FLOWTRAN on the UCS network. By 1985, 59 universities in the United States and Canada had used it. This represented about one-third of all the departments of chemical engineering.

Major deterrents to a wider use of FLOWTRAN were: (1) the inconvenience of having to submit card decks of input data in a batch mode to a remote batch machine, (2) the turn-around time for a run, which was typically one-to-two-days from the time the student submitted a card deck to the time the student received a print out, and (3) the cost of accessing the program on the UCS network. Typically a small-to-moderate size run would cost from $2.00 to $10.00. If access was from a remote job entry device, the cost would be more. Rarely did a student prepare an error-free input on the first, or even the second, try. Therefore, it might take a week to get a successful run. On the average, each university using FLOWTRAN was spending $1,500 per year.

In response to a request from CACHE, Monsanto Company announced on August 19, 1982 that load modules of FLOWTRAN on magnetic media would be made available to universities worldwide through CACHE for installation on in-house computers. The use of FLOWTRAN on the UCS network would be discontinued on January 15, 1986. By 1993,

FLOWTRAN load modules had been prepared for 14 different computer and operating system combinations, 190 (141 USA, 11 Canada, and 38 for 21 other foreign countries) load modules had been distributed worldwide, and approximately 15,000 copies of the three editions of *FLOWTRAN Simulation - An Introduction* had been sold.

Current Status of Steady-State Simulation

By the late 1980s, as discussed by Biegler (1989), computer-aided chemical process simulation had become virtually indispensable in process design and analysis. A large number of commercial simulators had become readily available and some of them, notably ASPEN PLUS of Aspen Technology, Inc., in Cambridge, Massachusetts, ChemCAD of Chem-Stations, Inc., in Houston, Texas, DESIGN II of ChemShare Corp. in Houston, Texas, HYSIM of Hyprotech Ltd. in Calgary, Alberta, Canada, and PRO/II of Simulation Sciences Inc., in Brea, California, were being licensed at low cost to universities. Compared to FLOWTRAN, some, most, or all of the latest versions of these simulators: (1) can be installed on a wider variety of computer/operating systems, including PCs with sufficient main and mass memory running under MS- or PC-DOS or one or more versions of Windows, (2) have much larger data banks, with more than 1000 chemical species, (3) have a much larger library of thermo-dynamic property models, including the ability to estimate K-values from equations of state, (4) have physical property models for entropy, transport properties, and other properties used to characterize materials, (5) have built-in tables for binary interaction parameters of activity coefficient correlations, (6) allow the user to specify any system of input and output units, (7) can generate graphs, (8) can handle solids, (9) have improved algorithms for converging multicomponent, multistage, separations, (10) have models for interlinked three-phase, and reactive distillation, (11) have models for kinetically controlled reactors, and (12) can compute simultaneous chemical and phase equilibrium by minimization of free energy. Thus, on April 4, 1994, Monsanto Company announced that it was their desire to bring to an orderly conclusion the CACHE FLOWTRAN project. Although FLOWTRAN can continue to be used for educational purposes on existing computers, no more load modules will be prepared or distributed. In 1993, at the AIChE Annual Meeting in St. Louis, the CACHE Corporation presented an award to Monsanto Company in recognition of its significant contribution to chemical engineering education. FLOWTRAN ushered in a major change in the way process design is taught to chemical engineers. More than any other individual, Robert H. Cavett of Monsanto was responsible for the successful development of sequential modular process simulation, which he implemented in FLOWTRAN. For this achievement, in 1987, he was the first recipient of the Computing Practice Award of the CAST Division of AIChE.

Optimization in FLOWTRAN

With the availability of FLOWTRAN from Monsanto in 1974, there was early interest by the late Professor Richard R. Hughes to also demonstrate and implement an optimization capability with process simulation and design. Professor Hughes had directed the development of the CHEOPS optimization package for Shell Development's process simulator and a similar capability was envisioned in the mid-1970s for FLOWTRAN. Fundamental work by Wilson and Beale in the mid-1960s led to Quadratic Approximation Programming (QAP), which was prototyped by Hughes and Isaacson in 1975. With QAP, reduced quadratic models for the objective function and constraints were constructed from the rigorous flowsheet models, and these reduced models were optimized to suggest a new base point. As a sequel to this work, A. L. Parker and Dick Hughes proposed an implementation of this optimization strategy to Monsanto, and a working version of QAP was demonstrated with an ammonia synthesis project using Monsanto's version of FLOWTRAN. Using about 65 simulation time equivalents on the ammonia process, this version took advantage of FLOWTRAN's data structures as well as the existing keyword interface. Still, this was the only problem solved by QAP on FLOWTRAN (Parker and Hughes, 1981).

Because the CACHE version of FLOWTRAN was available only through the UCS network, it took some time to implement Parker's approach for that version. As part of this exercise, a similarity between QAP and the evolving Successive Quadratic Programming (SQP) algorithm was noted, and this led to a simplified optimization algorithm which was termed Quadratic/Linear Approximation Programming (Q/LAP) (Biegler and Hughes, 1981). This approach was actually an SQP algorithm in disguise and improved the performance fourfold over the QAP strategy. However, developing an optimization approach proved expensive and clumsy on the UCS network and further development was done with a simpler prototype simulator, called SPAD, at the University of Wisconsin.

The QAP, Q/LAP, and virtually all previous flowsheet optimization approaches shared the same disadvantages. First, they required full flowsheet evaluations for objective and constraint function evaluations. This was expensive and also potentially unreliable; difficult intermediate points can lead to convergence failures. Second, these optimization approaches were implemented so that they needed to control the information flow in the overall process calculation. This led to a number of difficulties that deterred implementation of optimization strategies on existing process simulators.

However, the application of the SQP algorithm overcame both of these draw-backs. This algorithm can be interpreted both as the optimization of a quadratic model at each basepoint (iteration) as well as a (quasi-) Newton method applied to the optimality conditions of the process optimization problem. This second interpretation leads to the insight that both equality and inequality constraints could be converged simultaneously with a systematic improvement of the objective function. In particular, convergence of tear constraints, which often represented the least efficient part of a flowsheet simulation, could be incorporated into the formulation of the optimization problem and handled directly by the SQP algorithm. This insight further leads to a simplified implementation of an optimization algorithm to existing process simulators. Virtually all process simulators (and most prominently FLOWTRAN) have convergence blocks that appear as artificial units in the flowsheet calculation sequence. Upon convergence, a message is passed to the executive program of the simulator that signals the next calculation phase. This leads to the inclusion of sophisticated convergence algorithms largely independent of the architecture of the simulator and also allows the simple addition of an optimization capability - it is simply an extended convergence block. Moreover, this flexibility appears to have been anticipated by early SQP codes as well (e.g., VF02AD in the Harwell library). These were structured through reverse communication and thus allowed the information flow to be controlled by the user or a comprehensive modeling package. It appears that both recycle convergence blocks and the SQP code were made for each other. Not surprisingly, the insights of linking SQP with process simulators was also discovered independently at about the same time at the University of Illinois by Chen and Stadtherr (1985) and at Cambridge University by Kaijaluoto (1985).

At the University of Wisconsin and later at Carnegie-Mellon University, this optimization approach was demonstrated with the SPAD simulator by Biegler and Hughes (1982) and Ganesh and Biegler (1987) on several moderately-sized process flowsheets. However, SPAD proved to be limiting for further development and demonstration of flowsheet optimization algorithms. Fortunately, by 1984, FLOWTRAN load modules had been developed and distributed throughout academia for instructional use. A key feature of FLOWTRAN was its run-time compile architecture, which allowed a powerful FORTRAN insert capability both in specifying the problem and in linking user-written models. (This feature has also been used to advantage in the ASPEN simulator.) Moreover, the instructional documentation encouraged the development of these models by providing clear and ample information on data structures and internal utility routines. As a result, it was natural to reconsider FLOWTRAN as a candidate for demonstrating advanced flowsheet optimization strategies.

The SCOPT (Simultaneous Convergence and Optimization) package combined an improved SQP algorithm along with a tailored interface to the FLOWTRAN simulator. Written as a recycle convergence block with access to global decision variables and other FLOWTRAN data structures, it was easily implemented as a NEW BLOCK and required virtually no restructuring of the existing simulator. Moreover, existing FLOWTRAN features could be used directly in an efficient manner. Optimization problems were formulated through in-line FORTRAN and decision and dependent variables were identified through PUT statements, as detailed in the FLOWTRAN documentation. As a result, even a casual FLOWTRAN user could easily set up and optimize a given flowsheet.

The SCOPT package was written largely at Carnegie-Mellon with generous advice and assistance from the FLOWTRAN group at Monsanto. Between 1984 and 1986, the code was tested and documented, and more than a dozen test problems were developed and benchmarked. These included several unit and small loop optimizations, along with flowsheet optimization for ammonia synthesis, methanol synthesis, allyl chloride manufacture, and monochlorobenzene production. In addition, beta testing was carried out at a number of universities including the University of Pennsylvania, the University of Utah, and the University of Wisconsin. As a result, the CACHE FLOWTRAN simulator was fully extended with an optimization capability, and in 1986 the SCOPT block became an integral component of the FLOWTRAN package.

As part of FLOWTRAN, SCOPT has found considerable use in senior and graduate level courses in process simulation, design and optimization. Many have used this capability, for instance, in open-ended design projects. In addition, the SCOPT package has spawned a significant body of graduate research. At Carnegie-Mellon, the development of advanced optimization algorithms was done with SCOPT by Lang and Biegler (1987), along with the analysis and demonstration of simultaneous strategies for heat integration and flowsheet optimization by Lang, Biegler, and Grossmann (1988). Moreover, an early version of mixed integer nonlinear programming (MINLP) for process retrofits was also prototyped with SCOPT by Harsh, Saderne, and Biegler (1989). In addition, the SQP code from this package has been used separately to solve nonlinear programming problems in a variety of research domains. The SCOPT package has been adapted and modified for a number of commercial and academic simulation tools. For example, a modification of the SCOPT package forms the optimization capability in the SEPSIM process simulator of the Technical University of Denmark.

As FLOWTRAN has been superseded by more powerful commercial simulators, the classroom use of flowsheet optimization continues to evolve. While SQP still remains the workhorse of flowsheet optimization tools, numerous improvements have been made to this algorithm over the last decade. Work on incorporating SQP optimization strategies in commercial flowsheet simulators was started in the mid-80s. Most of these simulators, such as ASPEN PLUS, DESIGN/II, and PROCESS (later PRO/II), apply a structure similar to the one in FLOWTRAN. As with FLOWTRAN, probably the greatest advantages to these optimization approaches are the fast convergence of SQP (relatively few flowsheet evaluations are required) and the ease with which the optimization strategy can be implemented in an existing simulator. In fact, a further step was taken with the ProSim simulator, developed in France, where analytical derivatives were supplied for each module. SQP flowsheet optimization resulted in an order of magnitude improvement in performance with higher quality solutions!

Recently, there has also been strong development in equation-based process simulation, especially for optimization of process operations in real-time. A number of equation-based modeling tools (including SPEEDUP, DMO, MASSBAL, and NOVA) have been developed and all of these rely on SQP optimization strategies that are adapted to large-scale problems. Compared to the modular approaches of most sequential-modular simulators, as described above, equation-based strategies can be more than an order of magnitude faster, since all of the flowsheet equations are now converged simultaneously. However, equation-based simulators lack some of the intuitive structure of modular approaches, making them harder to construct and use. To preserve this modular structure and still effect simultaneous convergence of all the equations, the SQP strategy has recently been adapted (Schmid and Biegler, 1994) to deal with existing modules (e.g., columns and reactors) that converge simultaneously with the optimization problem and thus obtain the best of both worlds.

An important limitation of the flowsheet optimization method described above is that it can only find local optima. Therefore, good initializations are required to yield "good" (if not globally optimal) solutions. Recently, deterministic optimization algorithms have been developed that guarantee convergence to global optima for a large class of nonlinear problems. These strategies are somewhat more expensive than local tools and rely on combinatorial searches and convex relaxation in order to develop upper and lower bounds to the global solution. Often, they also incorporate local optimization solvers as well. A comprehensive review by Floudas and Grossmann (1994) summarizes their performance on structured process problems, such as pooling problems and heat exchanger networks. While, global optimization methods have yet to address problems of the size and generality of flowsheet optimization, progress in the last five years has been significant and is enhancing the industrial perception and application of these optimization tools. It is certain that global optimization and other future advances will be made in the development of flowsheet optimization algorithms and in the proper formulation of such problems. However, judging from the FLOWTRAN experience, it is crucial that these advances be aided by the development of widely distributed and user-friendly software. The resulting advice, support, and feedback are essential in identifying and tackling conceptual challenges and in opening up new areas for future research.

The Future of Process Simulation

Despite the significant progress that has been made in process simulation since the 1960s, users of computer-aided simulators are well aware of lingering limitations that still require remedies. As discussed by Seider, Brengel, and Widagdo (1991), current simulators are not yet utilizing the many new mathematical tools and approaches that are readily available. New numerical analysis tools such as homotopy continuation and the interval-Newton method with generalized bisection are needed to improve the robustness of nonlinear-equation solution, especially for multicomponent, multistage, equilibrium-based separation models, which for the more difficult problems fail to converge all too frequently. Currently, a simulator only searches for one solution. New tools, such as continuation and bifurcation, with graphical displays, are needed for assisting in the determination of solution multiplicity. Dynamic simulators are just beginning to emerge. They offer the opportunity to design more efficient and optimal control systems. Commercial rate-based separation operation models are becoming available and should be widely used in the near future. Membrane, adsorption, ion-exchange, chromatography, crystallization, and drying models need to be added to unit operations libraries. Equilibrium calculations need to be extended to handle solid phases, as well as liquid and gas phases.

References

Biegler, L. T., and R. R. Hughes (1981). *Chem. Eng. Progress*, **77**(4), 76.

Biegler, L. T., and R. R. Hughes (1982). *AIChE J.*, **28**, 994.

Biegler, L. T. (1989). *Chem. Eng. Progress*, **85**(10), 50-66.

Chen, H-S., and M. A. Stadtherr (1985). *AIChE J.*, **31**, 1843.

Clark, J. P., T. P. Kohler, and J. T. Sommerfeld (1980). Exercises in Process Simulation Using FLOWTRAN, Sec. Ed., *CACHE*.

Crowe, C. M., A. E. Hamielec, T. W. Hoffman, A. I. Johnson, D. R. Woods, and P. T. Shannon (1971). *Chemical Plant Simulation*, Prentice-Hall, New Jersey.

Floudas, C. A., and I. E. Grossmann (1994). *Proc. FOCAPD*, Snowmass, CO.

Ganesh, N., and L. T. Biegler (1987). *AIChE J.*, **33**, 282.

Harsh, M. G., P. Saderne, and L. T. Biegler (1989). *Comp. Chem. Eng.*, **13**, 947.

Henley, E. J., and E. M Rosen (1969). *Material and Energy Balance Computations*, Wiley, New York.

Hughes, R. R. (1982). CACHE Use of FLOWTRAN on UCS, Sec. Ed., *CACHE*.

Kaijaluoto, S., (1985). PhD Thesis, Cambridge University.

Lang, Y.-D., and L. T. Biegler (1987). *Comp. Chem. Eng.*, **11**, 143.

Lang, Y.-D., L. T. Biegler, and I. E. Grossmann (1988). *Comp. Chem. Eng.*, **12**, 311.

Parker, A. L., and R. R. Hughes (1981). *Comp. Chem. Eng.*, **5**, 123.

Schmid, C., and L. T. Biegler (1994). *Trans. I. Chem. E.*, **72**, 382.

Seader, J. D., W. D. Seider, and A. C. Pauls (1987). FLOWTRAN Simulation - An Introduction, Third Ed., *CACHE*.

Seider, W. D., D. D. Brengel, and S. Widagdo (1991). *AIChE J.*, **37**, 1-38.

OPTIMIZATION

Ignacio E. Grossmann
Carnegie Mellon University
Pittsburgh, PA 15213

Abstract

This chapter presents a general classification of optimization models, and discusses the role that several modelling systems can play in education for helping students to learn problem and model formulation, and to address relevant problems in chemical engineering.

Introduction

Chemical engineering is a very fertile area for optimization studies, since many problems can be formulated as mathematical models in which an objective function is to be minimized or maximized subject to a set of equality and/or inequality constraints. For example, the optimal operation of a chemical process can be modelled with an objective function of profit that is to be maximized and that accounts for the revenues of the product and cost of raw materials and utilities. The equality constraints (i.e., equations) correspond to the mass and heat balances, and the inequalities to process specifications (e.g. product purity) and physical limits (e.g. nonnegative flows, maximum pressure). In the past the teaching of optimization was largely an interesting theoretical exercise, but of limited application. The main emphasis was placed on one variable optimization problems and on linear programming. Most of the problems in chemical engineering, however, are multivariable and nonlinear, with constraints, and often require the use of discrete valued variables. Also, while optimization is a conceptually powerful framework, its impact has been somewhat limited in practice due to the lack of tools that facilitate the formulation and solution of these problems. In fact the common practice has been to rely on callable FORTRAN routines.

Recent years, however, have seen the development not only of new optimization algorithms, but also of powerful modelling systems. The important feature of these systems is that they provide an environment where the student need not be concerned with the details of providing the interfaces with various optimization codes (e.g. calling the appropriate routines, supplying the correct arguments for variables and functions and their derivatives, etc.). Instead, this environment automates the interfacing to a variety of algorithms that allow the student to *concentrate on the modelling of problems*, ultimately the main skill required for successful application of optimization in practice. More importantly, such an environment greatly reduces the time - up to one order of magnitude - required for the student to formulate and solve an optimization problem. This in turn allows the student to solve considerably more complex and interesting optimization problems than has been the case in the past.

We first give a brief overview of optimization models. We then introduce several modelling systems and educational materials that have proved to be useful in undergraduate and graduate education. Finally, we close with a discussion on future trends of the teaching of optimization.

Overview of Optimization

The use of optimization models in chemical engineering arises in a large variety of applications such as process design, process operation, process control, data fitting, and process analysis. Examples in process design include the synthesis of and sizing equipment for process flowsheets, separation sequences, or heat exchanger networks. Examples in process operations include production planning for refineries and production scheduling of batch processes. In process control, a good example is the determination of optimal temperature profiles in plug flow reactors. Data fitting involves both linear and nonlinear regression. An example of process analysis is the chemical equilibrium calculation through the minimization of the Gibbs free energy. All these problems have in common that they give rise to mathematical programming models involving continuous and discrete variables. These variables must be selected to satisfy equations and inequality constraints while at the same time optimizing a given objective function; the model is:

$$\begin{aligned} & \min f(x,y) \\ \text{s.t.}\quad & h(x,y) = 0 \\ & g(x,y) \le 0 \\ & x \in X,\ y \in Y \end{aligned} \qquad \text{(MP)}$$

The continuous variables are represented by x and the discrete variables by y, both with arbitrary dimensionality. The feasible region for the variables x and y is defined by the following constraints: $h(x,y) = 0$, equations describing the performance of a system (e.g., mass and energy balances, design equations); $g(x,y) \le 0$, inequalities related to specifications (e.g., minimum purity, maximum throughput), and by the sets X and Y. The former typically specify the ranges of values for the continuous variables (e.g., positive flows, minimum/maximum pressures for a chemical reactor), while the latter specifies the discrete choices (e.g. only 0-1 choices for say selection or not of a unit, or an integer number for number of plates). It should be noted that the model in (MP) can readily accommodate maximization problems (equivalent to minimizing the negative of the objective function), or problems involving greater than or equal to zero inequalities.

The mathematical model in (MP) is technically known as a "mathematical program" and an extensive body of literature exists on this subject (see Nemhauser et al., 1989, for a general review). From a practical standpoint, the important feature of a model MP is that it provides a powerful framework for modeling many optimization problems. Depending on the application at hand and the level of detail in the equations, the objective function and constraints can be given in explicit or implicit form. Explicit equations are often used for simplified models that are expressed with algebraic equations. In this case the evaluation of the model may be cheap, but may involve thousands of variables and constraints. Implicit equations arise when detailed and complex calculations must be performed in "black boxes", the case, for instance, with process simulator or differential reactor models. In these cases, the dimensionality for the optimization can be greatly reduced, but the evaluation of trial points for the optimization may be very expensive.

Formulating a given decision problem as a mathematical program requires three major assumptions: (a) that the criterion for optimal selection can be expressed through a single objective function, (b) that all the constraints must be exactly satisfied, and (c) that the parameters are deterministic in nature. Of course, these represent oversimplifications of real world problems. However, it should be noted that extensions based on the model in (MP) are available for addressing some of these issues. For instance, it is possible to relax the assumptions in (a) and (b) through multiobjective optimization methods, while the assumption in (c) can be relaxed through stochastic optimization methods.

If we assume no special structure to the problem in MP and to its various particular cases (e.g., only discrete or only continuous variables), then direct search techniques are often the easiest, but also the most time consuming, methods to apply. Here, either a systematic or random selection of trial points is chosen to evaluate and improve the objective function. Satisfaction of constraints can also be accomplished, but frequently with some difficulty (e.g., by using penalty functions). Perhaps the most popular direct search method that has emerged recently in chemical engineering is simulated annealing. This method is based on analogies with free energy minimization in statistical mechanics. This method is in principle easy to apply to problems with simple constraints, and is likely to find solutions that are close to the global optimum. However, aside from the fact that it often requires many thousands of function evaluations before the likely optimum is found, its performance tends to be highly dependent on the selection of parameters of the algorithm.

On the other hand, the most prevalent approach taken to date in optimization is to consider particular problem classes of MP, depending on the form of the objective function and constraints for which efficient solution methods can be derived to exploit special structures. The best known case of MP is the linear programming (LP) problem in which the objective function and all the constraints are linear, and all the variables are continuous:

$$\begin{aligned} & \min Z = c^T x \\ \text{s.t.} \quad & A x \leqq a \\ & x \geq 0 \end{aligned} \qquad \text{(LP)}$$

LP problems have the property that the optimum solution lies at a vertex of the feasible space. Also, any local solution corresponds to the global optimum. These problems have been successfully solved for many years with computer codes that are based on the simplex algorithm. Major changes that have taken place over the last ten years at the level of solution methods are the development of interior point algorithms that rely on nonlinear transformations and whose computational requirements are theoretically bounded by a polynomial expressed in terms of the problem size. Interestingly, this property is not shared by the simplex algorithm, which theoretically may require exponential time. Since this performance is rarely observed in practice, further significant advances have taken place for solving large scale problems with the simplex algorithm. With this algorithm, problems with up to 15,000 to 20,000 constraints can be solved quite efficiently; interior point methods tend to perform better in problems with up to 50,000 to 100,000 constraints.

The extension of the LP model that involves discrete variables is known as a mixed-integer linear program (MILP). The most common case is when the discrete variables have 0 or 1 values:

$$\begin{aligned} &\min Z = a^T y + c^T x \\ \text{s.t.} \quad &By + Ax \leqq b \qquad \text{(MILP)} \\ &y \in \{0,1\}^m \quad x \geq 0 \end{aligned}$$

This model greatly expands the capabilities of formulating real world problems, since one can include logical decisions with 0-1 variables, or, in general, account for discrete amounts. The most common method for MILP is the branch and bound search, which consists of solving a subset of LP subproblems while searching within a decision tree of the discrete variables. The other common approach relies on the use of cutting planes that attempt to make the MILP solvable as an LP with the addition of constraints.Because of the combinatorial problem that is introduced by the discrete variables, MILP problems have proved to be very hard to solve. In fact, theoretically, one can show that this class of problems is NP-complete; that is, there is no known algorithm for solving these problems in polynomial time. Nevertheless, recent advances based on combining branch and bound methods with cutting planes, and which have been coupled with advances in LP technology, are providing rigorous optimal solutions to problems that were regarded as unsolvable just ten years ago.

For the cases where all, or at least some, of the functions are nonlinear, and only continuous variables are involved, MP gives rise to nonlinear programming problems (NLP):

$$\begin{aligned} &\min f(x) \\ \text{s.t.} \quad &h(x) = 0 \qquad \text{(NLP)} \\ &g(x) \leq 0 \\ &x \in R^n \end{aligned}$$

For the case when the objective and constraint functions are differentiable, local optima can be defined by optimality conditions known as the Kuhn-Tucker conditions. These are perhaps the most common types of models that arise in chemical engineering. While ten years ago problems involving 100 variables for NLP were regarded as large, currently the solution of problems with several thousand variables is quite common. Reduced gradient and successive quadratic programming techniques, which can be derived by applying Newton's method to the Kuhn-Tucker conditions, have emerged as the major algorithms for NLP. The former method is better suited for problems with mostly linear constraints, while the latter is the method of choice for highly nonlinear problems. A limitation of these methods is that they are only guaranteed to converge to a local optimum. For problems that involve a convex objective function and a convex feasible region this is not a difficulty since these problems exhibit only one local optimum which therefore to the global optimum.

The extension of nonlinear programming for handling discrete variables yields a mixed-integer nonlinear programming (MINLP) problem that in its general form is identical to problem (MP). The more specific form that is normally assumed is linear in 0-1 variables and nonlinear in the continuous variables,

$$\begin{aligned} &\min Z = c^T y + f(x) \\ &B\, y + g(x) \leq 0 \\ &x \in X = \{x \mid x \in R^n, x^L \leq x \leq x^U\} \\ &y \in \{0,1\}^m \end{aligned} \qquad \text{(MINLP)}$$

MINLP problems were regarded as essentially unsolvable ten years ago. New algorithms such as the outer-approximation method and extensions of the Generalized Benders decomposition method have emerged as the major methods. These methods, which assume differentiability of the functions, consist of solving an alternating sequence of NLP subproblems and MILP master problems. The former optimize the continuous variables, while the latter optimize the discrete variables. As in the NLP case, global optimum solutions can only be guaranteed for convex problems. Solution of problems with typically up to 50 to 100 0-1 variables and 1000 continuous variables and constraints have been reported with these methods. Major difficulties encountered in MINLP include the ones encountered in MILP (combinatorial nature requiring large-scale computations) and in NLP (nonconvexities yielding local solutions).

Finally, it should also be noted that all the above methods assume that the problem in (MP) is expressed through algebraic equations. Very often, however, these models may involve differential equations as constraints. This gives rise to problems known as optimal control problems or as optimization of differential algebraic systems. The major approach that has emerged here is to transform these problems through discretization into nonlinear programming problems. The alternate approach is to solve the differential model in a routine which is then treated by the optimizer as a black box.

From the above review it is clear that there has been substantial progress in optimization algorithms over the last decade. These advances have in fact been accelerated by increased computational power and by the advent of powerful modeling systems, discussed in the next section.

Modelling Systems

Modelling systems have had a very large impact in practice, by making software systems in which optimization problems can be easily formulated in equation form readily accessible to many users of PC's, workstations and supercomputers. By separating the declarative part of a problem from the solution algorithms and their computer codes, these non-procedural modeling systems allow students to concentrate on the formulation of their models. These systems have typically reduced by at least one order of magnitude the time required to formulate optimization problems. For instance, in the past the equations for a model and any required analytical derivatives had to be supplied through subroutines or clumsy data files. This was not only a time consuming process, but also one that was prone to producing many errors. With current modeling systems, many of these problems have been virtually eliminated, greatly facilitating the formulation and solution of optimization problems.

We will now briefly review three modelling systems that have proved to be useful in education.

LINDO. This is an interactive program for solving LP and MILP problems. The simplex algorithm is used for the LP problem and the branch and bound method is used for the MILP. The program is available in PC and Macintosh versions, as well as for the UNIX operating system. LINDO is especially suitable at the undergraduate level due to the great ease of its use. When specifying an LP or MILP model, any name can be used for the variables. Also, there is no need to indicate the product of a coefficient times a variable with the * sign. Inequality signs are <= for less than or equal to, and => for greater than or equal to. There is also no need to explicitly specify non-negativity constraints on the variables, since all variables are assumed to have values greater than or equal to zero.

The problem formulation for a very small LP can be entered interactively as shown below (input in bold):

Enter model

MAX X1 + 2 X2

? ST

? X1 + X2 <= 3

? END

Examine model

LOOK ALL

```
[1] MAX    X1 + 2 X2
SUBJECT TO
  (2) X1 + X2 <=  3
END
```

Solve the LP

GO

Output from LINDO

```
LP OPTIMUM FOUND AT STEP    1
     OBJECTIVE FUNCTION VALUE
     1)   6.00000000
VARIABLE      VALUE          REDUCED COST
   X1          .000000          1.000000
   X2         3.000000           .000000
ROW   SLACK OR SURPLUS    DUAL PRICES
   2)          .000000          2.000000
```

The formulation can be edited and stored in a file. The model can also be specified in a file and accessed through the LINDO command TAKE. There are, of course, a number of other options which are explained in the program. The major advantages of LINDO are its great ease of use (no manual is needed) and its reliability. The instructor need spend almost no class time explaining the use of the program. A one page handout explaining the basics is sufficient. The only disadvantage of LINDO is that all equations must be entered in explicit form. Thus, one cannot use indexed variables and equations for specifying models in compact form. Therefore, LINDO is most suitable for small problems. The LINDO software and its book (Schrage, 1984) are distributed by LINDO Systems, Inc. The student version is available for problems with up to 100 constraints and 120 variables on the PC and 60 constraints and 120 variables on the Macintosh. Larger versions are also available.

GINO. This is an interactive program for solving NLP problems whose user interface is similar in nature to LINDO's. GINO uses a feasible path variant of the reduced gradient algorithm; the partial derivatives for the gradients are computed numerically by finite differences. The program is available in PC and Macintosh versions, as well as for the UNIX and VMS operating systems. GINO is suitable at the undergraduate level because of its great ease of use. Furthermore, GINO can also be used as an equation solver, and is also useful in courses such as thermodynamics and chemical reaction. Algebraic expressions used in the NLP can have arbitrary variable names, and the basic operator signs are as follows:

+ - Addition, subtraction

* Multiplication

/ Division

^ Exponentiation

LOG(X) natural logarithm

EXP(X) exponential function

= equal

<= less than or equal to

=> greater than or equal to

Each function in the formulation must end with a semicolon (;). The objective function is specified as MIN = or MAX = .

The problem formulation for the small NLP:

$$\min Z = x^2 + y^2$$
$$\text{s.t.} \quad 2x + y \geq 4$$
$$x \geq 0 \quad y \geq 0$$

can be entered interactively in GINO as shown below (input in bold):

Enter the model

```
MODEL:
? MIN= X ^ 2 + Y ^ 2 ;
? 2 * X + Y > 4 ;
? END
SLB X 0
SLB Y 0
```

Solve the NLP

```
GO
SOLUTION STATUS: OPTIMAL TO TOLERANCES. DUAL CONDITIONS: SATISFIED.
        OBJECTIVE FUNCTION VALUE
    1)      3.200000
 VARIABLE       VALUE          REDUCED COST
    X          1.600000          0.000000
    Y          0.800000          0.000000
   ROW   SLACK OR SURPLUS      PRICE
    2)        -0.000000         -1.600001
```

The formulation can be edited and stored in a file, or it can be entered through an input file which is accessed with the GINO command take. As with LINDO, the major advantages of GINO are its great ease of use (no manual is needed) and reliability. Also, as mentioned above, GINO can be used as a nonlinear equation solver. The disadvantage of GINO is that no indexed equations can be used. Therefore, this program is most suitable for small problems. The GINO software and its book (Liebman et al. 1986) are distributed by LINDO Systems, Inc. The student version is available for problems with up to 30 constraints (excluding simple lower and upper bounds) and 50 variables on both the PC and Macintosh. Larger versions are available.

GAMS. GAMS is an advanced modelling system which is not as easy to use as LINDO and GINO, but provides a flexible interface for the formulation of LP, MILP, NP and MINLP models and their solution with a variety of different algorithms. The models are supplied by the user in an input file (Filename.GMS) in the form of algebraic equations using a higher level language. GAMS then compiles the model and interfaces automatically with a "solver" (i.e., optimization algorithm). The compiled model, as well as the solution found by the solver, are then reported back to the user through an output file (Filename.LST on a PC and UNIX workstation, Filename.LIS on VMS).

In order to compile and execute the input file, the command is simply: GAMS filename. The GAMS input file is in general organized into the following sections:

a. Specification of indices and data

b. Listing of names and types of variables and equations (constraints and objective function)

c. Definition of the equations (constraints and objective function).

d. Specification of bounds, initial values and special options.

e. Call to the optimization solver.

While the format of the input files is not rigid, the syntax is. Also, there is a rather large number of keywords that provide flexibility for handling simple and complex models (all in equation form, either explicitly or with indices).The main solver types available in GAMS are as follows:

LP	linear programming
NLP	nonlinear programming
MIP	mixed-integer linear programming
RMIP	relaxed MILP; the integer variables are treated as continuous.
MINL	mixed-integer nonlinear programming

The optimization software available in the student edition of GAMS by Scientific Press is as follows:

LP	ZOOM, MINOS
MIP, RMIP	ZOOM
NLP	MINOS

ZOOM is a simplex and branch and bound code for LP and MILP, while MINOS is an infeasible path implementation of the reduced gradient method. In addition, there is a special student edition available through CACHE (see section below) which also includes the following software:

MINLP DICOPT++

DICOPT++ implements an augmented penalty variant of the outer-approximation method.

As an illustration of GAMS, consider the problem of assigning process streams to heat exchangers as described in pp. 409-410 of the text *Optimization of Chemical Processes* by Edgar and Himmelblau. The optimization problem is given by:

$$\begin{aligned} &\min Z = \sum_{i=1}^{n} \sum_{j=1}^{n} C_{ij} x_{ij} \\ &\text{s.t.} \sum_{i=1}^{n} x_{ij} = 1 \qquad j = 1..n \\ &\sum_{j=1}^{n} x_{ij} = 1 \qquad i = 1..n \\ &x_{ij} = 0,1 \qquad i = 1, n \;\; j = 1, n \end{aligned} \tag{AP}$$

which corresponds to the well known assignment problem. Here, i is the index for the n streams and j is the index for the n exchangers. The binary variable $x_{ij} = 1$ if stream i is assigned to exchanger j, and $x_{ij} = 0$ if it is not. The two equations simply state that every exchanger j must be assigned to one stream, and every stream i must be assigned to one exchanger.

The cost C_{ij} of assigning stream i to exchanger j is as follows:

Streams	*Exchangers* 1	2	3	4
A	94	1	54	68
B	74	10	88	82
C	73	88	8	76
D	11	74	81	21

We can formulate the above problem in GAMS almost in the form of the model in (AP) using index sets. The output file is shown below.

```
Test Problem
  4 *
  5 * Assignment problem for heat exchangers from pp.409-410 in
  6 * "Optimization of Chemical Processes" by Edgar and Himmelblau
  7 *
  8
  9  SETS
 10     I  streams      / A, B, C, D /
 11     J  exchangers   / 1*4 / ;
 12
 13  TABLE C(I,J)  Cost of assigning stream i to exchanger j
 14
 15          1    2    3    4
 16     A   94    1   54   68
 17     B   74   10   88   82
 18     C   73   88    8   76
 19     D   11   74   81   21 ;
 20
 21
 22  VARIABLES X(I,J), Z ;
 23  BINARY VARIABLES X(I,J);
 24
 25  EQUATIONS ASSI(J), ASSJ(I), OBJ;
 26
 27  ASSI(J)..  SUM( I, X(I,J) ) =E= 1;
 28  ASSJ(I)..  SUM( J, X(I,J) ) =E= 1;
 29  OBJ..     Z =E= SUM( (I,J), C(I,J)*X(I,J) ) ;
 30
 31  MODEL HEAT / ALL / ;
 32
 33  OPTION LIMROW = 0;
 34  OPTION LIMCOL = 0;
 35  OPTION MIP=ZOOM;
 36  OPTION SOLPRINT = OFF;
 37
 38  SOLVE HEAT USING MIP MINIMIZING Z;
 39
 40  DISPLAY X.L, Z.L ;

MODEL STATISTICS
BLOCKS OF EQUATIONS      3    SINGLE EQUATIONS       9
BLOCKS OF VARIABLES      2    SINGLE VARIABLES      17
NON ZERO ELEMENTS       49    DISCRETE VARIABLES    16
```

```
        S O L V E    S U M M A R Y
   MODEL   HEAT            OBJECTIVE Z
   TYPE    MIP             DIRECTION MINIMIZE
   SOLVER ZOOM             FROM LINE 38
**** SOLVER STATUS     1 NORMAL COMPLETION
**** MODEL STATUS      1 OPTIMAL
**** OBJECTIVE VALUE            97.0000
RESOURCE USAGE, LIMIT         0.390     1000.000
ITERATION COUNT, LIMIT        16        1000
            Iterations    Time
Initial LP         16     0.06
Heuristic           0     0.00
Branch and bound        0     0.00
Final LP            0     0.00

----     40 VARIABLE X.L
          1        2        3        4
A                                  1.000
B               1.000
C                        1.000
D      1.000

----     40 VARIABLE Z.L          =       97.000
```

Note that the SOLVE SUMMARY indicates that the optimum objective function value is Z = 97. Note also that the solution was obtained from the relaxed LP. This is not surprising, since it is well known that the assignment problem has a "unimodular" matrix and therefore the solutions for the x_{ij} are guaranteed to be 0-1 if we solve the problem as an LP.

It should be noted that a number of other modelling systems are available, including the AMPL system. AMPL is very similar in nature to GAMS. There are also chemical engineering modelling systems such as ASCEND and SPEED-UP which include capabilities for steady-state and dynamic simulation and nonlinear programming. SPEED-UP is distributed by Aspen Technology, while ASCEND can be obtained from Carnegie Mellon University. Both systems have capabilities for nonlinear programming.

Textbooks and Case Study

The two major textbooks available for teaching optimization in chemical engineering are:

Reklaitis, G.V., A. Ravindran and K. M. Ragsdell, *Engineering Optimization*, John Wiley (1983).

Edgar, T.F. and D.M. Himmelblau, *Optimization of Chemical Processes*, McGraw Hill (1988).

Both textbooks can be used at undergraduate and graduate levels. The text by Reklaitis et al. is more general and somewhat easier to follow. The book by Edgar and Himmelblau is more up to date and contains many examples relevant to chemical engineering. We have found that students like both textbooks, and that many of the exercises in them can be solved with LINDO, GINO or GAMS. Additional references appear in the bibliography.

In order to complement the above textbooks and to reinforce skills in problem formulation, CACHE has produced the case study:

Vol. 6: *Chemical Engineering Optimization Models with GAMS*, CACHE Design Case Study.

The case study is based on the GAMS modelling system and covers applications at various levels of complexity in the following areas:

Planning and scheduling of batch and continuous processes

Chemical and phase equilibrium

Design of heat exchanger networks, distillation columns, batch processes

Synthesis of reaction paths, heat exchanger networks, distillation sequences

Optimization of dynamic and distributed parameter models

This case study describes in some detail the formulation and solution of a total of 21 optimization problems that correspond to LP, MILP, NLP and MINLP models in the areas cited above. A special student version of GAMS for PC's is included that can handle problems with up to 1000 non-zero entries (up to 200 nonlinear) in the Jacobian and up to 20 discrete variables. The codes BDMLP, ZOOM, MINOS and DICOPT++ are included, as well as the *GAMS User's Guide*. Extensively documented GAMS input files are provided for each of these problems. Some of these input files are rather general, almost like self contained programs where problems can be specified with different data and dimensionality. This case study has been prepared by faculty and students at Carnegie Mellon University, Northwestern University and Princeton University. GAMS Development Corporation, the Licensing Technology Office at Stanford University, XMP Optimization Software and the Engineering Design Research Center at Carnegie Mellon have donated the computer software for this case study.

The case study is organized as follows. First a brief tutorial introduction to GAMS is given to illustrate at the simplest level some of its capabilities, and how to use the program. Also, some useful hints are included in this section. Next, the 21 problems listed in Table 1 are presented, with emphasis on the formulation of the problems. The GAMS input file for each problem is included (output listing files have been omitted except for the first problem because of space limitations), along with a discussion of results.

The GAMS input files of these problems are included in the diskette provided for this case study. A number of suggested exercises have also been included in the description of each problem. At the simplest level, they involve solving the same problem for different data; at the more complex level they involve modifying the formulations or explaining certain features in the model or algorithm. Several appendices are also included. The first one is a guide for the use of the MINLP optimization code DICOPT++, which is a non-standard option in the commercial version of GAMS. The second appendix shows how GAMS can be used to program an algorithm such as the Generalized Benders Decomposition. Finally, an appendix is included that provides some background on how to convert optimization problems with differential equations into nonlinear programming problems.

Table 1. Chemical Engineering Optimization Models with GAMS

1. LP model for refinery scheduling
2. LP model for production planning in multipurpose batch plants
3. LP transshipment model for predicting minimum utility consumption with constraints
4. LP/MILP model for reaction path synthesis
5. MILP model for the design of a chemical complex
6. MILP model for the scheduling of a multiproduct batch plant
7. MILP multiperiod model for the planning of chemical complexes
8. NLP model for power generation via fuel oil
9. NLP model for the optimization of an alkylation plant
10. NLP model for chemical equilibrium via minimization of Gibbs free energy
11. NLP model for phase and chemical equilibria via minimization of the Gibbs free energy
12. NLP model for global optimum search in the synthesis of heat exchanger networks
13. MINLP model for the selection of processes
14. MINLP model for the optimal design of multiproduct batch plants
15. MINLP model for simultaneous optimization in heat exchanger network synthesis
16. MINLP model for the synthesis of heat integrated distillation sequences
17. MINLP model for the synthesis of non-sharp distillation sequences
18. MINLP model for the optimal selection of feed tray in distillation columns
19. NLP differential/algebraic model for minimum time for a car to cover a fixed distance
20. NLP differential/algebraic model for optimal mixing of catalyst for packed bed reactor
21. NLP differential/algebraic model for parameter estimation in a batch reactor

Conclusions

Chemical engineering is a discipline in which many analysis, design and operations problems can be formulated as optimization models. Current courses and textbooks in chemical engineering optimization emphasize mostly theory and methods, and are restricted to rather small problems and applications. Although it is of course important to cover basic methods and the theory of optimization, it is also important to account for a number of new developments in this area. One of the most important developments is that new modelling systems such as LINDO,

GINO and GAMS offer the possibility to quickly model and solve a variety of optimization problems. Students can concentrate mostly on problem formulation, without having to spend too much effort on learning to use the software.

A second important development in optimization is that new and improved algorithms offer the possibility of solving much larger problems that could be handled previously. Linear programming problems involving thousands of constraints and thousands of variables can be readily solved. Similar trends are taking place for solution of mixed-integer linear programming problems. Another development is in the area of nonlinear programming, where currently one can solve problems involving several hundreds of equations and variables. Finally, important developments have also taken place in the solution of differential-algebraic systems and in the solution of mixed-integer nonlinear programming problems that, until recently, have received very little attention. There is a clear trend towards greatly improved capabilities for solving optimization models.

Given all of the above developments, there is an important educational need to reinforce the modelling skills of our students, to make use of both modern modelling systems and modern optimization algorithms, with which one can solve larger and more complex engineering problems. CACHE has produced a comprehensive case study to provide a set of chemical engineering problems to supplement optimization courses at both the undergraduate and graduate level. This case study is, of course, not meant to replace textbooks, but rather to provide additional material that will be useful for both instructors and students.

References

Bazaraa, M.S. and C.M. Shetty (1979). *Nonlinear Programming*, John Wiley.

Brooke, A., D. Kendrick and A. Meeraus (1988). *GAMS-A Users Guide*, Scientific Press.

Hillier, F.S. and G.J. Lieberman (1986). *Introduction to Operations Research*, Holden Day.

Liebman, J., L. Lasdon, L. Schrage and A. Warren (1986). *Modelling and Optimization with GINO*, Scientific Press.

Minoux, M. (1986). *Mathematical Programming: Theory and Algorithms*, John Wiley.

Nemhauser, G.L., A.H.G. Rinnoy Kan and M.J. Todd (eds). (1989). *Optimization, Handbook in Operations Research and Management Science*, Vol. 1, North Holland.

Schrage, L. (1984). *Linear Integer and Quadratic Programming with LINDO*, Scientific Press.

Williams, H.P. (1978). *Model Building in Mathematical Programming*, Wiley Interscience.

DESIGN CASE STUDIES

Ignacio Grossmann
Carnegie Mellon University
Pittsburgh, PA 15213

Manfred Morari
ETH Zentrum
Zurich CH-8092, Switzerland

Abstract

We discuss the role that design case studies play in undergraduate design courses and present a general overview of project assignment in design courses, and the evolution and trends of design case studies. A description of the CACHE design case studies is also given, followed by future trends.

Introduction

The design course is one of the more difficult courses to teach at the undergraduate level. A major problem has been the lack of design textbooks. From 1990 through 1995, only one new major design text, *Conceptual Design of Chemical Processes* by Douglas (1988), was published. This has meant that most faculty have had to put together course notes from a variety of different sources to keep up with new developments, design methodologies and computer tools. Another major problem in teaching design has been the lack of case studies that faculty can use as design projects. This means that faculty commonly spend a considerable amount of time to either get problems from industry or from the literature. CACHE, through its Process Design Case Studies Task Force, has tried to aid faculty with this problem.

The mission of the Process Design Case Studies Task Force is to promote the development of case studies for process design education to help instructors conduct design courses and provide examples for students. It is the chief objective of the CACHE Case Studies to demonstrate and elucidate the thought and decision process that is used in design, and to show how it can be supported by the use of computer tools for simulation, optimization and synthesis. Projects for the CACHE case studies have ranged from traditional to non-traditional processes, from grassroots to retrofit designs, and from large to small-scale problems. Task force activities include seeking and evaluating proposals for the development of case studies, reviewing case studies that are under development, and assisting in their final preparation, promoting recently developed case studies, acquiring design problem statements from industry, and organizing sessions and workshops at meetings dealing with process design education.

We give first a brief account of the common difficulties encountered in assigning and managing projects in the undergraduate design course, and then give an overview of the availability and evolution of design case studies. We then briefly describe the case studies that have been developed by CACHE. Finally, we discuss future plans and directions.

The Design Project

Teaching the design course is certainly a major challenge. In addition to covering lecture material, the instructor has to generate at least one design project. That generally means that the time spent in the course by both instructors and students is quite large, probably well beyond the credits earned. Furthermore, when you add the fact that difficulties may arise among the students in their groups, that they are not used to making decisions with incomplete information and that the grading of the project is somewhat subjective you have the perfect recipe for getting poor teaching ratings from students, which makes the teaching of the design course even more frustrating. While there are no magic recipes, one way of improving the teaching of the design course is by carefully selecting the design project and managing it in an efficient and meaningful manner for the students. Below we offer some general guidelines which clearly point to the usefulness of design case studies.

Firstly, there is the question as to whether it is better to assign one single problem to the whole class or a different one to each group of students. It may be a matter of personal preference, but we would recommend as a general rule to assign only one project. Unless the projects are relatively well defined or not very extensive, the work involved in gathering preliminary information, making sure the computer tools are applicable to the problem, preparing the teaching assistants, and keeping track of several projects simultaneously, is so great that it can become a very frustrating experience to assign different projects to different groups. Furthermore, even if only one project is given but it is sufficiently open ended, it will have multiple solutions, making it an ideal candidate for a design project. It is very interesting to see at the end how different alternatives are synthesized by the students.

The second major decision is whether the project should be a large design problem (e.g., design a complete flowsheet) or whether it should consist of solving perhaps two or more smaller problems. In the past, the practice has tended towards the first option. It has the advantage that students get to see what it takes to do a complete design. The drawback is that it exposes the students to only one problem which is what the second option seeks to avoid, as one can consider quite different applications with multiple projects. We personally feel that both options can be good experiences for students. The choice should be what the instructor believes he/she can do best.

The third major decision is how to integrate the design groups. In our experience what seems to work best is to let the students select their own partners within a specified deadline. Those who do not meet the deadline should then be assigned by the instructor. One can argue that this scheme has the disadvantage of creating unbalanced groups; presumably, some groups will be very strong, others will be weak. While this is generally true as far as the GPA is concerned, it is certainly not uncommon in design projects for some academically good students to do rather poorly and for some of the weaker students to do much better than expected. Furthermore, with our assignment scheme students feel more at ease because they make their own decisions about whom they will work with. Otherwise a common problem is that students can blame the instructor for making them work with students they do not get along with.

The fourth major decision is how to organize and manage the project. Here perhaps the most important rule is that a fair amount of work must be completed before the project starts. In particular, it is important not only to have a problem statement, but to have some rough preliminary calculations to avoid unpleasant surprises (e.g., no matter what design alternative is selected, it always leads to an economic loss). One also should check that the library has the literature that students are likely to need and whether it is possible to use the process simulator or other computer tools for the process. It is also important to involve teaching assistants or other faculty at this stage to make sure they get familiar with the project so that they can supervise the design groups more effectively. Clearly, if the instructor has access to a design case study, the task of performing the preliminary work is facilitated greatly. Another aspect that must be considered is very clear specification deadlines for students to submit their memos, progress reports, and final reports, right from the beginning of the course.

Finally, the last major decision is how to grade the projects. It is important that students recognize that a project is a team effort. They should also recognize that good performance is rewarded and that poor performance is unacceptable. One particular scheme that we have used is to assign 50% of the grade to the group effort (e.g., one common grade for the memos and final design reports). The remaining 50% of the grade is individual. It is a *subjective* grade (students are told about this) that is assigned by the project supervisor. The specific scheme that we have used in the past is to assign individual grades so that the average comes out to be the same as the group grade. Aside from using the input by the supervisor of that group, we also use evaluations by the students themselves. At the time each memo or report is submitted, we provide confidential evaluation forms in which students indicate the percent contribution of each group member and any special comments. In our experience, about 80% of the groups indicate the same percentage contribution for each student.

As seen above, assigning a project in a design course is a nontrivial matter, that can be facilitated with design case studies, discussed next.

Overview and Evolution of Design Case Studies

One of the first major efforts in generating design case studies culminated in the series of problems produced by Washington University in St. Louis, MO. These case studies were developed in the 1970s and tended to address interesting, but generally well defined, problems. In a similar vein, AIChE has been producing student contest problems annually which, while often challenging, have been somewhat restrictive in scope.

Recently there have been increasing demands for assigning new and different types of design projects. Firstly, there has been the trend of requiring that the design problems be open ended and ill-defined to bring them closer to "real world problems" in which students have to work in the form of teams, rather than individually. Secondly, given research advances in the area of process synthesis, there has been an increasing emphasis on the application of systematic techniques at the early stages of design, in which decisions normally have a large economic impact. In particular, greater emphasis is being placed in the generation and screening of alternatives (e.g., hierarchical decomposition) and in the efficient recovery of energy (i.e., pinch analysis). Thirdly, tools for process simulation have become commonplace, especially given the significant advances in computer performance and the introduction of graphical user interfaces that greatly simplify the use of these complex programs. It has been the changing nature of these three developments, and the need to provide a service to faculty teaching design, that have motivated the work of the Process Design Case Studies Task Force over the last ten years.

The first five volumes of the CACHE Design Case Studies are shown in Table 1.

Table 1. CACHE Design Case Studies.

Volume 1	Separation System for Recovery of Ethylene and Light Products from a Naphtha Pyrolysis Gas Steam
Volume 2	Design of an Ammonia Synthesis Plant
Volume 3	Design of an Ethanol Dehydration Plant
Volume 4	Alternative Fermentation Processes for Ethanol Production and Economic Analysis
Volume 5	Retrofit of a Heat Exchanger Network and Design of a Multiproduct Batch Plant

As will become apparent in the next section describing each of these volumes, the evolution of the CACHE case studies has proceeded as follows. The first three volumes were developed in the early 1980s and deal with the design and economic evaluation of large scale plants for manufacturing commodity chemicals (ethylene, ammonia, acetaldehyde). These design problems are fairly large in scope, and typically require groups of 4 to 6 students, although it is possible for an instructor to scale back the extent of these problems. Another common feature in these three case studies is the application of heuristics for most design decisions and of the pinch analysis for heat integration. Also, these three case studies illustrate extensive use of steady-state process simulators.

The fourth volume deals with the design of a fermentation process, to reflect the increasing interest in bioprocesses in the mid 1980s. The scope of the problem is similar to the first three, but it involves modification of a conventional simulator for handling nonstandard equipment and chemical compounds. Also, as with the first three volumes, special reactor models have to be developed.

The fifth volume represents a significant departure from the first four, in that it deals with two smaller problems that can be completed in 2-3 weeks and handled by groups of 2-3 students: retrofit of a heat exchanger network and design of a multiproduct batch plant. The intent here is to acquaint the student with some new trends that started to emerge in the late 1980s; the chemical industry was not building new large-scale plants, but retrofitting old ones, and it was concentrating on designs for manufacturing low volume specialty chemicals. Another interesting aspect of the fifth volume is that it assumes that the students will be working for a consulting firm, while the first four assume that the students will be working in divisions of large corporations. Although unintended at the time this volume was produced, it seems to reflect the emerging working environment for engineers in the 1990s.

One important aspect in the five volumes of the CACHE Design Case Studies is that they contain special material for the instructor to help him/her to organize the course. Also, since the case studies are generally well written and presented, they can serve as examples for students on how to write and communicate effectively.

Summaries of Case Studies

We include here brief descriptions of the first five volumes of the CACHE Design Case Studies; each of the case study problems originated in actual design courses and have been used extensively in the U.S. and overseas.

Volume 1. Separation System for Recovery of Ethylene and Light Products from a Naphtha Pyrolysis Gas Stream

The objective of this case study is the design of a separation train for recovery of seven product fractions from steam cracked naphtha involving about two dozen components. All stages of the design procedure, from preliminary calculations to detailed flowsheet calculations, are described. Emphasis is placed on the following steps: preliminary synthesis of the separation sequence, optimization of column pressure levels for energy cost minimization, and synthesis of a heat exchanger network. Distillation is used for all separation operations except for methane/hydrogen, for which a membrane separator is proposed. The distillation columns were designed using DESPAC, an interactive program developed at Carnegie Mellon University. Any other standard program could be used for this purpose.

Depending on the required detail and the availability of suitable distillation column design software, the case study is suitable as a one-term assignment for either a single student or a group of students. The published case study is based on the work of five students over a ten week period. The problem was posed by Dan Maisel from Exxon and the case study was prepared by Michael Lenncoff under the supervision of Ignacio Grossmann and Gary Blau of Carnegie Mellon University.

Volume 2. Design of an Ammonia Synthesis Plant

The objective of this case study is the design of an ammonia synthesis plant that uses hydrogen and nitrogen feedstocks from a coal gasification plant. All stages of the design procedure, from preliminary calculations to detailed flowsheet calculations, are described. Emphasis is placed on the following steps: screening of key flowsheet decisions (pressure of synthesis loop, ammonia recovery, synthesis of gas recycle, hydrogen recovery from purge stream), selection of reactor configuration, cost minimization, and synthesis of the heat exchanger network. The proposed design incorporates a medium-pressure synthesis loop with water absorption/distillation for ammonia recovery, and with membrane separation for hydrogen recovery. The process was designed with the simulator PROCESS from Simulation Sciences, and the ammonia reactor was designed with the special-purpose package QBED. A listing of this program is included in the case study.

Depending on the required detail and the availability of process simulation software, the case study is suitable as a one-term assignment for either a single student or a group of three students, while the final design report is based on the work of a group of five students. The problem statement was supplied by Philip A. Ruziska from Exxon Chemicals, and the case study was prepared by Stacy Bike under the supervision of Ignacio Grossmann from Carnegie Mellon University.

Volume 3. Preliminary Design of an Ethanol Dehydrogenation Plant

The objective of this case study is the preliminary design of an acetaldehyde synthesis process by ethanol dehydrogenation. The project covered all stages of the design procedure starting from consideration of qualitative aspects of the flowsheet and preliminary calculations to detailed process simulations and final economic evaluations. In this study, emphasis is placed on synthesizing a workable flowsheet and justifying its configuration, simulating and evaluating the design using a commercial process simulator, and deriving a heat recovery network for the final process. The main reaction in this process is the endothermic dehydrogenation of ethanol to acetaldehyde. However, under the specified reactor conditions, a number of byproducts are produced and their presence determines a number of interesting alternatives for separation.

Once these alternatives have been screened and a workable flowsheet has been synthesized, the study centers on the simulation of this flowsheet using PROCESS from Simsci, Inc. Here, some of the features, advantages, and limitations of this simulator are presented. Finally, the study concludes with a complementary presentation of this process simulated with the CACHE version of FLOWTRAN. While the aim of this study is not to provide a detailed comparison between PROCESS and FLOWTRAN, a useful description of the relative merits of these simulators can be readily observed.

This project is suitable for a one-term project by a five or six person team of senior design students. The results of two such teams are given in this study. This problem was posed by the Union Carbide Corporation and the case study was prepared under the supervision of L.T. Biegler of Carnegie Mellon University and the late R. R. Hughes of the University of Wisconsin.

Volume 4. Alternative Fermentation Processes for Ethanol Production

The objective of this case study is the preliminary design and economic evaluation of a fermentation process for the production of ethanol from a molasses feedstock. The intent is to expose the student to some non-traditional chemical engineering processes and to the expanding field of biotechnology. The scope of the study is such that groups of 2-3 students should be able to complete the design in about 30 days. The major focus of this design is the creation and rational development of a suitable process flowsheet, simulation of the flowsheet by the simulator FLOWTRAN, and economic evaluation and cost minimization of the final process.

The problem begins with the specification of the plant operating requirements. The type of fermentor to be used as well as plant operating conditions are left open. Suggested fermentors include batch, CSTR, and CSTR with cell recycle, as well as a novel extractive fermentor based on the use of hollow fiber membranes (HFEF). The choice of the fermentor will affect the nature of the flowsheet and lead to several design alternatives that the students will have to screen before arriving at a workable flowsheet that is ready for simulation. This case study includes a floppy disk with input files for the simulator FLOWTRAN and a program written in BASIC to evaluate the performance of CSTR fermentors.The problem statement was posed by Steven LeBlanc and Ronald L. Fournier and prepared under their supervision by the student Samer Naser at the University of Toledo.

Volume 5. Retrofit of a Heat Exchanger Network and Design of a Multiproduct Batch Plant

This volume contains two short design projects that can be developed by groups of 2-3 students in about two weeks. As opposed to the large projects that are commonly used in a design course, the objective of the case study is to expose students to a greater variety of problems and which are of current industrial significance.

The first problem deals with the retrofit of a heat exchanger network consisting of 8 exchangers with 5 hot and 3 cold processing streams as well as steam and cooling water. The layout of the network and areas of the exchangers are also given. The objective is to determine a retrofit design that can reduce the energy consumption within specified limits for the capital investment and payout times. This problem requires examination of alternatives for the level of energy recovery, matching of streams, addition of area, and removal or reassignment of existing exchangers and piping. This problem can be used to illustrate basic concepts of heat integration, as well as the application of computer software such as Target II, THEN, MAGNETS and RESHEX.

The second problem deals with the design of a batch processing plant that has to manufacture 4 different products, all of which require 5 similar processing steps (reaction, product recovery, purification, crystallization and centrifuge). An important aspect of this problem is that the production schedule and inventory must be anticipated at the design stage. Furthermore, this problem also requires analyzing alternatives for merging processing tasks into single units, and using parallel units with and without intermediate storage. The use of Gantt charts is emphasized to examine some of these alternatives. The case study also includes two sets of homework problems with solutions that can be used to provide the basic background for the two problems. This case study has been prepared by the students Richard Koehler and Brenda Raich at Carnegie Mellon University under the supervision of Ignacio Grossmann, who developed the problem statements and educational material.

Concluding Remarks and Future Directions

Though no two design problems are alike, there is a general logical sequence of basic steps which lead to a good design. It is the chief objective of the CACHE Case Studies to demonstrate and elucidate this thought and decision process. The CACHE Case Studies are different from final student or industrial project reports in that they not only present one final solution, but show the whole solution procedure leading from the problem statement to the final solution(s) in an organized manner. This chapter has discussed general issues involved in projects for the undergraduate design course. The role of design case studies to facilitate this task has been discussed. Finally the first five volumes of the CACHE Process Design Case Studies have been described.

It should be noted that the sixth volume of this series, *Chemical Engineering Optimization*

Models with GAMS, is devoted to optimization. This case study is described in the preceding chapter and would normally be used in an optimization course, although parts of it might also be incorporated into a design course, if optimization is taught as part of it. Also, a new case study on process operations (data reconciliation in a refinery) is under development at McMaster University and will be available soon.

As for future trends, we believe that the design course will continue to be the major capstone course in chemical engineering. It is the only course in the curriculum in which the students put together and apply all the material they have learned in chemical engineering. It is a course in which students have to make decisions, often with limited or incomplete information. It is also a course in which students have to learn how to work and get along with their group members. Finally, it is a course in which students learn the importance of written communication as a complement to the technical work.

It should also be noted that the teaching of the design course will be facilitated in the future with at least two new design textbooks. *Chemical Process Design* by Robin Smith (1996) has just been published, while *Systematic Methods for Chemical Process Design* by L. Biegler, I. Grossmann, and A. Westerberg is due to appear in 1997.

The design project, however, will continue to play a central role in the design course. In order for the project to fulfill a relevant and timely role, it is important that future case studies address nonconventional processes (e.g., electronics) and pressing industrial problems (e.g., environmental). It is also important that the projects be directed towards operational issues. The case studies should incorporate the use of new computer tools and environments, and reflect new trends in industrial practice. We hope that CACHE can continue to promote the creation and distribution of such case studies, and encourage faculty who can develop good design problems to participate with CACHE in this endeavor.

PROCESS CONTROL

Yaman Arkun
Georgia Institute of Technology
Atlanta, GA 30332

Carlos E. Garcia
Shell Oil Company
Deer Park, TX 77536

Abstract

This document summarizes CACHE's involvement in process control education during the last 25 years. A historical review of past activities is given; current trends are pointed out and a futuristic perspective on chemical process control education is provided.

Introduction

We review and put into perspective CACHE's involvement in process control education over the last 25 years, hoping that this information will prove useful for chemical engineering educators. As in essentially all of its activities, CACHE's role here has been primarily to facilitate the introduction of computer aids and define new educational directions in process control. In doing so, CACHE has tried to delineate innovative ways to use computer tools to develop creative teaching media which will help to improve the learning skills of our students.

Process control in the chemical engineering curriculum occupies a unique place. Firstly, its boundaries are not as rigid or well-defined as in some other courses. It requires a working knowledge and synthesis of many diverse and interdisciplinary concepts ranging from basic principles of modeling to controller design. Secondly, it studies techniques and computer algorithms which are used in the chemical process industry in connection with real-time monitoring and digital control. Therefore, in addition to the basics, some exposure to computer implementation is also important. Considering the computational challenges faced during modeling, dynamic analysis and controller synthesis, and the technological advances made in industrial computer control practice, it is not surprising that process control education has witnessed increasing use of computer hardware and software over the years. This transformation was accompanied by interesting debates on the appropriate course coverage to bridge the gap between academic training and industrial practice (Edgar, 1990). In what follows, we review CACHE's role in this evolution by describing its major activities and the educational materials it has introduced for control courses. Finally we will express our views on the current and future directions of process control education.

CACHE's involvement in process control education dates back to the early 1970s. In these earlier days of its establishment, with the help of many faculty from different universities, CACHE developed 150 computer programs and modular instruction materials (Edgar et al., 1988) for core chemical engineering courses, including process control. These FORTRAN programs included the following control modules: roots ofthe characteristic equation, bode plots of transfer functions, control of a pure timedelay, simulation of tank level control, and frequency response data via pulse testing. They became quite popular and were reprinted three times. These programs relieved students from repetitive and tedious calculations and allowed them to spend more time in creative problem solving. It was also the first time that interactive computer graphics had been introduced into the chemical engineering curriculum. At the time (1976), the first author was in graduate school performing TA duties in an undergraduate process control course. He remembers vividly the excitement caused by the introduction of root locus and bode plots. CACHE was also involved in preparation of instructional process control modules published by AIChE in five volumes (AIChEMI, 1981-1986), and which are still available.

Recognizing the importance of interactive graphics, CACHE formed a task force on computer graphics in the late seventies. Its main goal was to develop an awareness of the power and pedagogical value of interactive graphics in chemical engineering education. Its activities included publication of tutorials and position papers providing information on interactive graphics devices, and recommended hardware configurations and potential applications (Carnahan et al., 1978; Reklaitis et al., 1983). From these studies, process control emerged as a prime candidate to benefit from the use of interactive graphics tools. In particular, graphical tools for controller tuning, such as Nyquist plots, became more widely used as teaching tools. Several departments began development of their own dynamic simulators and design tools with graphical interfaces.

A CACHE survey carried out in the summer of 1981 revealed that while access to graphics hardware in chemical engineering departments was growing, educational use was still quite limited by the availability of general purpose software, and hardware consisted mostly of older monochrome Textronix devices. In retrospect, the picture has changed drastically in the last decade. Starting with Program CC (Thompson, 1985) and continuing with MATLAB, general purpose commercial packages have taken the place of much in-house control software in recent years. At the same time, rapidly growing computer technology has allowed use of interactive graphics on many different hardware platforms, including PCs, Macs and UNIX workstations, all at affordable prices. Today, commercial packages such as MATLAB are used in quite a few chemical engineering departments (Kantor, 1992). More will be said about the current status of these tools in process control education later on.

One of the most significant commitments of CACHE to process control education has been in the area of real-time computing. Motivated by the lack of educational material in this area in the 1970s, the Real-Time Computing Task Force published eight introductory monographs (Mellichamp, 1977) that covered basics of real-time programming and architectures, prototype laboratory experiments, computer interfaces and application software, and data acquisition and control algorithms. These monographs, which later became a textbook (Mellichamp, 1983), were in high demand for teaching real-time systems and are still being used in some digital process control courses. In parallel with CACHE activities, integration of real-time computing into process control teaching had already started in the late seventies in quite a few chemical engineering departments (Morari and Ray, 1979).

Faced with the pervasive use of personal computers in the1980s, CACHE, following its traditional interest in real-time computing, took a new initiative, "microcomputer applications in the laboratory". The mission was to assist chemical engineering departments in developing undergraduate laboratory experiments involving on-line computer applications. The motivation was to give a picture of the wide array of classical and modern engineering concepts that have been successfully built into computer-aided laboratory experiments, of the many very different hardware and software approaches, and of the pedagogical objectives to be fulfilled when the computer was used for control purposes. The specific CACHE contributions included a *Survey* of U.S. and foreign chemical engineering departments (Cinar and Mellichamp, 1986), an *Anthology* (Cinar and Mellichamp, 1988), and a publication describing Real-Time Task Force activities (Arkun, et al., 1989). The *Anthology* contains descriptions of twenty one experiments covering the areas of thermodynamics, fluid dynamics, heat and mass transfer, chemical and biochemical reactors, process dynamics, and control. In process control education this document should help, especially faculty who are developing on-line computer control applications for the first time. There is no doubt that teaching process control with this type of laboratory instruction is very desirable, as the students are introduced to real problems, and acquire valuable hands-on experience with control hardware and different data acquisition and control practices.

In the 1990s, CACHE has been involved with two major products, the Model Predictive Control Toolbox (CACHE-Tools) (Morari, 1993) and PICLES (Cooper, 1993). While CACHE-Tools have been developed as a CACHE task force activity, PICLES was created at the University of Connecticut by Douglas Cooper with CACHE assistance in testing and disseminating the program. PICLES (Process Identification and Control Laboratory Experiment Simulator) is a training simulator that provides access to several simulated processes with real world behaviors; processes include gravity drained tanks, a heat exchanger, a pumped tank and a binary distillation column. It also includes "mystery processes", with models hidden from the students, in which input and output data are displayed for subsequent identification and control studies. In all these processes, PICLES demonstrates the significance of nonlinearites, time delays, noisy measurements, inverse and integrating open-loop responses, and multivariable interactions. The available controllers include different PID algorithms, Smith predictor, feed forward control and a decoupler. PICLES provides a color graphic display of process equipment with some animation, historical data of inputs and outputs, and controller templates (see Fig. 1). The program runs on IBM-compatible personal computers. It can be incorporated easily into existing undergraduate process control courses either by assigning homework around the processes it provides or by treating it as a simulated experiment in the control laboratory.

The commitment to bring new and proven industrial control practices into academic training is best reflected in the recently introduced software, CACHE-Tools. This is a collection of MATLAB functions developed for the analysis and design of model predictive control (MPC) systems. MATLAB runs on almost all personal computers and engineering workstations and is available at a significant educational discount. MPC was conceived in the 1970s primarily for industry. It can be now safely be argued that it is the most widely used multivariable control algorithm in the chemical process industries. MPC is a novel technique for feedback control that makes explicit use of a process model. It uses the model to predict the effect of control

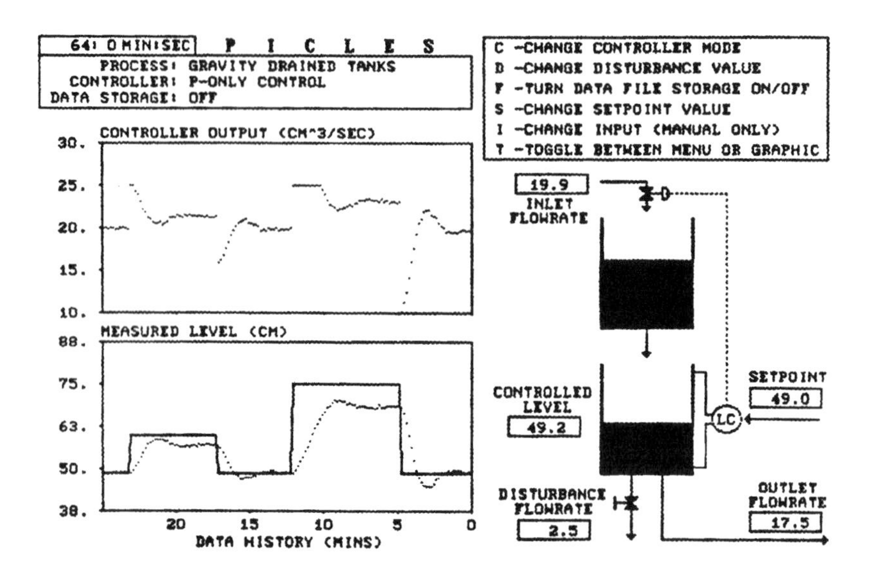

Figure 1. PICLES - Gravity-Drained Tanks under P-Only control shows increaing offset as set point moves further from design value.

actions and to estimate disturbances (see Fig. 2). The control actions are calculated from an on-line optimization. The fact that a process model is involved makes MPC quite useful for teaching, since the interesting interplay between dynamic modeling and control can be taught and explored in a design setting that has immediate industrial applicability. We can now appreciate better the significance of different types of process models for control purposes. Also, with MPC, real and difficult process problems such as multivariable interactions, delays, measurable disturbances, an, most importantly, constraints on both process inputs and outputs, can be handled directly.

CACHE-Tools is intended for both the classroom and the practicing engineer. It can help in communicating the concepts of MPC to students in an introductory course. At the same time it is sophisticated enough to use in higher level control courses and also to train practicing engineers. Recent undergraduate textbooks (Seborg, et al., 1989; Deshpande and Ash, 1988) include some introductory material on MPC. However, more tutorial teaching material is needed, and we expect the monograph (Morari, et al., 1994) will be valuable in this regard.

Finally, CACHE has been quite active in organizing process control conferences. CPC (Chemical Process Control) has become the premier international conference series bringing together participants from academia and industry. Although these conferences are primarily research oriented, they have impacted process control education by providing a forum for discussing industrial viewpoints and new educational paradigms (Downs and Doss, 1991).

CACHE's major activities in process control in the last 25 years is summarized in Fig. 3.

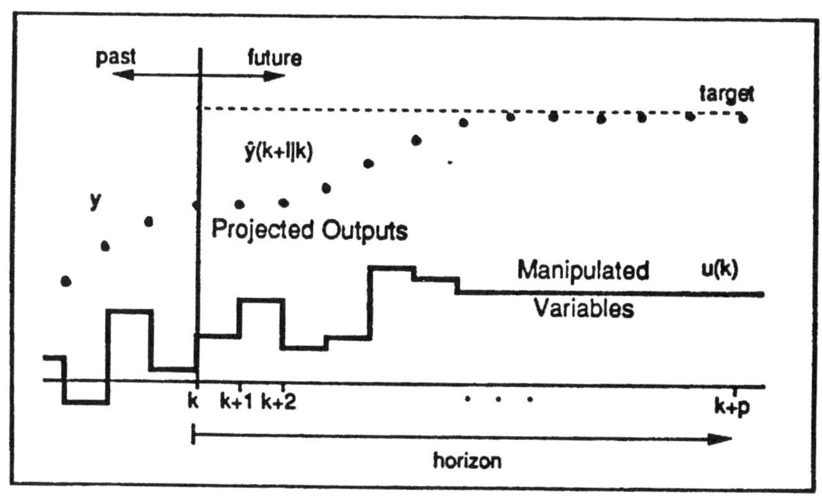

•Use model to predict the future changes of controlled variables.

•Correct the prediction using the disturbance effect estimate.

•Calculate manipulated variable moves to:

min[{prediction-setpoint}+{control effort}] subject to constraints on inputs

Figure 2. Model Predictive Control.

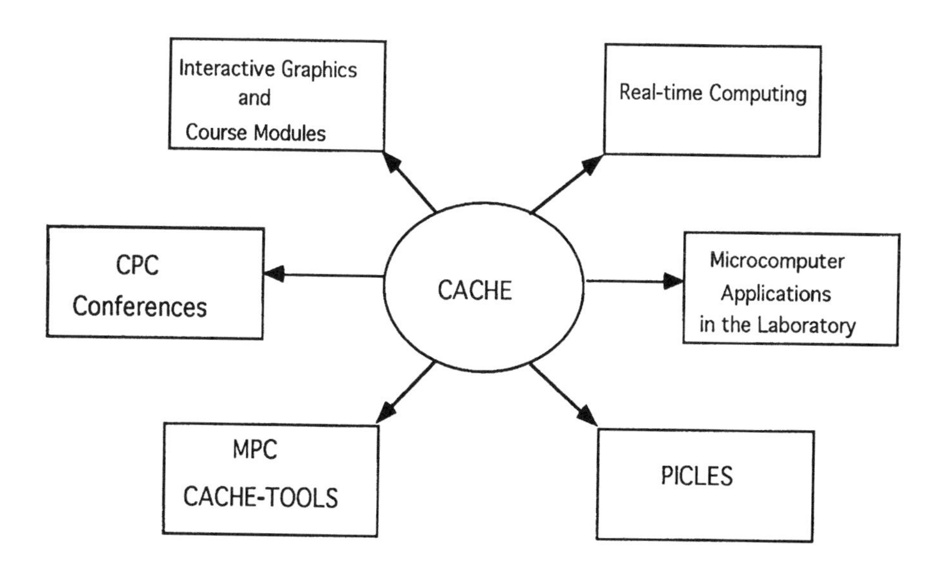

Figure 3. CACHE activities in process control.

Present Status and Future Trends

Today there seem to be two dominant trends in the development of computer aided educational materials. One is to create tool boxes around commercially available general purpose software. The other is the case study approach. For example, MPC CACHE-Tools has been designed around MATLAB, which is readily accessible and offers many useful capabilities, including symbolic manipulation, identification, control system analysis and design, digital signal processing, statistical analysis, and optimization. Any of these features can be used in connection with CACHE-Tools, if needed. In addition, a separate software package, SIMULINK, has added a nonlinear dynamic simulation and modeling environment to MATLAB's analysis environment. With the help of SIMULINK, different process case studies can be developed and interfaced with MPC or any other control algorithm. In fact, today quite a few departments are in the process of creating these types of case study modules. The first author has used SIMULINK control modules as part of a design project in an undergraduate control course in the last three or so years (Fig. 4). In the future, it would be valuable to document and distribute a library of such case studies to process control instructors. Finally, we should draw attention to the efforts of our colleagues devoted to process control education independent of CACHE activities. There have been numerous valuable contributions in the creation of educational materials, including software development, some of which is documented in Seborg, Edgar and Mellichamp (1989). Among these, in particular, CONSYD from Wisconsin, and UC-Online from Berkeley have contributed significantly to the teaching of process control.

Most of the changes that have occurred in process control education (and practice) in the last 25 years can be attributed to the introduction of faster and cheaper microcomputers. This has forced process control education to shift in emphasis toward "softer" chemical process operation improvement and away from the traditional "hard" instrumentation emphasis. Although theoretical course subjects have not changed in any significant way, we have seen more coverage of digital control with laboratory experiments, introduction of new computer algorithms like MPC, and some exposure to statistical process control. Basically, process control education has gone through a period of evaluation of new topics and teaching techniques in response to rapidly changing computer hardware and software developments, and new advances in control research and industrial practices.

We anticipate that the next twenty five years will be even more revolutionary. With increased implementation of advanced control systems and techniques in industry and with increased investment in sophisticated process control equipment, engineers need to be better educated in maximizing the use and value of control technology in manufacturing operations. This requires the plant engineer to have a thorough understanding of control technology and of the processes and the skills needed to successfully combine them to operate a plant profitably. Currently, this task is delegated to the "control engineer" or to vendors who, because of their lack of day to day process exposure, are limited in identifying opportunities. To achieve this, we see a long term challenge to shift the way engineers are educated away from steady-state design of unit operations and more into how plants are operated in real time. It is conceivable that every unit operations course could cover dynamic transient concepts and education on how these units are operated. The process control course will then become more broad and dedicated to teach all technology that drives those processes to satisfy operating objectives: control theory and algorithms, advisory and information systems, optimization, etc. Laboratory work

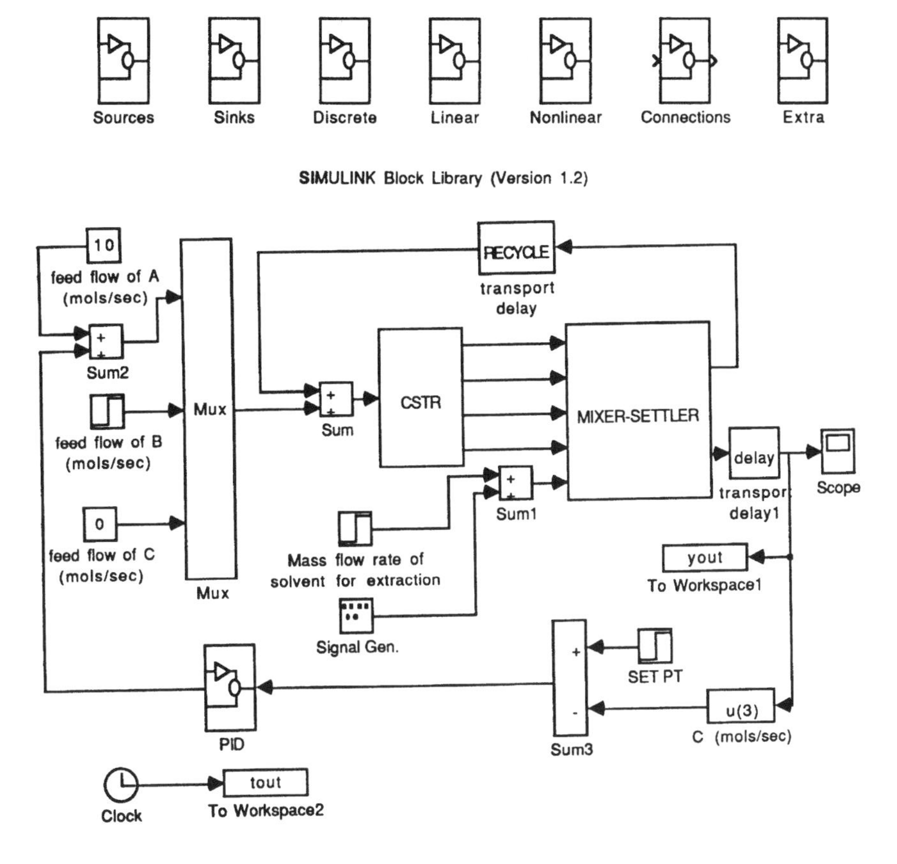

Figure 4. SIMULINK Module - Control of a reactor system with recycle.

will be used to emphasize the integration of these skills with process operation through control hardware and software.

Because of these trends, educational aids for control should help engineers learn the process dynamics as well as the control principles. Therefore, there will be a need for dynamic simulators with computer interfaces like those used in industrial operations. We also need educational aids that help in learning the mathematical concepts, the development of control strategies and control algorithm design, and the art of evaluating whether control systems are helping meet operating objectives. Therefore, topics such as process monitoring, statistical analysis and performance evaluation, model identification and parameter estimation, and large scale computing and optimization will need to be integrated with traditional process control education and covered in more depth in undergraduate courses. At the same time, in laboratories we expect to see more advanced instrumentation with friendly real-time interfaces, increased computational power and data analysis capabilities, advanced multivariable control algorithms, and less traditional experiments. Laboratory experience should prepare students for correctly using control skills on processes.

A most important area which will impact the way we teach process control in the future is dynamic modeling. Models are an excellent way to learn process dynamics and of course are the main requirement for control design and implementation. Systematic development and utilization of the appropriate models for different control purposes and analysis of effects of modeling inaccuracies on controller performance will find a permanent place in control education, as we learn more from research findings in these areas.

Also, as we broaden the impact of process control in operations, two challenging topics that have been haunting us for quite some time in education will increase in importance. These are integration of control with process design, and plant wide control (as opposed to unit operations control). Despite the need expressed by industry (Downs and Doss, 1991) and increased coverage in recent textbooks (Seborg, et al., 1989; Stephanopoulos, 1985), it is not clear how to teach these concepts and the demanding design tasks involved in the most efficient way. There must be more activity that sheds light in this direction, if process control education is to make a big impact on chemical engineering practice.

In summary, efficiencies in computing and the continued development of improved modeling and control tools better integrate this technology with plant operations and educational needs to reflect that trend. We see computer-based educational aids in control as being the key in teaching this fundamental technology to engineers, and expect CACHE to have a continued and major emphasis on process control education in the future.

References

AIChEMI Modular Instruction Series; published by AIChE: New York, 1981-86: Series A, Process Control; T. F. Edgar, Ed.

Arkun,Y., A. Cinar, and D. A. Mellichamp (1989). *Laboratory Applications of Microcomputers*. CACHE News, Spring.

Carnahan, B., R. S. H. Mah, and H. S. Fogler (1978). *Computer Graphics in Chemical Engineering Education*. CACHE: Cambridge, MA, June.

Cinar, A. and D. A. Mellichamp (1986). *Survey: Experiments Involving On-Line Computer Applications in Chemical Engineering Undergraduate Laboratories*. CACHE: Austin, TX, Sept.

Cinar, A. and D. A. Mellichamp (1988). *On-Line Computer Applications in the Undergraduate Chemical Engineering Laboratory: A CACHE Anthology*. CACHE: Austin, TX, Sept.

Computer Programs for Chemical Engineering Education. Sterling Swift Publishing: Austin, TX, (1972).

Cooper, D. J. (1993). PICLES: The Process Identification and Control Laboratory Experiment Simulator. CACHE News, Fall.

Deshpande, P. B., and R. H. Ash (1988). *Computer Process Control with Advanced Control Applications*, 2nd Ed., ISA.

Downs, J. J., and J. E. Doss (1991). The View from North American Industry, *CPC IV Proc.*, Y. Arkun and W.H. Ray, Eds., AIChE.

Edgar, T. F. (1990). Process Control Education in the Year 2000. *Chem. Eng. Edu*, Spring, 72-77.

Edgar, T. F., R. S. H. Mah, G. V. Reklaitis, and D. M. Himmelblau (1988). Computer Aids in Chemical Engineering Education. *CHEMTECH*, **18**, 277-283.

Kantor, J. C. (1992). Matrix Oriented Computation Using Matlab, CACHE News.

MATLAB, The MathWorks, Inc., South Natick, MA.

Mellichamp, D. A. (ed.) (1977). *CACHE Monograph Series in Real-Time Computing*. CACHE: Cambridge, MA.

Mellichamp, D. A. (ed.) (1983). *Real-Time Computing*. Van Nostrand-Reinhold: New York.

Morari, M. (1993). CACHE News.

Morari, M and W. H. Ray (1979). The Integration of Real-Time Computing into Process Control Technology, *Chem. End. Edu.*, 160-162.

Morari, M., C. E. Garcia, J. H. Lee, and D. M. Prett (1994). *Model Predictive Control*. Prentice Hall.

Reklaitis, G. V., R. S. H. Mah, and T. F. Edgar (1983). *Computer Graphics in ChE Curriculum*. CACHE: Salt Lake City.

Seborg, D. E., T. F. Edgar, and D. A. Mellichamp (1989). *Process Dynamics and Control*. John Wiley & Sons.

Stephanopoulos, G. (1985). *Chemical Process Control*. Prentice Hall.

Thompson, P. T. (1985). *Program CC*. System Technology, Inc., Hawthorne, CA.

LABORATORY AUTOMATION AND REAL-TIME COMPUTING

Duncan A. Mellichamp
University of California
Santa Barbara, CA 93106

Babu Joseph
Washington University
St. Louis, MO 63130

Abstract

We discuss the history and CACHE contributions to the field of laboratory automation and real-time computing. The evolution of the field through the last few decades is examined, and notable historical milestones are mentioned. The chapter concludes with a brief look at the future of the field.

Introduction

The use of computers for data acquisition and control in the chemical engineering laboratory was very quickly viewed by university faculty as having great potential for application in both teaching and research. Indeed, the early days of CACHE coincide with the first commercial developments of minicomputers, the innovation that would make possible practical university applications in this area. Thus, it should be no surprise that one of CACHE's first new activities, following the initial focus on large-scale computing applications, was on "real-time computing," i.e., the use of digital computers for data acquisition and control and for special purpose dynamic simulation (provided primarily via analog computers prior to that time).

This chapter looks back over those early beginnings in the late 1960s, the development of the minicomputer-based applications that occurred during the 1970s, the microcomputer revolution of the 1980's, and finally where the field appears to be going in the 1990s. The entire area of laboratory automation can now be fairly characterized as representing a mature field, i.e., one in which the emphasis is on utilizing an inexpensive, well understood, and rapidly expanding technology (rather than on trying to understand a brand new technology, acquire it, and bend it to purpose).

One of CACHE's most significant publication projects is represented by the CACHE Monograph Series in real-time computing, eight volumes written by early specialists in the field and edited by Mellichamp (1977-1979). These provided a major impetus at the time for prospective new users to begin work in this new area. Clear evidence of the relative maturity of the field just one decade later is provided by the CACHE *Anthology of Applications in On-Line Automation*, edited by Mellichamp and Cinar (1988) and containing nearly two dozen specific examples of undergraduate laboratory projects utilizing the computer.

Early Developments (1960s)

Pioneering academic workers in this area came primarily from one of two groups: (1) those having experience with industrial digital computer control projects, and/or (2) those with a background in analog hybrid computing. During the early 1960s industrial control projects had progressed from the first supervisory control projects (in which a large-scale digital computer was connected to analog instrumentation for the purpose of optimizing the placement of controller set-points) to so-called direct digital control systems (in which similarly large computers converted process measurements directly to digital form and manipulated the process actuators directly, as well). In either case, the computer systems used to carry out industrial control were quite large, costing on the order of hundreds of thousands of dollars; nevertheless academic appetites were greatly whetted by experiences with such systems, particularly those individuals interested in carrying out control research or in monitoring complex laboratory processes.

Once reliable (solid-state) and relatively inexpensive computers became available, monitoring and control facilities were established in chemical engineering departments at several universities where access to necessary funding was found. One cites, in particular, the early work of Grant Fisher at Alberta and Roger Schmitz, then at Illinois, both able to build facilities around the IBM 1800, a machine of modest capabilities (by today's standards) adapted from the earlier stand-alone Model 1620. Other academics came to laboratory real-time computing via a less expensive route during this period, by utilizing the digital computer part of an analog hybrid computer — the sort used for simulation purposes in a number of departments in the early 1960s. One can cite Cecil Smith at Louisiana State as one who followed this path.

By the end of this decade, a number of chemical engineering faculty were utilizing any available computer, e.g., an IBM 1800 linked from another building or across campus, as at Santa Barbara, facilities shared with analytical chemists, or whatever. All of the facilities of this period suffered from high costs and continued low reliability and from hardware and software limitations that, in retrospect, seem overwhelming by today's measures. On the other hand, some significant research was performed; most often this involved dedicated technicians and graduate students who practically lived with the systems.

The Age of the Minicomputer (1970s)

The first practical minicomputer, the DEC PDP-8, appeared commercially in the mid-1960s. This 12-bit, single-accumulator machine was purchased and used by a relatively large number of laboratory chemists but, for some unexplained reason, the PDP-8 was largely ignored by chemical engineers. However, the Data General NOVA, with its larger 16-bit word size and more advanced multiple-accumulator architecture, was introduced around 1970 and did catch on with a number of innovators. Among the first were Joe Wright at McMaster and Duncan Mellichamp at U.C. Santa Barbara. Wright was involved in the early development of

Data General's Real-Time Operating System (RTOS), the first in a series of minicomputer software developments that allowed programs for laboratory experiments to be written in an organized (structured) manner.

When DEC later released the more powerful PDP-11 with its own real-time operating system, a second group of chemical engineering faculty chose to build on that system. Ed Freeh and John Heibel at Ohio State were workers in this group. Both the early DG and DEC systems supported real-time processing functions — analog-to-digital and digital-to-analog converters, time-keeping and interrupt-service routines directly accessible to the user/programmer, etc. During the early 1970s Hewlett-Packard, a traditional major player in laboratory instrumentation, introduced a line of special purpose computers for process monitoring and control, as did several other minor players who have long since disappeared into the economic black hole of technical successes but business failures.

Soon after 1970, the CACHE Real-Time Task Force was formed and began its early efforts to define the field and to provide guidance to those faculty looking for ways to utilize these new computer capabilities. At that time, at least four different manufacturers were providing computers roughly equivalent in terms of capability but totally incompatible with each other; the small chemical engineering user community was badly splintered as a result. Thus CACHE's first efforts to provide guidance in this area were significantly complicated by the lack of standardization in the hardware area. Early efforts of the new Real-Time Task Force attempted to focus on two "demonstration projects," the interfacing and use of a digital computer with (1) a gas chromatograph and (2) a laboratory-scale distillation column. Beyond several descriptive overviews and surveys, no specific CACHE products came out of this early period, although several workshops dealing with the principles of this new field were held under CACHE auspices.

Just as chemical engineering real-time users tended to fall into groups depending on their original choice of hardware — the PDP-11 users, the DG Nova/Eclipse users, and the rest (mostly IBM 1800 and HP stand-alone digital instrumentation systems) — the choice of software for user application programs was split, as well. The earliest applications on the smaller machines were forced to utilize assembly language as the programming medium because of the small amount of available main memory (initially 8 Kbytes). However, as soon as memory prices dropped (mid-70s), most users followed the IBM 1800 approach of using real-time versions of the FORTRAN language. One exception involved the use of various versions of real-time Basic, an interpreted language that was/is particularly useful in an educational format. During this period several academic projects led to published descriptions of undergraduate laboratory experiments suitable for demonstrations. For example, an NSF supported instructional project at UC Santa Barbara illustrated the use of a digital computer for the most important operations — analog data acquisition, binary input, output and logic, process control, etc. — with applications to three bench-scale undergraduate laboratory experiments. Documentation of this work was distributed to all CACHE participating schools at the close of the project.

The members of the NSF-required overview panel for the Santa Barbara project were (no accident here) virtually contiguous with the CACHE Real-Time Task Force. At one of their early meetings, this group discussed the idea of putting together a set of written materials that would deal with the many key principles and issues in this new area. Out of these discussions (and many future meetings of the group!), the CACHE Monograph Series in real-time computing, a set of eight volumes, was edited by Mellichamp and published by CACHE between 1977 and 1979. This series, in somewhat updated form, was published by Van Nostrand Reinhold as a professional text (Mellichamp, 1983). The book was chosen almost immediately as an "alternate selection" by one of the major computer book clubs. However, the chief importance of these materials was as a relatively early primer that helped many new users become established in the area.

Two stumbling blocks prevented many departments from introducing computer control in their laboratories during this period. One was the high cost of the hardware; a PDP-11 system with real-time hardware peripherals sold for about $20,000 in the late 1970s. The second problem was the high overhead in terms of effort required by faculty to become familiar enough with the hardware and software that it could be effectively used in a classroom or laboratory. A few chemical engineering faculty, nevertheless, came to real-time computing through research that required computer-based data acquisition and control facilities and through the efforts of their graduate students.

Change occurred rapidly following the introduction of the microprocessor. Industrially, control instrumentation hardware began to be replaced with microprocessor-based distributed computer control systems. These systems, however, still carried a heavy price tag due to their orientation toward large-scale continuous processing plants. The situation for smaller-scale academic research and instructional applications changed radically with the introduction of microprocessor-based personal computers, the topic of the next section.

The Microcomputer Revolution (1980s)

The late 1970s saw the introduction of the first microcomputers. While no different conceptually than their minicomputer predecessors, microcomputers possessed distinct advantages of scale and cost. Thus, many of the trends in laboratory automation one notes today did not come from the minicomputer applications of the 1970s, i.e., large, user-written programs (e.g., in FORTRAN), but rather reflect the microcomputer philosophy of very large main memory (> 1 Mbyte of RAM) used in conjunction with commercial software specifically tailored for data acquisition and control and for ease of user interactions.

The Apple II computer was introduced in 1977, the brainchild of two computer hackers working out of a garage. This microcomputer in a new package represented an astonishing step forward in that for the first time the average professional user could afford to purchase a computer, set it up, and run it without much prior knowledge of hardware and software. It was never really intended to be used for real-time computing applications in the laboratory. It did not even have a FORTRAN compiler, but it did come with a built-in Basic interpreter.

However, in addition to its ease of use, the designers of this revolutionary machine provided it with an open architecture that quickly led to the birth of a subsidiary multibillion dollar industry: making peripherals that extended the power of the basic system. Within a very short time period, new start-up companies began to manufacture real-time peripherals for the Apple II. As with the computer itself, these boards were priced low enough that one could set up a computerized data acquisition and control system at prices starting below $5000.

Many universities took advantage of these reduced costs and increased ease of use to equip their control laboratories with personal computer based systems. Joseph and Elliot (1984, 1985) describe an undergraduate process control laboratory built using bench-scale units interfaced to Apple II computers. One of the chemical engineering faculty members who embraced this new microprocessor technology was Peter Rony at Virginia Polytechnic Institute and University located in Blacksburg, Virginia. Under his leadership CACHE organized workshops on Microcomputer Interfacing for interested chemical engineering faculty.

IBM introduced their first personal computer in 1981, providing yet another breakthrough in cost and performance. Again, although this computer was intended to be a hobby and recreational machine, it found wide applications in professional and other areas, including the laboratory. Academic users desiring flexibility and low-cost preferred a data acquisition system built around these low-cost personal computers rather than around a dedicated industrial grade computerized Data Acquisition and Control (DAC) system. The latter provided performance and ease of use only with a large price tag and lack of flexibility. With each generation of microprocessor development, the increasing power of these personal systems accompanied by ever decreasing costs led to rapid popularity among academic scientists and engineers.

Simultaneously, another revolution was taking place in the software available for data acquisition and control applications. The need to write code in assembly language was replaced, first by Basic interpreter-based applications programming capability. With a suddenly large installed base of microcomputer users, many software companies were established that catered specifically to the Data Acquisition and Control market. Prominent among them was a company begun by Fred Putnam, a chemical engineering professor at MIT, now one of the leading suppliers of PC-based DAC software to industry (Laboratory Technologies, Inc.). Software developments have kept pace with the increasing power available as each successive generation of microprocessors has been commercialized.

With the availability of real-time control software, it became possible to introduce concepts of distributed computer control in undergraduate chemical engineering courses. Alan Foss at the University of California, Berkeley, developed a software package for his process control course which is currently available to other chemical engineering departments under license (Foss, 1989). The book by Joseph (1988) provides a good summary of the technology that was available in the 1980s for laboratory automation. The CACHE Anthology of Applications in On-Line Laboratory Automation (Mellichamp and Cinar, 1988) provides a number of illustrations of chemical engineering undergraduate laboratory experiments that were developed during this period.

Laboratory Automation Today (1990s)

Today, it is not surprising to walk into an undergraduate chemical engineering laboratory or a graduate research laboratory and see many of the experiments interfaced to microcomputer-based DAC systems. Computers are used routinely for acquiring data, processing it, and reporting the results. The software running the actual data acquisition and control functions hides much of the detail, freeing the user to focus on the experiment and the data rather than how it is acquired. These menu-driven packages provide "visual programming," usually a way of connecting blocks or icons into an instrumentation block diagram. They offer on-line help to set up the DAC application so that a DAC system can be up and running in a few hours instead of the days, or even months, required in the past.

A typical laboratory data acquisition system today will be built around a desk top microcomputer package containing an Intel Pentium, IBM/Motorola Power PC, or equivalent class microprocessor with a few real-time peripherals, usually including industry standard 12-bit converters. A single card handling 16 channels of analog input, 4 channels of analog output, and 16 channels of binary input/output can be purchased for under $600. One probably will pay somewhat more for the software that is needed to set up the DAC system. DAC computers now can be easily networked to share data among multiple machines and to do remote data acquisition. Real-time software packages allow true distributed computing using these networked PC's. Interestingly, this concept is now used in industry to construct low cost distributed control systems. Thus the functionalities now available to the academic user are in no way restricted, either by hardware or by software, vis-à-vis industrial applications systems.

A key recent development on the DAC scene is the arrival of visual programming. Companies like National Instruments, Hewlett Packard, Iconics, Data Tranlation and Keithly-Metrabyte now offer software (LabView, LabWindows, HPVEE, Genesis, etc.) which further remove the programmer from the low-level assembly language and high-level FORTRAN programming of the past. These applications allow the user to construct a DAC program by connecting blocks together visually, in much the same way one would construct a functional block diagram in which each block provides certain capabilities such as analog read or write, data manipulation, data storage, or a simple controller function. Such applications quickly and easily allow the user to provide the DAC program a graphical front-end for communication with the operator — complete with indicators, strip-chart recorders, buttons, switches, and the like — all of which will be maintained in real-time during runs, including the provision of animation features, changing colors, etc. Thus the DAC programmer now constructs complete applications without the need to type-in long programs of code with the inevitable endless series of bugs and faults that must be removed before useful runs can be initiated. Since the "block diagram" now becomes both documentation and run-time program, a single session at the computer can encompass programming, running, debugging, and documenting "program" and results. This single incredible step has revolutionized DAC programming, making the writing and altering of programs extraordinarily easy and fast.

Another relatively recent development has been the increased use of realistic simulation exercises to teach principles of control. For example, the UC-ONLINE package from UC/Berkeley provides software that can simulate the start-up and control of distillation units, steam generation systems, inventory control and other processes. Purdue University has released a number of simulated "process modules" that students can exercise to obtain realistic plant data.

A second interesting addition has been the increased use of control design packages such as MATLAB (Mathworks, Inc.), MATRIX_X (Integrated Systems, Inc.), and ACSL (MGA, Inc.), which reduce the time required to design and test new concepts and ideas. All of these systems are not yet fully integrated to real-time peripherals but soon will be. MATLAB is already provided with DACs capabilities, and in work conducted at the Ecole Polytechnique in Lausanne (Gillet et al, 1993), a version of MATLAB was interfaced to LabView, thus providing process interfacing flexibility with high-level analytical capabilities.

Finally, there has been a recent development in the field of software for the steady-state design of chemical plants and processes, e.g., ASPEN (AspenTech), HYSIM (Hyprotech), Chemcad (Chemstations) and PRO II (Simulation Sciences). At least two of these companies have announced the availability of dynamic simulators which offer the prospect of students having the capability of working with realistic plant-scale facilities and of investigating the effect of choosing alternative control structures, control algorithms, or controller settings on plant operability and profitability. These entirely realistic studies offer the advantage of providing industrial-level problems without having to provide inordinately expensive facilities (then only at pilot scale) in university laboratories.

The Future

The computer has now become an integral part of the chemical engineering laboratory both in data acquisition and in data processing. The power of both hardware and software continues to grow at an ever increasing pace. Certainly these developments are going to impact even more on the future chemical engineering laboratory.

We expect to see more and more realistic simulation of both laboratory-scale equipment as well as plant wide simulations, so that students will be able to experiment with new concepts and ideas in the laboratory — assembling and building, then collecting data on simulated experimental setups. Data analysis and simulation tools will enable students to match their models with the data that they collect. Similarly they will be able to experience a plant start up and the procedure of bringing it under control from the computer console.

It is likely that sensors of the future will have their own computers to do self diagnosis, calibration, and error detection. Interfacing will become synonymous with data communication. These changes generally will continue to reduce the time and effort needed to set up DAC systems.

The control laboratory will most likely utilize interfaces similar to those found in industrial DCS installations. Thus students will be able to configure, implement, and test control system designs on processes similar to the ones they will encounter when they go into practice. The availability of advanced artificial-intelligence-based software integrated with control system design and simulation packages should allow students to develop and test complex control system designs in short laboratory sessions.

In summary, the future looks quite exciting for laboratory automation. The features mentioned here are on the verge of reality even now.

References

Aspen Technology, Inc., *Aspen Plus* Software, 10 Canal Park, Cambridge, MA 02141

Chemstations, Inc., *Chemcad* Software, 10375 Richmond Ave, Suite 1225, Houston, TX 77042

Data Translation, *DT-VEE* Software, 100 Locke Dr., Marlboro, MA 01752

Foss, A. (1987). *UC-ONLINE*: Berkeley's Multiloop Computer Control Program, *Chemical Engineering Education*, Summer.

Gillet, D., C. Salzmann, R. Longchamp and D. Bonvin (1993). A Methodology for Development of Scientific Teachware with Application to Adaptive Control, *American Control Conference*, San Francisco.

Hewlett Packard, *HP-VEE* Software, 1501 Page Mill Road, Palo Alto, CA 94304

Hyprotech Inc., *Hysim* Software, Alberta, Canada

Iconics, Inc., *Genesis Control Series*, 100 Foxborough Blvd., Foxborough, MA 02035.

Integrated Systems, Inc., *MATRIXX*, Santa Clara, CA 95054

Joseph, B. (1988). *Real-Time Personal Computing for Data Acquisition and Control*, Prentice Hall, New Jersey.

Joseph, B. and D. Elliott (1984). Experiments in Process Control Using Personal Computers, *Chemical Engineering Education*, **18**(4).

Joseph, B., D. Millard and D. Elliott (1985). Experiments in Temperature Measurement and Control Using Microcomputers, *Control Systems*, **5**(3).

Keithly-Metrabyte, *ViewDAC* Software, 440 Myles Standish Blvd, Taunton, MA 02780

Laboratory Technologies Inc., *LabTech* Software, 400 Research Dr., Wilmington, MA 01887

Mathworks, Inc., *MATLAB* Software, Cochituate Place, 24 Prime Parkway, Natick, MA 01760

Mellichamp, D.A., Ed. (1977-1979). Monograph Series in *Real-Time Computing*, Vol's I-VIII, *CACHE*, Austin, TX.

Mellichamp, D.A., Ed. (1983). *Real-time Computing with Applications to Data Acquisition and Control*, Van Nostrand Reinhold, New York.

Mellichamp, D.A., and A. Cinar, Ed's. (1988). On-Line Computer Applications in the Undergraduate Chemical Engineering Laboratory: A CACHE Anthology, *CACHE*, Austin, TX.

MGA, *ACSL* Software, 200 Baker Ave., Concord, MA 01742.

National Instruments, *LabWindows* Software, 6504 Bridge Point Parkway, Austin, TX 78730.

CACHE AND THE USE OF COMPUTER NETWORKS FOR EDUCATION

Peter R. Rony
Virginia Polytechnic Institute
Blacksburg, VA 24061

Abstract

Though CACHE was a pioneer in the use of electronic mail in chemical engineering education, it did not play a role in the broader networking community. The promise of worldwide networking and communication of digital text, image, animation, audio, and video files has arrived with World Wide Web (WWW). Internet software – Mosaic, Netscape, HGopher, WS_FTP, and others – for a Windows platform is described.

CACHE Contributions to Chemical Engineering Electronic Communications

Within the field of chemical engineering, CACHE preceded AIChE and other chemical engineering organizations in championing electronic communications, *viz*., electronic mail. CACHE, to use a familiar cliché, kept the torch lit within chemical engineering; CACHE, to use an unfamiliar cliché, was probably the first chemical engineering organization to pave its minuscule portion of the information superhighway with a bit of asphalt, which rapidly developed potholes. The role of CACHE in the broader networking community within science and engineering was negligible.

Phase 1. The CACHE National Electronic Mail Experiment: ITT Compmail+ ™

To quote the April 9, 1985 CACHE Electronic Mail Task Force proposal (Rony, 1985):

> "The objectives of the CACHE National Electronic Mail Experiment are: (1) to catalyze the creation of a widely used national chemical engineering electronic mail network between academic, industrial, and government sites, (2) to issue a report that documents concrete examples of how to use electronic mail, identifies potential chemical engineering applications, and lists typical costs, (3) to extend such a network to our international colleagues, (4) to publicize the use of electronic mail in CACHE, CAST, AIChE, and other chemical engineering organizations, (5) to identify important uses for electronic mail in chemical engineering, (6) to perform limited tests of alternative mail services, and (7) to publish an article on electronic mail in *Chemical Engineering Progress*."

Initial publicity for the CACHE National Electronic Mail Experiment appeared in *CACHE News* (September 1984) and at the CACHE Reception, 1984 AIChE Annual Meeting

(San Francisco, November 1984). The participation of CACHE in electronic communications within chemical engineering can be subdivided into at least four phases:

The first CACHE experiment in electronic communications was conducted during 1984-85 (Rony, Hale and Wright, 1986a). Though members of the chemical engineering community were invited to participate, few colleagues did so, other than members of the CACHE Board of Trustees, each of whom was given both a Compmail+ account and instructions how to use it. As of January 31, 1986, CACHE sponsored 55 Compmail+ accounts, only 20 of which were ever used. The experiment was not successful. The most successful use of Compmail+ was the transfer of files for *CAST Communications*. By April 1986, the experiment was concluded.

Phase 2. The CACHE National Electronic Mail Experiment: BITNET

Phase 2 of the CACHE National Electronic Mail Experiment started with the publication of excerpts from the *Science* article, "Computer Networking for Scientists," (Rony, Hale and Wright, 1986b) in the April 1986 issue of *CACHE News* (Rony, 1986). Several steps were recommended for chemical engineering departments, including the submission of faculty userids to the CACHE Electronic Mail Task Force. For historical purposes, it is appropriate to note that the initial list of BITNET userids (Rony, 1986) included eighteen names at fifteen universities and one corporation in four countries. Technion - Israel Institute of Technology - was the leader, with three faculty represented.

Late in 1984, when the CACHE National Electronic Mail Experiment was first proposed, John Seinfeld (Cal. Tech.) suggested that the Electronic Mail Task Force find ways to involve students. The result was the CACHE Computer Networking Student Contest, which was first announced in the September 1986 issue of *CACHE News* (CACHE, 1986). Unfortunately, little interest was generated, and the minimum requirement of at least 4 student-chapter entries was not met.

Phase 3. The CACHE National Electronic Mail Experiment: IBM Grand at Louisiana State

A serious attempt was made by IBM and Louisiana State University to promote wide-area networking within chemical engineering through the testing of experimental IBM network server software called GRAND. Status reports, announcements, a user's manual, and userid lists were made available by CACHE through CACHE News, the CACHE reception at Annual AIChE meetings, and by direct mail to the CACHE representatives in chemical engineering departments (Rony, 1987; Rony, Wright and Rawson, 1987a, 1987b; Reible, 1987, 1988; Cutlip, 1987; Reible and Rony, 1988). Unfortunately, the IBM/LSU GRAND experiment was not successful; GRAND was infrequently used by chemical engineers and was eventually abandoned.

Phase 4. Creation and Maintenance of BITNET/Internet Userid Lists

Though not successful as a wide-area network file server for the chemical engineering community, the IBM/LSU GRAND experiment did succeed in promoting substantial interest in BITNET among chemical engineering faculty and departments. Perhaps the most significant contribution to *CACHE News* by the task force was the two-part "A User's Guide to Electronic Mail" (AAS, 1989, 1990), which was published through the permission and kindness of the American Astronomical Society (AAS). Concerning electronic communications, it is worth noting how forward looking the AAS was with its 1989 *Electronic Mail Directory*, a distinct contrast to the extreme conservatism (which continued through 1993) of the AIChE.

Starting in fall 1989, the focus of the CACHE Electronic Mail Task force shifted to the creation and maintenance of BITNET/Internet userid lists for both chemical engineering faculty and chemical engineering departments. The Fall 1989 issue of *CACHE News* reported on the validation of BITNET userids, including the significant contribution of Robert Brodkey of Ohio State University to this effort (Rony, 1989b).

During Phase 4, the only constant about the task force identity was change: the task force name changed from the "Electronic Mail Task Force" at its creation in 1985 to the "Electronic Communications Task Force" in 1989 and finally to the "Electronic Networking Task Force" in 1990. The use of the term *networking* marked a degree of maturity in the task force's perception of its technological niche. The task force chairman (Rony) would like to acknowledge his task force colleagues during the first four phases of task-force activities: Joe Wright, John Hale, Norman Rawson, Robert Brodkey, John Hassler, and Wayne Crowl. All were influential in identifying the task force mission statement, vision, and key results.

John Hassler, in the fall 1989 issue of *CACHE News*, contributed comments on UUNET (Unix Users Net), sometimes known as USENET, which was probably the second nationwide network after ARPANET. Having done so, he inquired, "Don't ChE types have anything to discuss?" (Rony, 1989b) Richard Zollars established a discussion group, IFPHEN-L, for those interested in interfacial phenomena. At the invitation of the task force, he published an article entitled "E-mail Discussion Groups: A List Owner's Perspective." Zollars (1990) wrote: "At the moment, I receive approximately one to two requests per week for subscriptions and only about one message per month. Thus, the time that I spend administering the mail list is less than an hour every month. I would prefer more business ..."

Phase 5 (see below) was previewed by a March 6, 1989 report to the CACHE Executive Committee by the Electronic Communications Task Force. From an article by Susan Winitsh, the following were noted (Rony, 1989a; Wintsch, 1989):

> "A new paradigm is emerging: the *national collaboratory*, a framework in which scholars across the nation interact as if they were across the hall from one another, and alternatively called 'geographically-distributed problem solving' by Peter Denning."
>
> "Another scenario: An accelerated version of the age-old scholarly process of acquiring and disseminating knowledge and information."

Additional observations and conclusions were (Rony, 1989a):

> "Most (if not all) recent chemical engineering efforts to develop ChE-specific bulletin boards and wide-area network file servers - to state the matter bluntly - have not been successful, nor is there any reason to believe that they will become successful during the next 12 months."
>
> "Most chemical engineers, so far, have not adapted to the emerging style of rapid, interactive, electronic communications and collaboration. We probably are not the only academic discipline in which such a situation exists."

Phase 5. Chemical Engineering Educators on Internet

Phase 5 will succeed for CACHE when the organization, or its members, actively use the broader electronic capabilities of the Internet, not just electronic mail. Such capabilities are discussed in "Current Practice" later in this article. Several contributions to *CACHE News* (Rony, 1991; Kim, 1992a, 1992b), and a proposal by a CACHE trustee (Kantor, 1993), predicted a significant shift in task force interests from those in Phase 4.

Anonymous FTP servers were the subject of Sangtae Kim's article in the Fall 1992 issue of *CACHE News* (Kim, 1992a).

> "There are now literally thousands of anonymous FTP servers in the USA and beyond, containing a veritable smorgasbord of software packages: plotting routines, 3-D graphics and visualization packages, spreadsheets, symbolic algebra programs, compilers, desktop publishing software, and yes, even some nifty games. The only way to keep track of them is with the help of (what else?) anonymous FTP servers!"

At the summer 1993 Mt. Crested Butte CACHE trustees meeting, Jeffrey Kantor proposed that CACHE support the electronic distribution of its software and documentation via the Internet (Kantor, 1993).

> "The types of information amenable to electronic distribution include: (a) case studies, (b) educational software, (c) phone book information on CACHE contributors, (d) articles from past issues of *CACHE News*, (e) meeting information and announcements, and (f) links to other sources of information."

> "The initial task of establishing a Gopher and FTP site is straightforward and would be easily accomplished. The more difficult issues deal with the copyright and ownership of freely-accessible information ..."

Sangtae and Jeffrey were on target. FTP, Gopher, name servers, and other Internet software and protocols - especially the World Wide Web (WWW) - represent the expanded world of electronic networking for chemical engineering educators.

Current Practice: The Wide, Wide World of Internet

What is Internet?

According to the AAS, in *A User's Guide to Electronic Mail* (Part 1) (AAS, 1989):

> "The Defense Advanced Research Project Agency (D)ARPA network was initially set up by the U.S. Department of Defense in 1969. It is now a part of the ARPA Internet, which uses *TCP/IP* (Transmission Control Protocol/Internet Protocol) communications and includes over 30,000 hosts (1987) and more than 570 networks in several domains: **COM** (commercial organizations), **EDU** (educational/research organizations), **GOV** (civilian government organizations), **MIL** (Department of Defense), **ORG** (other organizations), **NET** (network resources)."

> "Most Internet network sites that astronomers communicate with will be in the **EDU** domain (universities, national observatories). There are additional do-

mains for countries outside the USA, e.g., **UK** (United Kingdom) and **AU** (Australia). Internet includes some transcontinental and transatlantic satellite links (SATNET). Typical delivery time on Internet is a few minutes."

"In Internet, individual computers are assigned numerical addresses within a hierarchical system, with the first number in the address being the number of the individual network on Internet. For example, 4.0.0.0 is SATNET, 10.0.0.0 is the ARPA network, 128.112.0.0 is the Princeton network, and 128.112.24.2 is an individual machine at Princeton. These addresses are mapped against alphanumeric addresses via host tables. Thus, the machine 128.112.24.2 corresponds to pupgg.princeton.edu."

"Internet is the fastest growing of the United States networks and presently is supported by DARPA, the National Science Foundation, NASA, the Department of Energy, and the United States Geological Service. NSF has the mandate to support national networking for the scientific research community ..."

As additional examples, the author's Internet nodes are the IBM mainframe at Virginia Tech., VTVM1.CC.VT.EDU (128.173.4.1) and a Hewlett-Packard server, VT.EDU. In February 1994, he installed an IBM ValuePoint 486DX, which functions as an Internet node named RONY.CHE.VT.EDU (128.173.164.0), in an undergraduate controls laboratory.

What are the Basic Applications of Internet?

A superb, early text on the subject of Internet on Unix platforms is *The Whole Internet* by Ed Krol (Krol, 1992). In subdividing the world of Internet, important categories for the individual user include (a) computer platform, (b) communications hardware, and (c) Internet software. For category (c), Krol identified the following Internet applications and created the following individual chapters for each: Remote Login, Moving Files, Electronic Mail, Network News, Finding Software, Finding Someone, Tunneling Through the Internet, Searching Indexed Databases, Hypertext Spanning the Internet, and Other Applications.

For readers who either have Unix workstations or plan to use Unix for Internet communications and desire a comprehensive perspective on most key aspects of Internet, Krol's book is excellent (Krol, 1992). Rather than re-invent the wheel in this page-limited chapter, the author has decided to ignore Unix-workstation-based Internet and instead to focus on an alternative, more popular platform - Windows - for Internet applications. Current Internet software development for Windows is in a rapid state of flux; it is an exciting time for chemical engineering colleagues who use IBM PC machines or clones.

Our discussion of Internet for Windows starts with Winsock, which is the key to all other Internet applications software. The focus then shifts to three interesting Internet software packages for Windows, namely, WS_FTP, HGopher, and Mosaic. Space is not available to discuss the remaining Windows Internet applications software: Trumpet, QWS3270 Telnet, Eudora, WFTPD, Ping, Finger, and Archie. Krol's outstanding book is extensively excerpted in this chapter (Krol, 1992).

The Standard TCP/IP Windows Interface (TRUMPET WINSOCK Software)

According to Harry Kriz in his useful electronic document, "Windows and TCP/IP for Internet Access," (Kriz, 1994):

> "'Winsock' is the buzzword that dominates discussion about TCP/IP and Windows ... applications ... In order to get these applications working, there are only a few things that an end-user needs to know about Winsock and how it supports Windows applications.
>
> "In layman's terminology, the term 'Winsock' refers to a technical specification that describes a standard interface between a Windows TCP/IP application (such as a Gopher client) and the underlying TCP/IP protocol stack that takes care of transporting data on a TCP/IP network such as the Internet. When I invoke a program such as HGopher, it calls procedures from the WINSOCK.DLL dynamic link library. These procedures in turn invoke procedures in the drivers supplied with the TCP/IP protocol stack. The TCP/IP drivers communicate with the computer's Ethernet card through the packet driver. For serial line communications, the TCP/IP drivers communicate with a SLIP driver to enable network communications through the serial port.
>
> "The WINSOCK.DLL file is not a generic file that can be used on any system. Each vendor of a TCP/IP protocol stack supplies a proprietary WINSOCK.DLL that works only with that vendor's TCP/IP stack.
>
> "The advantage to the developer of a Winsock-compliant application program is that the application will work with any vendor's Winsock implementation ... It is this aspect of the Winsock standard that has resulted in the blossoming of Winsock-compliant shareware applications since the summer of 1993."

The dominant Winsock was Peter Tattam's *Trumpet Winsock*, which included a TCP/IP protocol stack and basic clients such as Telnet, FTP, Ping, and Archie. For a shareware fee of $20, version 1.0A was available by anonymous FTP as files twsk10a.zip (February 3, 1994) and winapps.zip (November 30, 1993) by anonymous FTP from FTP.UTAS.EDU.AU in subdirectory /PC/TRUMPET/WINSOCK or by Gopher from INFO.UTAS.EDU.AU under menu item Utas FTP Archive (Kriz, 1994).

What is a Packet Driver?

In order to run Trumpet Winsock, you must either have a serial communications port, which permits use of a serial-link interface protocol (SLIP), or preferably an Ethernet hardware communications card, which requires the use of a packet driver for both the card and for WINPKT.COM, a virtual packet driver interface for Windows 3.1.

A *packet driver* is "a small piece of software that sits between your network card (e.g., Etherlink III) and your TCP program. This driver provides a standard interface that many programs can use in a manner analogous to BIOS calls using software interrupts." (Tattam, 1994). For the 3Comm Etherlink III interface card, two lines in the DOS file, AUTOEXEC.BAT, do the trick:

```
3C509 0x60
WINPKT 0x60
```

0x60 is a software interrupt vector with a hexadecimal code of 60H. Other types of packet drivers can be obtained from the Crynwr Packet Driver Collection available over Internet.

Moving Files: Anonymous FTP (WS_FTP Software)

One of the jewels in the Internet application suite is a program that allows you to use anonymous FTP to transfer files from another Internet node to your node.

> "The essential idea of anonymous FTP is that user accounts and passwords are not required. The user's account is replaced by the generic name account, *anonymous*." (Kim, 1992a).

> "The File Transfer Protocol was originally a Unix utility used for interactively transferring files. High-quality, public-domain clients and servers are available for most computing platforms. The advantages of FTP are its interactivity and the ability to transfer binary files. The main disadvantage is that the required interactive access to the Internet is universal. FTP is probably the most widely used mechanism for software distribution on the Internet" (Kantor, 1993).

With DOS-based anonymous FTP, a knowledge of FTP commands - e.g., **ftp, get, put, quit, bye, cd, binary, ascii, dir, ls, lcd, mput, mget, close** - is required. With Windows-based FTP, and specifically with John Junod's **WS_FTP** software (Junod, 1994), intuitive mouse point-and-click operations, menus, and other graphical-user-interface (GUI) features substantially simplify the task of transferring files over the Internet. The detailed treatment of Unix instructions provided in Chapter 6 of reference (Krol, 1992) is no longer required for the Windows platform. FTP client **WS_FTP** is available as file ws_ftp.zip (February 9, 1994) by anonymous FTP from FTP.USMA.EDU in subdirectory /PUB/MSDOS/WINSOCK.FILES (Kriz, 1994).

Tunneling Through the Internet: Gopher (HGopher Software)

> "Gopher, or more accurately, 'the Internet Gopher,' allows you to browse for resources using menus. When you find something you like, you can read or access it through the Gopher without having to worry about domain names, IP addresses, changing programs, etc. For example, if you want to access the on-line library catalog at the University of California, rather than looking up the address and telnetting to it, you find an entry in a menu and select it. The Gopher then 'goes fer' it.

> "The big advantage of Gopher isn't so much that you don't have to look up the address or name of resources, or that you don't have to use several commands to get what you want. The real cleverness is that it lets you browse through the Internet's resources, regardless of their type, like you might browse through your local library with books, filmstrips, and phonograph records on the same subject grouped together ...

> "... Think of the pre-Gopher Internet as a set of public libraries without card catalogs and libraries. To find something, you have to wander aimlessly until you stumble on something interesting. This kind of library isn't very useful, unless you already know in great detail what you want to find, and where you're likely to find it. A Gopher server is like hiring a librarian, who creates a card catalog

> subject index ... Unfortunately, Gopher services did not hire highly trained librarians. There's no standard subject list, like the Library of Congress Subject Headings, used on Gophers to organize things ... Gopher does not allow you to access anything that couldn't be made available by other means. There are no specially formatted 'Gopher resources' out there for you to access, in the sense that there are FTP archives or white pages directories ... Gopher knows which application (telnet, FTP, white pages, etc.) to use to get a particular thing you are interested in and does it for you. Each type of resource is handled a bit differently. However, they are all handled in an intuitive manner consistent with the feel of the Gopher client you are using" (Krol, 1992).

Martyn Hampson's *HGopher*, Version 2.4, is currently the best Gopher+ client for Windows 3.1 and Winsock (Hampson, 1994). It is available by anonymous FTP as file hgopher2.4.zip from BOOMBOX.MICRO.UMN.EDU in subdirectory /PUB/GOPHER/WINDOWS. Mr. Hampson suggests that you donate $10.00 to your favorite charity if you like *HGopher* (Kriz, 1994). [*Editor's note:* The appearance of "search engines" such as *Yahoo* and AltaVista during 1995 and 1996 has made Gopher largely obsolete.]

Hypertext Spanning the Internet: World-Wide Web (Mosaic Software)

The *World-Wide Web (WWW)* is based upon hypertext, a software technology first made popular on the Macintosh platform.

> "Hypertext is a method of presenting information where selected words in the text can be 'expanded' at any time to provide other information about the word. that is, these words are links to other documents, which may be text, files, pictures, anything" (Krol, 1992).

Krol has done an effective job of distinguishing between WWW and Gopher; rather than rephrase his comparison, we quote it directly (Krol, 1992).

> "What is WWW about? It's an attempt to organize all the information on the Internet, plus whatever local information you want, as a set of hypertext documents. You traverse the network by moving from one document to another via 'links.' ... Your home page is the hypertext document you see when you first enter the Web.
>
> "... While there are a lot of similarities, the Web and Gopher differ in several ways. First, the Web is based on hypertext documents, and is structured by links between pages of hypertext. There are no rules about which documents can point where - a link can point to anything that the creator finds interesting ... The Gopher just isn't as flexible. Its presentation is based on individual resources and servers. When you're looking at an FTP resource, this may not make much of a difference; in either case, you'll see a list of files. But the Gopher doesn't know anything about what's inside of files; it doesn't have a concept of a 'link' between something interesting on one server, and something related somewhere else.
>
> "Second, the Web does a much better job of providing a uniform interface to different kinds of services. Providing a uniform interface is also one of the Gopher's goals; but the hypertext model allows the Web to go much further. What does this mean in practice? For one thing, there are really only two Web com-

> mands: follow a link ... and perform a search ... No matter what kind of resource you're using, these two commands are all you need. With Gopher, the interface tends to change according to the resource you're using."

The markup language used by the World-Wide Web is called *HTML*, or *HyperText Markup Language*. *A Beginner's Guide to HTML* describes the minimal HTML document, including titles, headings, and paragraphs; linking to other documents, including the uniform resource locator (URL) and anchors to specific sections; additional markup tags for lists, preformatted text, extended quotes, character formatting, and special characters; in-line images; external images; troubleshooting; and ends with a comprehensive example (NCSA, 1993).

NCSA *Mosaic* for Windows is available by anonymous FTP server as file mos_20a1.zip from NCSA's FTP server, FTP.NCSA.UIUC.EDU in subdirectory PC/MOSAIC (Wilson and Mittelhauser, 1994). Reference (NCSA, 1993), on HTML, is available from PUBS@NCSA.UIUC.EDU. [*Editor's note:* A commercial browser, *Netscape*, available for virtually all hardware platforms, replaced Mosaic as the dominant Web browser during 1995 and 1996. A new Microsoft browser, *Explorer*, targeted for use with Windows on PC compatibles, was introduced in 1996, and has achieved significant success in the marketplace.]

Other Sources for Winsock and Internet Information

It is appropriate to conclude this section of the chapter with a final mention of Harry Kriz's, *Windows and TCP/IP for Internet Access* (Kriz, 1994). Harry (1) identified key Windows/Winsock software and described his experience with each briefly, (2) identified for each software package the author, version, license arrangement, file name/date, and availability at a specific anonymous FTP or Gopher site, and (3) provided readers with suggestions for other sources of Winsock information. For additional information, contact Harry M. Kriz, University Libraries, Virginia Polytechnic Institute & State University, Blacksburg, VA 24061-0434. Email: hmkriz @ vt.edu. In 1996, Harry received about 20,000 WWW "hits" per month.

Internet reference books were a growth industry in 1994. Consider *The Windows Internet Tour Guide* by Michael Fraase (Fraase, 1994); *Using the Internet* by Tolhurst, Pike, Blanton, and Harris (Tolhurst et. al., 1994); *Navigating the Internet* by Mark Gibbs and Richard Smith (Gibbs and Smith, 1993); and *The Internet Directory* by Eric Braun (Braun, 1994).

The Future: Internet and Chemical Engineering Education

How will Internet and networking affect chemical engineering education? The author suspects that Internet has not had the impact within both chemical engineering education and research that it has already had in other fields – e.g., computer science, physics, astronomy, electrical engineering – that are more attuned to the computer networking revolution. Our national society, AIChE, has been slow to participate in Internet. Where on the information superhighway have been our profession's leaders? Perhaps riding bicycles on side streets.

A revolution in the convenience of Internet software is now occurring. Command-line oriented DOS and Unix Internet clients are rapidly giving way to Unix, Macintosh, and Windows graphical user interface (GUI) clients. Internet users on the Windows platform now have the convenience of software such as Trumpet Winsock, Trumpet, Mosaic, HGopher, WS_FTP,

WS_PING, WFTPD, Eudora, Netscape, and others. Internet usage has become fun, as opposed to being tedious under DOS and under Unix before X-Window. Most of the new Internet Windows/Winsock software is either freeware, shareware, or requires a nominal license fee of under $50.

Internet users no longer need to participate in the Unix culture in order to enjoy the benefits of Internet; Unix workstations are no longer required, nor is there a need to remember numerous Unix commands. Internet software on both the Macintosh and Windows platforms is already excellent and will soon be better. With the expected merging of platforms, e.g., as promoted by the PowerPC chip vendors, many more users will be able to select the platform "identity" of their PC without purchasing entirely different hardware. Internet applications software will be outstanding on any platform that a user selects.

When we consider the potential impact of Internet on chemical engineering education, it is appropriate to consider this issue from the point of view of our customers, students. There are far more questions than answers. Some of the questions are provocative.

Will student access to the Internet be direct or indirect? In other words, will students (a) own their personal computers and use them in their rooms to access the Internet anytime during the day? (b) use personal computers networked to the Internet only in university computer laboratories? (c) depend upon intermediaries - e.g., chemical engineering faculty - to utilize Internet on their behalf?

Who will pay the costs of access to Internet? Will students (a) pay a monthly charge for high-speed modem or local-area network access to the Internet directly from their own rooms? (b) pay a semester laboratory charge for access to the Internet through university laboratories? (c) depend upon the chemical engineering department to pay for access to the Internet?

What will be the trade-off between computer storage (memory) costs for digital information and the convenience of Internet access for such information? Will the convenience and low cost of write-once or read/write optical memory be such that downloaded files will be stored optically rather than on more expensive magnetic disk? Will the convenience, speed, accessibility, and low cost of Internet be such that neither optical nor magnetic local storage are required at all? In other words, have the CD-R and CD-ROM already been supplanted by the Internet? Is mass local storage at an individual personal computer required anymore?

What types of information will be available to students and faculty through the Internet? Will changes in educational paradigms occur as a direct consequence of the Internet? How will the Internet affect the creation, testing, marketing, distribution, sale, and use of textbooks? How is the Internet already affecting the creation, testing, marketing, distribution, sale, and use of computer data (programs, images, video, audio, text, presentations, executables, and so forth)? How will the Internet affect the need for a student to physically attend lectures on a university campus?

Recall the article by Susan Winitsh (Wintsch, 1989; Rony, 1989b) in which it was noted that a new paradigm is emerging:

> "the national collaboratory, a framework in which scholars across the nation interact as if they were across the hall from one another."

Will the Internet allow several chemical engineering departments, including both faculty and students, to collaborate during a semester in a semester course that is jointly offered by all of the departments? Will we have an "educational collaboratory?" How will the Internet affect the need for a chemical engineering department to maintain departmental faculty competence to teach all required courses in the curriculum? Once the Internet becomes widely available to students with fast service, will a chemical engineering department need as many teaching faculty as it has in the past?

In chemical engineering education, what are the educational values of a printed textbook, of an electronic textbook the pages of which can be printed on demand, of personal contact with a faculty member, of impersonal contact with a faculty member in a lecture with large class size, of personal contact with graduate teaching assistants, of chemical engineering applications software, of homework assignment evaluation, of taking handwritten notes in a lecture, and so forth?

Greater attention should be paid to electronic information as a valuable resource. Chemical engineering authors should be careful to archive files of their manuscripts no matter what application software generated them. Handwritten answer books for chemical engineering textbooks should be scanned, compressed, archived, and made available to faculty in electronic form. Entire textbooks should be made available as electronic documents that can be communicated over the Internet under controlled conditions.

Page limitations prohibit speculation on most of the above questions. However, one speculation is offered.

Speculation: The Marketing of Chemical Engineering Textbooks

Assume that you desire to create a new chemical engineering textbook. Your primary obstacle will not be the creation of the textbook (it probably is already 50% finished based upon your lecture notes over the past several years) nor the creation of camera-ready copy (you are already proficient at desktop publishing). The primary obstacle will be the *marketing* of the textbook. You need to convince a major technical publishing company - e.g., McGraw-Hill, Prentice-Hall, Wiley, or several others in chemical engineering - that it should publish your new textbook. The likelihood is that none of them will do so. Why not? They already each have their in-house example of a textbook in your field, and do not need to add a second (or third) title.

The major publisher may have your textbook proposal reviewed. All such reviews will be subjected to the prevailing dogma of current experts in the field. Proposals that either deviate significantly, or do not present any change from current dogma and paradigms, will likely be rejected. You may feel completely dependent today on the marketing strengths of the major publishers of chemical engineering textbooks. If no publisher is interested in your proposal, is your wonderful work dead in the water?

Perhaps not. Enter the Internet. Make the assumption that Internet nodes are widely available at all major chemical engineering departments and that, at long last, most ChE faculty are both computer-literate and Internet-literate. Does such a situation permit a new and different marketing paradigm for chemical engineering textbook authors? Certainly. It is a new marketing paradigm that does not depend upon any of the major publishers, as we currently know them.

Consider software shareware and freeware. In each case, copyright control does exist and is held by one or a few individuals (or, on occasion, companies). Yet the marketing (distribution) mechanism is by the Internet or by CD-ROM. Potential users have the opportunity to test the software before deciding to use it. If it is freeware, no royalty payment is required. If it is shareware, a specified royalty payment is requested. An example is Peter Tattam's Trumpet Winsock, which requires a $20 license fee after 30 days of use.

The identification and announcement of shareware licensing fees already means that the Internet is widely used for the marketing of software. This precedent being established, it should be possible to use the Internet for a productive educational and scholarly purpose, namely, the marketing of "textbook shareware" to universities. The broader ethics and value of marketing textbook shareware via the Internet is identical, in the author's opinion, to the ethics and value of marketing computer shareware. Universities can easily police the unauthorized use of textbook shareware in a course.

Consider a textbook that is normally marketed by a major publisher in printed form for a list price of $60. Add state tax to the $60. Assume a royalty payment to the author of 10% of the wholesale price of $45, or $4.50 royalty per book.

Now consider the same textbook marketed and distributed in electronic form - for example, Adobe Acrobat Portable Document Format (PDF) - over the Internet for a single-user license (royalty) fee of $4.50. The recipient of the electronic textbook has the option of printing it or retaining most pages in electronic form. The cost of such printing could be $5 to $20. The result of this alternative paradigm is that the textbook author receives an identical royalty, but students pay less than half the cost for a textbook that is not quite as convenient to use and certainly not as permanent.

The trade-off? Textbook cost to student versus convenience, form, and permanence. Will this paradigm occur? Certainly. Will it eventually replace the printed textbook publisher? Who knows?

For the textbook author there are several advantages associated with the new paradigm. No longer will the major textbook publishers have veto power over the publication of a new textbook. License (royalty) fees per electronic textbook could well be higher than for a textbook printed by traditional publisher. The textbook could be revised annually. Copyright control of the textbook could remain in the hands of the author(s). The textbook could contain substantial color, audio, and video by virtue of its electronic form. A new marketing channel, the Internet, makes this all possible.

When you think of the Internet, think of *marketing*. Whether or not the rules of the Internet permit such marketing today, it is clear that the needs of the information superhighway will change such that marketing will become a (perhaps *the*) major component in the future. And when you think of marketing, think of *entrepreneurship*. Internet potentially is liberating to textbook authors, just as it has been for computer shareware authors.

Jeffrey Kantor is certainly correct in his assessment of the Internet situation (Kantor, 1993). The time has come for CACHE to support the electronic distribution of documentation, and perhaps selected software, via the Internet. FTP, Gopher, WWW, Netscape, WAIS and other services can all be seriously considered today. Anticipated problems with copyright issues or the possible loss of revenue for software are non-issues, in the author's opinion. There is sufficient electronic information that can be freely disseminated to anyone, not just to chemical engineers, to create a viable CACHE.ORG server. Software that has commercial value or copyright sensitivity can be distributed by more traditional, controllable channels (e.g., diskettes, CD-ROMs). The emergence of CD-Rs and CD-ROMs and low-cost, high-capacity removable magnetic diskettes (e.g., Iomega's Zip "floppies") eliminate the 1.44-MB barrier of high-density diskettes for program storage.

Addendum

This manuscript was originally prepared in 1994. As of late 1996, substantial changes have occurred in the computer and networking landscape. The author would like to connect some points made in the paper to current trends in the computing environment:

1. Within chemical engineering education, the use of the Internet - WWW, email, and FTP - has become both obvious and pervasive. The change in computing perspective, away from laboratory-based PCs, may well become a "paradigm shift" in higher education by the end of the century.

2. Netscape and Microsoft Explorer have replaced Mosaic as the dominant WWW browsers.

3. New WWW data types – real audio, VRML (Virtual Reality Markup Language), and perhaps others – have appeared.

4. The programming language, *Java*, introduced in early 1996 by Sun Microsystems, now represents a potential threat to Microsoft's dominance. The author hopes that Netscape Corporation and Sun MicroSystems succeed in their revolutionary transformation of Internet usage.

5. The author's "Speculation: The Marketing of Chemical Engineering Textbooks" has led directly to an entrepreneurial activity to promote such a possibility. The author has applied for a Federal trademark at the USPTO for the service mark, Sharebook (TM), and also has acquired the domain name, sharebook.com. A sharebook site at URL http://www.sharebook.com is currently under construction in late1996.

6. A new class of computer hardware, the "network computer" or "NC", is about to emerge on the computing scene. It represents one answer to a question posed in the original manuscript: Is local optical or magnetic storage required on every computer?

7. The dominant software applications in 1996 are word processing, email, the World Wide Web, the spreadsheet, and presentation software.

8. The 1980s ethical concerns about "marketing" on the Internet have disappeared by 1996. The dominant domain is already *.COM, which has supplanted the *.EDU and *.ORG domains in number of servers by perhaps a factor of ten to one hundred.

9. Secure, perhaps encrypted, electronic "digital cash" should arrive on the network

computing scene sometime during 1997. The potential for entrepreneurial activity on the Internet will become compelling.

10. The original manuscript statement, "When you think of the Internet, think of marketing," remains accurate as of late 1996.
11. The late 1990s is an exciting time for young, energetic, computer-savvy, and market-savvy chemical engineering professionals.
12. Because of the impact of the Internet, the future of the traditional academic department in higher education as we knew it during the 1980s is questionable during the early decades of the new millennium.
13. Because of its exponentially increasing popularity, the Internet within the United States may develop frequent instances of "network gridlock" by 1997. Already, local Internet gridlock instances are being discussed using weather as a metaphor. Internet "caches" will become commonplace by the end of the millennium.
14. The historical examples of the early failures of CACHE Corporation projects – which were designed to stimulate the use of email – provide a useful lesson for the future. As Santayana stated, "Those who cannot remember the past are condemned to repeat it."
15. The questions that were asked in 1994 concerning the potential impact of the Internet on chemical engineering education remain valid.
16. With improving price/performance ratios for both hardware and software, it is much more likely during the late 1990s that a chemical engineering student will have, during the undergraduate years, both (a) his/her own multimedia computer and (b) either a modem or Ethernet link to an Internet server. In other words, student access to the Internet will be direct during the late 1990s.
17. We started with the desktop PC during the late 1970s and early 1980s, and ended the millennium with the desktop PC, the laptop PC, and the NC. The author suggests that, as the technology wheel turns, the next innovation will become the KISSPC, namely, the keep-it-simple-stupid-PC. Users may ultimately revolt against the unstated expectation that a computer user must become a one-person computer center. Software has become too complex, bloated, and buggy.

References

American Astronomical Society (1989). A User's Guide to Electronic Mail (Part 1), *CACHE News*, **29**, 22-26.

American Astronomical Society (1990). A User's Guide to Electronic Mail (Part 2), *CACHE News*, **30**, 9-17.

Braun, Eric. *The Internet Directory*, Fawcett Columbine, NY, 1994.

CACHE Electronic Mail Task Force (1986). CACHE Computer Networking Student Contest, *CACHE News*, **23**, 3-5.

Cutlip, Michael D. GRAND Software: Visit to City University of New York Computer Center, *CACHE News*, **25**, 35 (Fall 1987).

Fraase, Michael. *The Windows Internet Tour Guide: Cruising the Internet the Easy Way*, Ventana Press, Chapel Hill, NC, 1994.

Gibbs, Mark, and Richard Smith, *Navigating the Internet*, SAMS Publishing, Carmel, IN, 1993.

Hampson, Martyn. *A Gopher Client for Windows 3.1, Version 2.4*, 1994.

Junod, John. *Windows Sockets FTP Client Application WS_FTP, Version 94.02.08*, February 8, 1994.

Kantor, Jeffrey C. *CACHE on the Internet*, Proposal to CACHE trustees, July 22, 1993.

Kim, Sangtae (1992a). Anonymous FTP Servers, *CACHE News*, **35**, 10-11.

Kim, Sangtae (1992b). Name Servers: or Electronic Mail Made Easy, *CACHE News*, **35**, 12.

Kriz, Harry M. Windows and TCP/IP for Internet Access, Release 3, February 9, 1994; electronic document available from hmkriz@vt.edu.

Krol, Ed. *The Whole Internet: User's Guide & Catalog*, O' Reilly & Associates, Inc. (103 Morris Street, Suite A, Sebastopol CA 95472), 1992.

NCSA, *A Beginner's Guide to HTML*, 1993.

Reible, D. D. (1987). CACHE Bulletin Board Service, *CACHE News*, **25**, 30-31. Note: This article was mistakenly attributed to Peter R. Rony.

Reible, Danny D (1988). LSU/CACHE Bulletin Board: General Information and User's Manual, *CACHE News*, **26**, 24-26.

Reible, Danny, and Peter Rony (1988). Status of GRAND Wide-Area Network File Server at Louisiana State University, *CACHE News*, **26**, 16-17.

Rony, Peter R (1985). The CACHE National Electronic Mail Experiment, CACHE Communications Task Force Proposal, April 9.

Rony, Peter R (1986). The CACHE National Electronic Mail Experiment: Part 2. Bitnet, Arpanet, and NSFNet, *CACHE News*, **22**, 10-12.

Rony, Peter R (1987a). Progress Toward a Wide-Area Network for Chemical Engineers, *CACHE News*, **25**, 36-38.

Rony, Peter R (1988). Standard List of Bitnet Userid Nicknames, *CACHE News*, **26**, 17-18.

Rony, Peter R (1989a). *Electronic Communications Task Force*, Memorandum to CACHE Executive Committee, March 6.

Rony, Peter R (1989b). Electronic Mail Task Force, *CACHE News*, **29**, 17-18.

Rony, Peter (1991). NSF Electronic Proposal Submission Project: A Report, *CACHE News*, **32**, 5-6.

Rony, Peter R., J.C. Hale, and J.D. Wright (1986a). The CACHE National Electronic Mail Experiment: Part 1. Compmail, *CACHE News*, **22**, 9-10.

Rony, Peter R., J.C. Hale, and J.D. Wright (1986b). Computer Networking for Scientists. *CACHE News*, **22**; comments on article from *Science* (February 26).

Rony, Peter, J. Wright, and N. Rawson (1987a). *Bitnet Wide-Area Network File Server for Chemical Engineering*, Memorandum sent to chemical engineering departments, June 20.

Rony, Peter, J. Wright, and N. Rawson (1987b). Bitnet User Identification Numbers, *CACHE News*, **25**, 28-29.

Tattam, Peter R (1994). *Trumpet Winsock, Version 1.0*, Trumpet Software International.

Tolhurst, William A., Mary Ann Pike, Keith A. Blanton, and John R. Harris (1994). *Using the Internet, Special Edition*, QUE Corporation, Indianapolis, IN.

Wintsch, Susan (1989). Toward a National Research and Education Network, National Science Foundation MOSAIC, 20 (4), 32-42.

Wilson, Chris, and Jon Mittelhauser (1994). *NCSA Mosaic for Microsoft Windows, Installation and Configuration Guide*.

Zollars, Richard L (1990). E-mail Discussion Groups: A List Owner's Perspective, *CACHE News*, **31**, 36-37.

INTELLIGENT SYSTEMS IN PROCESS OPERATIONS, DESIGN AND SAFETY

Steven R. McVey and James F. Davis
Ohio State University
Columbus, OH 43210

Venkat Venkatasubramanian
Purdue University
West Lafayette, IN 47907

Abstract

Many of the problems in process operations, design and safety are ill-structured and share certain generic characteristics that make the traditional approaches to automation very difficult if not impossible. The domain of intelligent systems offers a set of powerful concepts and techniques that are indispensable in addressing the challenges faced in process engineering. In particular, intelligent systems are poised to play a central role in the emerging paradigm of computer integrated manufacturing (CIM) applied to chemical process industries. Partial CIM implementations are quite prevalent and have resulted in the anticipated benefits of improved efficiency, better product quality, and lower operating costs.

In this paper, we trace the development of intelligent systems through the needs for their presence and their defining characteristics, examine a number of applications in process operations, design and safety, and articulate the value added.

Introduction

Over the recent decade artificial intelligence (AI) has emerged as an important problem-solving paradigm in process systems engineering. Intelligent systems is a branch of artificial intelligence with an interface to some applications area such as process engineering. While it is difficult to define or quantify precisely what intelligence is, whether it is natural or artificial, it is, however, useful to have some working definition of artificial intelligence and intelligent systems. In this sense, AI may be defined as the study of mental faculties through the use of computational models (Charniak and McDermott, 1985). Intelligent systems are defined in a similar vein as systems that exhibit the characteristics we associate with intelligence in human behavior - understanding language, learning, reasoning, solving problems, and so on (Barr and F.A. Feigenbaum, 1989). In recent practice, however, the term "intelligent systems" has come to be used to refer to a wide variety of artificial intelligence-based methodologies that include knowledge-based systems (a.k.a. expert systems), neural networks, genetic algorithms, fuzzy logic, machine learning, qualitative simulation, and others. In this paper, we review the origins, evolution and the applications of intelligent systems in the context of process systems engineering such as in process operations, process and product design and process safety.

The origins of intelligent systems research and development in process systems engineering can be traced to the late 1970's and early 1980's. A landmark is FOCAPD'83 (Conference on Foundations of Computer-Aided Process Design, held in 1983) where the keynote speaker proposed the application of expert systems in process operations. The CACHE/CAST sponsored conference sparked an interest that resulted in many systems-oriented researchers in academia initiating investigations. Early industrial implementations, while holding promise, foundered because of unmet performance expectations, difficult maintenance, high cost of specialized computers and software, lack of appropriate interfaces, and lack of integration. To reflect changes in perspective, the term "expert system" evolved to "knowledge-based system". This terminology attempted to capture a perspective that "expert systems" were not as much "expert" as they were viable and practical approaches that draw upon and encode expert knowledge and experience. The CACHE monograph series *Artificial Intelligence in Process Systems Engineering*, (Stephanopoulos and Davis, 1990-1992) provides details on knowledge-based system development through four monographs covering (1) an overview (Stephanopoulos, 1990), (2) rule-based systems (Davis and Gandikota, 1990), (3) knowledge representations (Ungar and Venkatasubramanian, 1990) and (4) object-oriented programming (Forsythe et al., 1992).

While the early research focus was on expert systems, recent breakthroughs in computer technologies brought neural networks into practical consideration with a resulting research shift toward this approach. While knowledge-based systems took advantage of expert knowledge and experience, neural networks provided the mechanisms for identifying the inherent correlations in process operating data. As was the case with knowledge-based systems, neural networks experienced an overshoot in expectations followed by a reconciliation of advantages and limitations. The combined experiences with these technologies have resulted in a much more rational stance that there is no single, uniform solution to modeling the behaviors of chemical processes. In fact, together these two technologies help harness information from both the process and the expert.

The current view recognizes that the artificial intelligence techniques while useful, must be considered as components of hybrid systems (Antsaklis and Passino, 1993; Rao et al., 1993). AI approaches such as knowledge-based systems, neural networks, genetic algorithms, etc., have to be integrated with more traditional techniques based on mathematical models, statistical tools and so on for a successful approach to computer-aided problem-solving in process systems engineering. The term "intelligent system" attempts to recognize this realization while also noting the inclusion of approaches that enable the integrated system to "automate tasks not covered by traditional numerical algorithms (Stephanopoulos and Han, 1994)."

Process Operations, Design and Safety

Many of the problems in process operations, process and product design and process safety are ill-structured and share certain generic characteristics that make the traditional approaches to automation very difficult if not impossible (Venkatasubramanian, 1994). These may be summarized as the following list of needs that one faces in many practical situations:

- *To reason with incomplete and/or uncertain information about the process or the product(s).* For example, in fault diagnosis, the sensor data may be uncertain due to noise and incomplete as all the relevant information may not be available. In hazard and operability analysis, one trys to identify abnormal situations based on qualitative information. These necessitate the use of qualitative modeling and reasoning.
- *To understand, and hence represent, process behavior at different levels of detail depending on the context.* For example, in process design it is important to be able to reason in a hierarchical manner. The understanding of the cause and effect interactions in a complex process system requires the generation of explanations hierarchically at different levels of detail.
- *To make assumptions about a process when modeling or describing it.* One then has to ensure the validity and consistency of these assumptions under dynamical conditions during the course of the problem-solving process. This also leads to adaptation when the underlying assumptions change. Again, this is a commonly found situation in design and model development.
- *To integrate different tasks and solution approaches.* This requires integrating different problem-solving paradigms, knowledge representation schemes, and search techniques. This is an important requirement for the development of an approach to integrate planning, scheduling, supervisory control, diagnosis and regulatory control.
- *To keep the role of the human in the decision-making process active and engaged.* This is crucial for the success of the system in many applications such as fault diagnosis.

Thus, it can be seen that the automation of process operations, design and safety require approaches that address such needs and demands. These are the challenges that intelligent systems can address. We will elaborate on this by describing five important characteristics of an intelligent system. These by no means represent an exhaustive characterization of system intelligence, but are intended to provide some measures of the extended value brought about by intelligent system approaches.

Qualitative Modeling

Inspired by the successes of CIM in discrete parts manufacturing, attempts have been made to generalize this concept to chemical process industries (Williams, 1989, Venkatasubramanian and Stanley, 1994, Macchietto, 1994). The application of CIM to the chemical process industries introduces challenges that do not exist for discrete parts manufacture (Madono and Umeda, 1994). These challenges result from complex chemical and physical processes and their interactions. Although mathematical models for these processes may exist, they are often inadequate for describing or predicting many phenomena of critical interest in the efficient operation of a plant. For example, complex chemical, kinetic and transport effects are often ill-understood quantitatively. As a result, mathematical models that approximate behaviors may be insufficient for activities such as control, planning, design or diagnosis. Other physical processes such as corrosion, fouling, mechanical wear, equipment failure and fatigue occur by poorly-understood mechanisms and are thus very difficult to model mathematically. Likewise, design selection decisions like sieve tray spacing, or unit configuration do not lend themselves to mathematical models and optimization.

Experienced plant operating personnel or experienced designers, on the other hand, are quite adept at making appropriate decisions regarding complex situations through qualitative considerations. These decisions are based on qualitative models derived from information gained directly from observed behaviors of a process operation (historical) and from experience with similar process components (design/predictive).

Intelligent system approaches provide the capability for encoding various forms of qualitative models, thereby extending modeling capacity beyond mathematical description. This qualitative modeling aspect is a primary distinction between intelligent systems and other, strictly mathematical approaches. The immediate value of an intelligent system (component) is apparent when the underlying model is describing (or predicting) useful plant phenomena and/or assimilating information into active decisions that cannot be modeled effectively using mathematical representations. The breadth of modeling is brought out by recent work on modeling languages such as DESIGN-KIT (Stephanopoulos et al., 1987), MODEL.LA (Stephanopoulos, 1990), and Functional Representation (Chandrasekaran, 1994; Goodaker and Davis, 1994).

Diagnosis is certainly an activity that lends itself to qualitative models since failure situations are very difficult to model mathematically. As examples, CATCRACKER (Ramesh et al., 1992) and CATDEX (Venkatasubramanian, 1988) are knowledge-based systems for diagnosing problems in fluid catalytic cracking units. The complexity of multiphase chemical reactions, fluid mechanics, thermal effects, and mass transport makes it nearly impossible to develop mathematical models of failure conditions that are of sufficient numerical resolution. However, qualitative models in the form of cause-and-effect relationships and that use qualitative interpretations of the process data are quite adequate in isolating failure conditions.

Similarly, pattern recognition neural networks make it possible to formulate black box models that map between process data and diagnostic conditions based on past occurrences (e.g., Leonard and Kramer, 1993; Davis and Wang, 1993; Vaidhyanathan and Venkatasubramanian, 1992). If there is sufficient operating data available, these models can be quite adequate for diagnosis when mathematical models are not. In process design, knowledge-based system approaches are very effective for modeling selection decisions, designing alternatives when constraints fail and managing the use of mathematical simulation. Qualitative models, based on experience, provide near optimized selections and effectively manage the complexity of the design problem.

Process hazards review and safety analysis, such as the HAZOP analysis, is another area where industrial practitioners have long used qualitative modeling and reasoning extensively to identify abnormal process situations. Recent work on this problem has demonstrated that the intelligent systems approach can be used successfully to automate HAZOP analysis of large-scale process P&IDs with significant gains in efficiency, accuracy and documentation (Venkatasubramanian and Vaidhyanathan, 1994). The success stems from the judicious use of qualitative modeling and object-oriented knowledge representation techniques.

The value of any intelligent system component first lies in its underlying model. It must be a model of some aspect of the process or decision making that cannot be modeled adequately using mathematics. Ill-conceived intelligent systems attempt to use qualitative models when mathematical models exist and are adequate or when there is not a good knowledge source to build an adequate model.

Cognition

An advantage of intelligent systems is that the techniques can be used not only to model process behaviors qualitatively but they can also simultaneously be used to model decision-making processes. Reasoning processes can take the form of simple mappings such as those found in neural networks or table look-up approaches or more complex processes that make use of specific knowledge organizations and inference strategies such as hierarchical classification, fault tree diagnosis, constraint-based reasoning, model-based reasoning, case-based reasoning, and so on.

Regardless of the approach, there is an added level of value associated with cognition. Cognition is the ability of an intelligent system to provide output that is at a symbol level, directly understandable to the user without further deliberation or interpretation. Cognition then is a measure of a system's ability to both explain its reasoning and communicate with the user in an understandable form.

Cognition can be broken down into two types, both of which may be present in an intelligent system. A cognitive system is one that: (i) provides conclusions in an interpreted, qualitative form that requires no further deliberation, and/or (ii) maintains this qualitative form throughout its manipulation and representation.

A system that gives results in an understandable form has, in a sense, performed the interpretation that otherwise would be left to the user. Cognition is therefore a characteristic of practical value in that results can be understood and used directly. In contrast, a numerical simulation that provides its results as a table of numbers or a graphics presentation leaves a great deal of analysis and interpretation to the person receiving the output.

An important practical implication of the second type of cognition is explanation capability. The reasons a system reaches a specific conclusion are readily understood when a qualitative representation is used. Although such systems may not be programmed to produce an explanation, the transparency of their qualitative representation facilitates understanding of their reasoning.

We can see the elements of cognition in any intelligent system component. For example, neural networks are well suited to process monitoring, diagnosis, sensor validation and data interpretation where pattern recognition plays a role. Generally, these systems classify data patterns into categories that are defined during a training phase (e.g., Tsen et al., 1994; Whiteley and Davis, 1994; Bakshi and Stephanopoulos, 1994; Kavuri and Venkatasubramanian, 1993 and 1994). QI-Map is a specific application example that illustrates the cognitive nature of neural networks (Whiteley and Davis, 1994; Davis and Wang, 1993). Implemented in ART2 (Carpenter and Grossberg, 1987), QI-Map has been used for data interpretation, monitoring and detection, and diagnosis. Trained using process data, the network maps from multisensor data input to appropriate labels that provide interpretations for immediate use. Training is supervised and the source of knowledge for the appropriate labels is expertise. For detection, the labels are "normal" and "not normal"; for data interpretation, there are many possible labels such as skewed, pulsing, cycling, tail, etc.; for diagnosis, the labels are specific root causes such as "plugged line," "contaminated feed," and so on. All of these labels are directly understandable without further deliberation. Neural net systems, however, are unable to explain their interpretations and would require additional symbolic-reasoning layers or modules to provide for that capability.

Similarly, the cognitive nature of fuzzy reasoning approaches derives from the assignment of qualitative, linguistic labels to numerical data based on membership functions (e.g., Roffel and Chin, 1991; Huang and Fan, 1993; Yamashita et al., 1988). Promising applications of fuzzy reasoning have been seen in the area of process control. Controller actions appear as the consequence of qualitative rules that are invoked by the presence of certain "fuzzy" conditions that are so labeled. As implied previously, the labels can be whatever is needed to be understandable and appropriate for taking action. Unlike neural networks, process data are not used directly to establish the membership functions. Rather they are constructed based on expertise with the process and the data. This additional emphasis on expert knowledge in the construction of the membership functions provides a capacity for explaining interpretations that neural networks do not have.

Knowledge-based systems also can exhibit one or both types of cognition. In table look-up, systems map directly from input data to conclusions by explicitly enumerating all input-output combinations (Saraiva and Stephanopoulos, 1994). The output is in an understandable form but there is no additional information inherent in the system to explain the conclusions. Alternatively, knowledge-based systems that are based on knowledge organizations to represent fragments of information and then search to piece information together provide a greater capability for explanation (Davis, 1994).

All intelligent system components offer cognition of the first type. Cognition together with qualitative modeling therefore define the basic value of any intelligent system. Cognition of the second type represents a higher degree of functionality, since explanation is possible. Systems exhibiting cognition of the second type involve a greater emphasis on expert knowledge.

Deliberation to Achieve Generalization

Deliberation to achieve generalization is associated with a significantly greater level of system performance. It refers to the ability of a system to make use of a representation and manipulation by some search mechanism (deliberation) to reach correct conclusions without having explicitly enumerated all the possible relationships between attribute combinations and conclusions (generalization). A very significant emphasis on various forms of expert knowledge is required, but the benefits are that the system can reach conclusions about situations that were not explicitly covered in the construction of the system and that may not have been previously considered.

As an illustration, consider a knowledge-based system whose representation consists of a hierarchy of nodes representing plant systems and subsystems. Each node contains evaluation information that results in the node being accepted or rejected based on the presence of some attributes. Acceptance of a given node results in the search continuing to its successors, each of which is constructed similarly. In constructing the representation, one defines each node separately with its own attribute of focus and then links it to other nodes. Provided the nodes are defined and linked together appropriately, the knowledge-based system will be able to deliberate on all combinations of attributes, even though the combinations are not explicitly enumerated. As the number of nodes and attributes increases, the capacity of representation and search becomes enormous.

In contrast, consider a spreadsheet package that provides a macro command, which executes a search for an input string in the left-hand column of a table and returns the string in the corresponding right-hand column. Usually called "table-lookup," this procedure produces the same results as a "deliberative" knowledge-based system, if the pre-enumeration is complete. The drawback is that all possible attribute combinations must be entered into the table along with the corresponding conclusions. Even simple problem domains that require a relatively few number of attributes and conclusions are not amenable to solution via table-lookup.

Good examples of applications that benefit from deliberation and generalization are fault diagnostic knowledge-based systems such as FALCON (Rowan, 1988, 1992), PX (Prasad et al., 1993), Shell KB (Kumar and Davis, 1994), CATCRACKER (Ramesh et al., 1992), MODEX2 (Venkatasubramanian and Rich, 1988), and Faultfinder (Hunt et al., 1993). All of these systems emphasize knowledge bases, various representations and corresponding inference mechanisms that make it possible to accommodate and react to a very wide range of possible operating situations without having to pre-enumerate every combination. These representations are usually some form of directed graph in which nodes contain separate pieces of qualitative knowledge about process faults. Different inference mechanisms are used to traverse the representations, examine the nodes and draw conclusions.

FALCON is a system that relies on a representation that combines qualitative knowledge about the process with mathematical simulations. Generalization occurs through selective simulation of various normal and failure conditions coupled with appropriate interpretation. PX, Shell KB, and CATCRACKER are all systems based on hierarchical classification. Hierarchical classification leads to generalization by decomposing the process into a hierarchical set of subcategories, i.e., subsystems and/or fault categories and then pre-enumerating the evaluation patterns for each. Generalization occurs as a result of the interactions between the subcategories as the diagnosis proceeds. In addition to its establish-refine mechanism, the Shell KB system utilizes a technique called hypothesis queuing (HQ), which prioritizes malfunction hypotheses based on their likelihood in a given situation and directs the search accordingly. HQ extends the existing deliberation mechanism and thus enhances the performance of the system.

Faultfinder, MODEX2, and HAZOP (Catino et al., 1991) are examples of systems that achieve generalization by linking device models together according to the topology of the process. By propagating effects forward and backward from unit model to unit model, the system is able to generate process behaviors without pre-enumerating all possible symptom patterns.

An intelligent system designed to achieve generalization represents a significant increase in the value of the system. This level of system functionality builds upon the elements of modeling and cognition to provide the added capacity of reaching correct conclusions about situations that have not been considered previously. Various intelligent systems for diagnosis, hazard identification, design and planning have demonstrated this capability.

Adaptation

Adaptation (synonymous with learning) refers to the ability of a system to make changes to its representation in response to new information through deliberation. This definition allows us to differentiate between a deliberative adaptive system and say, a model identification problem, which is merely a parameterized model that automatically adapts to new process data. For example, systems such as neural adaptive, model-predictive, and dynamic matrix controllers update model parameters or network weights as process data are received so that the controllers use the latest model of the process. Although these systems do, in a sense, adapt by changing parameter values, no deliberation, no cognition and no qualitative modeling are involved. As a result, these systems, while important components of an integrated intelligent system are not considered "intelligent system" components in themselves.

Unlike the identification techniques mentioned above, systems that deliberatively adapt by updating their own knowledge representations are scarce. There is current work on adaptation of process monitoring systems (e.g., Davis et al., 1994; Bakshi and Stephanopoulos, 1994). It is important for a monitoring system to incorporate new patterns in its representation as they occur. At present, these systems are able to partially automate the adaptation process but elements of supervision are still required. There are only a few examples of work on learning of knowledge-based systems in process engineering (e.g., Modi et al., 1993, Rich and Venkatasubramanian, 1989).

Autonomy

Autonomy refers to the ability of a system to achieve a fixed high-level goal through appropriate accomplishment and/or revision of subgoals without needing outside intervention. It represents an ultimate capability for an intelligent system. The ability of a system to perform its task autonomously depends greatly on how well it copes with uncertain information and changing circumstances. Autonomy is generally considered to have a greater role in large, distributed systems containing several component modules that must communicate with each other effectively in order for the system to attain its goal (Antsaklis and Passino, 1993). The absence of human intervention places great demands on integrated intelligent systems, which must draw upon all the previously mentioned characteristics in order to effect autonomy. In particular, the notion of cognition must be extended to encompass the communication among subsystems. In addition to producing results that are understandable to humans, these systems also manage information that is passed from one module to another. The inputs and outputs must at some point appear in forms that are meaningful to other modules, intelligent or otherwise.

Systems pointed toward autonomy are at the early stages of development. Redmill et al. (1994) have considered autonomous control of rocket engines in unmanned space vehicles. Marchio and Davis (1994) have prototyped a procedure management system for plant start-ups and cyclical operations. Bhatnagar et al. (1990) have developed a system for failure action management in nuclear power plants. Each of these systems exhibit aspects of autonomy through sophisticated forms of deliberation. Full implementation of CIM systems today is confounded by difficulties in integrating lower-level supervisory and control execution activities with the higher-level planning and scheduling functions. As a consequence, conventional agents, i.e., humans, are currently used to fill the "gap." Autonomous systems, by definition, cannot rely on this intervention. Adaptation is of major importance to maintaining autonomy, as is generalization. These must exhibit a high degree of development in order to successfully automate human problem solving and decision making tasks. Clearly, the achievement of autonomy is an extremely challenging objective that will become increasingly relevant as work towards integrating intelligent systems continues.

Conclusions

In this paper we have attempted to define the value added by intelligent systems in terms of five characteristics: qualitative modeling, cognition, deliberation to achieve generalization, adaptation, and autonomy. These incorporate major aspects of "intelligent" functionality observed in the different techniques and applications that have appeared thus far. Qualitative modeling represents an essential component of cognition and generalization and underlies all aspects of system intelligence. The benefits of these characteristics culminate in autonomy, which embodies all of them to some degree. As each of these is "built into" a system, the improved performance brought by the system to the particular application increases.

Current research is generally focused on development of the qualitative, generalization, and cognitive aspects of intelligent applications. In the past few years, however, research efforts have also begun to emphasize integration and, in particular, the challenges presented by integrated intelligent systems. Adaptation techniques are in the very early stages of development. Autonomy of large integrated intelligent systems that can effectively manage complex interactions of disparate information and techniques remains a goal that has yet to be realized.

Acknowledgments

We would like to thank Dr. B. Chandrasekaran (Computer and Information Science) and Dr. U. Ozguner (Electrical Engineering) for their helpful participation in our discussions on intelligent systems.

References

Antsaklis, P. and K. Passino (1993). Introduction to intelligent control systems with high degrees of autonomy, in *An Introduction to Intelligent and Autonomous Control*, P. Antsaklis and K. Passino (eds.), Kluwer Academic Publishers, Boston.

Barr, A. and F.A.Feigenbaum (1989). *The Handbook of Artificial Intelligence, Vol. 1*, William Kaufman Pub.

Bhatnagar, R., D. Miller, B. Hajek, and J. Stasenko (1990). An integrated operator advisor system for plant monitoring, procedure management, and diagnosis. *Nuclear Tech.*, **89**, 281-317.

Bakshi, B. and G. Stephanopoulos (1994). Representation of process trends - III. Multiscale extraction of trends from process data. *Computers Chem. Engng.*, **18**, 4, 267-302.

Bakshi, B. and G. Stephanopoulos (1994). Representation of process trends - IV. Induction of real-time patterns from operating data for diagnosis and supervisory control. *Computers Chem. Engng.*, **18**, 4, 303-332.

Calandranis, J., G. Stephanopoulos and S. Nunokawa (1990). DiAD-Kit/Boiler: on-line performance monitoring and diagnosis. *Chem. Engng. Prog.*, **86**, 1, 60-68.

Carpenter, G. and S. Grossberg (1987). ART2: self-organization of stable category recognition codes for analog input patterns. *Applied Optics*, **26**, 4919-4930.

Catino, C., S. Grantham and L. Ungar (1991). Automatic generation of qualitative models of chemical process units. *Computers Chem. Engng.*, **15**, 8, 583-599.

Chandrasekaran, B. (1994). Functional representation and causal processes, in *Advances in Computers*, 38, Academic Press, 73-143.

Charniak and McDermott (1985). *Introduction to Artificial Intelligence*. Addison-Wesley Pub.

Cott, B. and S. Macchietto (1989). An integrated approach to computer-aided operation of batch chemical plants. *Computers Chem. Engng.*, **13**, 11/12, 1263-1271.

Crooks, C. A., K. Kuriyan and S. Macchietto (1992). Integration of batch plant design, automation, and operation software tools. *Computers Chem. Engng.*, **16S**, S289-S296.

Davis, J. (1994). On-line knowledge -based systems in process operations: the critical importance of structure on integration. *Proceedings IFAC Symposium on Advanced Control of Chemical Processes, ADCHEM'94*, Kyoto, Japan.

Davis, J. and M. Gandikota (1990). Rule-based systems in chemical engineering, *CACHE Monograph Series, AI, in Process Systems Engineering. vol. II*, G. Stephanopoulos and J. Davis (eds.) CACHE.

Davis, J. and C. Wang (1993). Adaptive resonance theory for on-line detection and sensor validation: A practical view. *AIChE Annual Meeting*, St. Louis MO, November 1993.

Davis, J., C. Wang and J. Whiteley (1994). Real-time knowledge-based interpretation and adaptation of process data. *Proceedings Fifth International Symposium on Process Systems Engineering*, Seoul, Korea, 1153-1159.

Forsythe, R., S. Prickett, and M. Mavrovouniotis (1992). An introduction to object-oriented programming in process engineering, *CACHE Monograph Series, AI in Process Systems Engineering. vol. IV*, G. Stephanopoulos and J. Davis (eds.) CACHE.

Goodaker, A. and J. Davis (1994). Sharable engineering knowledge databases in support of HAZOP analysis of process plants. *AIChE Annual Meeting*, San Francisco, November 1994.

Huang, Y. and L. Fan (1993). A fuzzy-logic-based approach to building efficient fuzzy rule-based expert systems. *Computers Chem. Engng.*, **17**, 2, 181-192.

Hunt, A., B. Kelly, J. Mullhi, F. Lees, and A. Rushton (1993). The propagation of faults in process plants: 6, overview of and modeling for, fault tree synthesis. *Reliability Engng. and Proc. Safety*, **39**, 2, 173-194.

Hunt, A., B. Kelly, J. Mullhi, F. Lees, and A. Rushton (1993). The propagation of faults in process plants: 7, divider and header units in fault tree synthesis. *Reliability Engng. and Proc. Safety*, **39**, 2, 195-209.

Hunt, A., B. Kelly, J. Mullhi, F. Lees, and A. Rushton (1993). The propagation of faults in process plants: 8, control systems in fault tree synthesis. *Reliability Engng. and Proc. Safety*, **39**, 2, 211-227.

Hunt, A., B. Kelly, J. Mullhi, F. Lees, and A. Rushton (1993). The propagation of faults in process plants: 9, trip systems in fault tree synthesis. *Reliability Engng. and Proc. Safety*, **39**, 2, 229-241.

Hunt, A., B. Kelly, J. Mullhi, F. Lees, and A. Rushton (1993). The propagation of faults in process plants: 10, fault tree synthesis - 2. *Reliability Engng. and Proc. Safety*, **39**, 2, 243-250.

Kramer, M. and R. Mah (1994). Model-based monitoring. *Proceedings of the Second International Conf. on FOCAPO*, Crested Butte CO, CACHE/Elsevier, 45-68.

Kumar, S. and J. Davis (1994). Knowledge-based problem solving in after-cycle, root cause diagnosis of batch operations. submitted to *Industrial and Chemistry Research*.

Kavuri, S. and V. Venkatasubramanian (1993). Using fuzzy clustering with ellipsoidal units in neural networks for robust fault classification. *Computers Chem. Engng.*, **17**, 8, 765-784.

Kavuri, S. and V. Venkatasubramanian (1994). Neural Network Decomposition Strategies for Large Scale Fault Diagnosis.*Special Issue of Intl. J. Control*, (Ed.) Morari and Morris, **59**, 3, 767-792.

Leonard, J. and M. Kramer (1993). Diagnosing dynamic faults using modular neural nets. *IEEE Expert*, **8**, 2, 44-53.

Macchietto, S. (1994). Bridging the gap - integration of design, scheduling, and process control. *Proceedings of the Second International Conf. on FOCAPO*, Crested Butte, CO, CACHE/Elsevier, 207-230.

Madono, H. and T. Umeda (1994). Outlook of process CIM in Japanese chemical industries. *Proceedings of the Second International Conf. on FOCAPO*, Crested Butte, CO, CACHE/Elsevier, 163-177.

Marchio, J. and J. Davis (1994). A generic knowledge-based architecture for batch process diagnostic advisory systems. paper presented at *AIChE National Spring Meeting*, 1994.

Modi, A., A. Newell, D. Steier, and A. Westerberg (1993). Building a chemical process design system within Soar: part 2 - learning issues. to be published in *Computers Chem. Engng.*

Prasad, P. and J. Davis (1993). A framework for knowledge-based diagnosis in process operations, chapter in *An Introduction to Intelligent and Autonomous Control*, P. Antsaklis and K. Passino (eds.), Kluwer Academic Publishers, Boston.

Prasad, P., J. Davis, H. Gehrhardt and B. Feay (1993). Generic nature of task-based framework for plant-wide diagnosis. in revision *Computers Chem. Engng.*

Prasad, P., Y. Jirapinyo and J. Davis (1993). Fault trees for knowledge acquisition and hierarchical classification for diagnosis. *IEEE Expert*, July 1993.

Psichogios, D. and L. Ungar (1992). A hybrid neural network - first principles approach to process modeling. *AIChE Journal*, **38**, 10, 1499-1511.

Ramesh, T., J. Davis and G. Schwenzer (1992). Knowledge-based diagnostic systems for continuous process operations based on the task framework. *Computers Chem. Engng.*, **16**, 2, 109-127.

Rao, M., Q. Wang and J. Cha (1993). *Integrated Distributed Intelligent Systems in Manufacturing*. Chapman & Hall, London.

Redmill, K., U. Ozguner, J. Musgrave, W. Merrill (1994). Design of intelligent hierarchical controllers for a space shuttle vehicle. *Proceeding 1993 IEEE International Symposium on Intelligent Control*, 64-69.

Rich, S. H. and V. Venkatasubramanian (1989). Causality-based Failure driven Learning in Diagnostic Expert Systems, *AIChE Journal*, **35,** 6, 943-950.

Roffel, B. and P. Chin (1991). Fuzzy control of a polymerization reactor. *Hydrocarbon Processing.*, **70**, 6, 47-49.

Rowan, D. (1988). AI enhances on-line fault diagnosis. *InTech*, **35**, 5, 52-55.

Rowan, D. (1992). Beyond Falcon: Industrial applications of knowledge-based systems. *Proceedings IFAC On-line Fault Detection and Supervision in the Chemical Process Industries*, 215-217.

Saraiva, P. and G. Stephanopoulos (1994). Data-driven learning frameworks for continuous process improvement. *Proceedings Fifth International Symposium on Process Systems Engineering*, Seoul, Korea, 1275-1281.

Stephanopoulos, G. (1990). Brief overview of AI and its role in process systems engineering, *CACHE Monograph Series, AI in Process Systems Engineering. vol. I*, G. Stephanopoulos and J. Davis (eds.) CACHE.

Stephanopoulos, G. and C. Han (1994). Intelligent systems in process engineering: a review. *Proceedings Fifth International Symposium on Process Systems Engineering*, Seoul, Korea, 1339-1366.

Stephanopoulos, G., G. Henning, and H. Leone (1990). MODEL.LA. A modeling language for process engineering - I. The formal framework. *Computers Chem. Engng.*, **14**, 8, 813-846.

Stephanopoulos, G., G. Henning, and H. Leone (1990). MODEL.LA. A modeling language for process engineering - II. Multifaceted modeling of processing steps. *Computers Chem. Engng.*, **14**, 8, 847-869.

Stephanopoulos, G., J. Johnston, T. Kriticos, R. Lakshmanan, M. Mavrovouniotis, and C. Siletti (1987). DESIGN-KIT: an object-oriented environment for process engineering. *Computers Chem. Engng.*, **11**, 6, 655-674.

Tsen, A., S. Jang, D. Wong and B. Joseph (1994). Predictive control of quality in batch polymerization using artificial neural network models. *Proceedings Fifth International Symposium on Process Systems Engineering*, Seoul, Korea, 899-911.

Ungar, L. and V. Venkatasubramanian (1990). Knowledge representation, *CACHE Monograph Series, AI in Process Systems Engineering. vol. III*, G. Stephanopoulos and J. Davis (eds.) CACHE.

Vaidhyanathan, R. and V. Venkatasubramanian (1992). Representing and diagnosing dynamic process data using neural networks. *Engng. Appl. of Artif. Intell.*, **5**, 11-21.

Venkatasubramanian, V. (1988). CATDEX: An expert system for diagnosis of a catalytic cracking operation. *CACHE Case-Studies Series,* Knowledge-Based Systems in Process Engineering, G. Stephanopoulos (ed.), vol. II, CACHE.

Venkatasubramanian, V. (1994). Towards Integrated Process Supervision: Current Trends and Future Directions, Keynote Address, *Second IFAC International Conference on AI/KBS in Process Control,* University of Lund, Lund, Sweden.

Venkatasubramanian, V. and S. Rich (1988). An object-oriented two-tier architecture for integrating compiled and deep-level knowledge for process diagnosis. *Computers Chem. Engng.*, **12**, 9, 903-921.

Venkatasubramanian, V. and G. Stanley (1994). Integration of process monitoring, diagnosis and control: issues and emerging trends. *Proceedings of the Second International Conf. on FOCAPO*, Crested Butte CO, CACHE/Elsevier, 179-206.

Venkatasubramanian, V. and R. Vaidhyanathan (1994). A Knowledge-Based Framework for Automating HAZOP Analysis, *AIChE Journal*, **40**, 3, 496-505.

Whiteley, J. and J. Davis (1994). A similarity-based approach to interpretation of sensor data using adaptive resonance theory. *Computers Chem. Engng.*, **18**, 7, 637-661.

Williams, T. (ed.) (1989). *A Reference Model for Computer Integrated Manufacturing (CIM).* Instrument Society of America, Research Triangle Park NC.

Yamashita, Y., S. Matsumoto and M. Suzuki (1988). Start-up of a catalytic reactor by fuzzy controller. *Journal of Chem. Engng. of Japan*, **21**, 3, 277-282.

LANGUAGES AND PROGRAMMING PARADIGMS

George Stephanopoulos and Chonghun Han
Massachusetts Institute of Technology
Cambridge, MA 02139

Abstract

Are programming and computer science relevant to the education and professional needs of the vast majority of chemical engineers? In this chapter we explore this question in the presence of shifting paradigms in programming languages and software design. After a brief history in the evolution of programming languages, high-level, chemical engineering-oriented languages are discussed as the emerging programming paradigm, and certain conclusions are drawn on the importance of computer science as a pivotal element in engineering education and practice.

Introduction

The explosive growth in the use of the digital computer is a natural phase in the saga of the Second Industrial Revolution. If the First Industrial Revolution in 18th century England ushered the world into an era characterized by machines that extended, multiplied and leveraged the humans' *physical capabilities*, the Second, currently in progress, is based on machines that extend, multiple and leverage the humans' *mental abilities*. The thinking man, homo sapiens, has returned to its Platonic roots where, "*all virtue is one thing, knowledge*." Using the power of modern computer science and technology, software systems are continuously developed to; preserve knowledge for it is perishable, clone it for it is scarce, make it precise for it is often vague, centralize it for it is dispersed, and make it portable for it is difficult to distribute. The implications are staggering and have already manifested themselves, reaching the most remote corners of the earth and inner sancta of our private lives.

Central to this development has been the story of the continuously evolving programming languages and programming paradigms. From FORTRAN 0 (in 1954) to FORTRAN 90, from ALGOL 58 (in 1958) to BASIC and PL/1, and through ALGOL 68, to Pascal and C, from ALGOL 60 through SIMULA I to Smalltalk 80 and C++, from LISP through ZetaLisp to Common LISP and CLOS, the programming languages have been evolving like the natural languages of humans; to fit the evolving needs in representing the world and articulating ideas. From the 0, 1 alphabet to modern modeling languages, computer scientists and engineers continue to enrich and formalize the way humans communicate with a computer. There is no end to this process, as there is no end to the evolution of the human languages.

The expressive power of a programming language influences the breadth and depth of computer-aided human solutions and their communication to others, very much as the progress of human civilization moved in tandem with the growing expressiveness and precision of human languages. The following corollary is a little discussed understatement, and, for some, a trivial redundancy: *limited grasp of programming languages limits the sophistication of computer-aided engineering solutions*. If this is true, then the theoretical foundations of a computer language are as central to engineering education as physics, mathematics, and chemistry.

As the growing pervasiveness of computers in present-day life demands from humans increased familiarity and higher-level communication skills with machines, the above corollary becomes all too important. It suggests that only natural language-like communication will allow humans to fully exploit the capabilities of a digital computer; *not in executing prepackaged programs, but in becoming an integral part of and extending the human ability*. A student of chemical engineering should not be forced to express his/her ideas in 0 and 1's, FORTRAN statements, or any other linguistic straight jacket. Instead, he/she should be able to communicate naturally with machines, using the language of chemical engineering science and practice. Therefore, the natural form of computer programming for a chemical engineer is worlds apart from the current paradigm, which is full of intellectually stifling coding and, at times unbearably, painful debugging.

The evolution of programming languages has been followed by a shift in the programming paradigm, i.e., the process of designing and developing software systems. From the unstructured and convoluted spaghetti-like designs of earlier times (which can still be seen today in commercially successful software products), to modular and structured-programming products, to easily reusable and maintainable object-oriented codes, the process of designing a software system has been largely streamlined and rationalized. Today, the development of a complex software system is not in the realm of a creative "hacker," but instead, is the result of teamwork, and resembles more the production line of a modern manufacturing plant, whose units are not constrained by geographic location.

In the midst of all these developments, what is the role that programming languages and programming paradigms can play in the education of a chemical engineer? How do they affect the way that a professional chemical engineer carries out his/her work; product/process research, development or engineering? In the following sections of this chapter we will try to discuss the above issues. Specifically, the next section will provide a brief overview of the historical evolution of programming languages, and will attempt to underline the foundations which may constitute an educational component in undergraduate engineering curricula. The third section will focus on "modeling languages," a natural evolution of programming languages that seems to serve better present and anticipated future needs for chemical and other types of engineers. The fourth section introduces the task-oriented languages which complement the modeling languages to form the paradigm of future programming. The shift in software design paradigms will be discussed in the fifth section. Section six will attempt to draw some conclusions, which may have an effect on the computer-aided education and practice of chemical engineers.

Programming Languages

The history of the evolution of programming languages is a fascinating subject that transgresses the boundaries of interests of computer scientists, and offers useful insights on the interplay of intellectual contributions, commercial interests, and human inertia. The encyclopedia edited by Wexelblat (1981) offers a rich panorama of this history. The book by Sebesta (1993) provides the reader with a shorter version of this history, and an educational exposition to programming languages. For a formal exposition to the theory of programming languages, the book by Meyer (1990) is an exceptional source. The first digital computers were constructed in the 1940s in response to the needs of scientific applications, which involved the solution of large and complex numerical problems. *PlanKalcul* was the first high-level programming language. A fairly rich language, it was composed by the German scientist Konrad Zuse in 1945, but was not published until 1972 (Zuse, 1972). Although Zuse wrote many programs, using PlanKalkul, and executed them on Z4, a computer he constructed with electromechanical relays, his work did not have any impact on the subsequent development of programming languages, since it was not widely known for a long time.

FORTRAN - The Paradigm of Imperative Programming

The decisive step forward was the development of FORTRAN (FORmula TRANslator), in response to the new capabilities offered by the IBM 704 computer. The road to the formulation of FORTRAN had been paved by a number of earlier and parallel research efforts, such as: John Mauchly's interpreted *ShortCode* (Wexelblat, 1981), developed in 1949 and run on BINAC and UNIVAC I computers; John Backus' *Speedcoding* interpreter with the ability to process floating-point data; Grace Hopper's development of advanced compiling systems for the UNIVAC computers; Alick Glennie's *Autocode* compiler for the Manchester Mark I computer.

The development of FORTRAN 0 began in 1954 (IBM, 1954) and was implemented the following year. Subsequent modifications led to FORTRAN I, whose compiler was released in 1957. FORTRAN II was introduced the following year, it was succeeded by FORTRAN III in 1959, and reached maturity with FORTRAN IV in 1962. FORTRAN IV, one of the most widely used programming languages, became the standard for an explosive growth in scientific computing during the period 1962 to 1978, when it was succeeded by FORTRAN 77, and which in turn was followed by FORTRAN 90 (ANSI, 1990). The transformation of FORTRAN over a period of 25 years is remarkable. Becoming the standard language for scientific computing, it was continuously under pressure to adapt and include new features, which were required by the expanded needs of its users. Other programming languages, developed during the period 1960-1990, had many inherent design advantages over FORTRAN, but none could replace it, since FORTRAN kept adapting to include the most attractive of the missing features.

FORTRAN is also important from another point of view. It represents the prototypical paradigm of *Imperative Programming Languages*. The imperative style of programming is characterized by the presence of constructs describing commands issued to a machine (Meyer, 1990). The construct that makes a programming language most prominently imperative is the *instruction*. Thus, programs written in an imperative language are sequences of instructions, with each instruction describing a set of actions to be performed. Each instruction changes the state of the program, by modifying the values of variables and defining the sequence of subsequent instructions to be executed.

The Gap Between Mathematics and Imperative Languages

Mathematical notation is *referentially transparent* (Meyer, 1990). For example, the relationships, x = 1 and x + z < 10, imply that, 1 + z < 10. Although it is made explicit, referential transparency is extremely important in mathematical reasoning, since all FORmula manipulations (arithmetic and logical) rely on it. Imperative programming languages, like FORTRAN, on the other hand, violate referential transparency, because they permit side effects on the variables of a program. For example, consider the following sequence of FORTRAN instructions, defining a function, F(x): Instruction-i; y=y+1 Instruction-(i+1); F=y+x Instruction-(i+2); End. Although called function, F(x) is not a function in the mathematical sense. It does not produce a result computed from its arguments, but changes the state of the program, by modifying the value of the variable y every time that the function is called. As a consequence, referential transparency among different expressions is lost. *Aliasing*, the access of a given object through more than one name, is another feature of the imperative programming languages that generates a gap between mathematics and the encoded computer programs.

ALGOL and Its Offsprings - More Imperative Programming Languages

The Association for Computing Machinery (ACM) in the U.S. and the Society for Applied Mathematics and Mechanics (GAMM) in Germany, decided, in 1958, to join efforts and produce a universal, machine-independent programming language, which would be as close as possible to "standard mathematical notation." Thus, ALGOL 58 (ALGOrithmic Language) came into existence, formalizing the concept of data type, adding compound statements, expanding the number of array dimensions, allowing nested conditional statements, and a number of other highly desirable features. In 1960, a six-day meeting in Paris produced ALGOL 60, which was the first language to be described in detailed form, using the Backus-Naur syntax (Naur, 1960) that is still widely used. ALGOL 60 became one of the most pivotal contributions in the evolution of programming languages, introducing the concept of "Block structure," implementing recursion, and allowing the passing of parameters to subprograms by value or by name. Its direct descendant, ALGOL 68, introduced the revolutionary innovations of "user-defined data types," and "dynamic arrays," and thus led to the design of two new offsprings with significant influence; Pascal, and C. Although Pascal and C did not introduce any significant innovations, they both became far more popular than ALGOL 68. Pascal (Wirth, 1971) became the language of choice for the teaching of programming, while C evolved into the language of choice for the professional systems programmers. C became the first truly portable language, and in addition to ALGOL 68, it owes a lot of its character to languages such as, CPL, BCPL, and B. Ada is also a descendant of ALGOL 68, being based on Pascal.

Data Abstraction - From SIMULA to Smalltalk and C++

SIMULA I is another language that draws its ancestry from ALGOL 60. It was designed and implemented in 1964 to support system simulation applications. Its successor, SIMULA 67 (Dahl and Nygaard, 1967) introduced one of the most pivotal innovations, the *class* construct, which became the basis for data abstraction and the subsequent maturation of object-oriented programming ideas. A "class" was a construct, which packaged into one entity both (i) the data structure that defined the attributes of the class, and (ii) the routines that manipulated the data structure.

The "class" concept of SIMULA 67 became the pivotal element for the development of *Dynabook* (Kay, 1969), a computer program that used the idea of a "desk-top" on a graphic screen, to simulate the working environment of future computer users. Thus, the concept of "windows," currently visible in every personal computer, was born. Moving to Xerox, Kay led a group effort, whose objective was to create a language that would support the highly iconic and graphic paradigm of the man-machine interaction, and build a computer which would deliver it. Smalltalk-72, -74, -76, -78 represent the early versions of a revolutionary programming language, which took on its matured form in Smalltalk-80.

All program units in Smalltalk are *objects*, and can be viewed as instances of abstract data types, called *classes*. The Smalltalk classes are very similar to those of SIMULA 67. Each class contains a set of attributes, which define the data structure of the class, and a series of procedures, called *methods*, which operate on the attributes. During the execution of a Smalltalk program the objects "call and ask" other objects to carry out specific tasks. The calling object does not need to know "how" the receiver of its message will carry out the task, thus leading to complete encapsulation. Smalltalk introduced two innovative ideas, *inheritance* and *dynamic typing*, which, along with the concept of class, formed the basis for what it has come to be known as, *object-oriented programming* (OOP). The notion of inheritance allows the straightforward extension of existing classes to new ones with specialized features. Thus, a set of existing classes can be continuously extended with new classes by making full utilization of existing libraries in a very efficient manner. This reusability and extensibility is at the core of vast improvements in software development efficiency, which has been realized in recent years.

Although object-oriented programming has become and will remain for the next decade the dominant paradigm, Smalltalk is not the language of choice. Instead, C++, an evolutionary object-oriented version of C seems to become the standard. The first steps towards C++ were taken by Bjarne Stroustrup at Bell Laboratories in 1980 (Stroustrup, 1983). He introduced a new data type, called "class," which captured the essential character of classes as it was designed by SIMULA67 and Smalltalk. Additional modifications of C brought in inheritance, but not dynamic typing. Between 1986 and 1994, C++ continued to evolve and today, commercial C++ versions includes; multiple inheritance, and a growing number of product-specific classes which encapsulate elements of graphic interfaces (e.g. windows, menus, buttons, tables, lists, icons, spreadsheets, graphs, etc.), procedures for solving numerical problems, interfaces to databases, and many others. The reusability and extensibility of C++ has made it the replacement of C for software engineering. The growth rate on the number of C++ classes within specific programming environments has been phenomenal. Small companies appear everyday, marketing C++ classes for specific market needs.

C++ is a successful hybrid of imperative and object-oriented programming. Through evolution has moved the programming to the next phase, i.e. the OOP. It seems that evolution is a safer road to commercial success than revolution, a feature attempted by Smalltalk.

Borland's Turbo Pascal is another example of an imperative language's evolution to include object-oriented ideas. How far behind is Object-FORTRAN?

The universal acceptance of the OOP paradigm led to a fury of commercial activity during the last decade, which led to a number of OOP languages, like; CLOS, ACTOR, Eiffel, Objective-C and others. The unavoidable standardization has recently produced a commercial streamlining, with C++ taking the dominant position and Smalltalk proving its endurance through a smaller circle of programming artists.

LISP - The Paradigm of Functional Programming

Imperative languages are "polluted by side effects, assignments, explicit sequencing, and other repugnant features" (Meyer, 1990), all of which have led to a serious gap between the language of mathematics and that of the corresponding imperative examples, e.g. FORTRAN, Pascal, and C. LISP is an example of a programming language which proves that all the above weaknesses are not necessary evils. Other examples are FP and MIRANDA, data flow languages, such as LUCID, or logic programming languages, such as PROLOG.

LISP (LISt Processing) was developed by John McCarthy in 1958 and was aimed at supporting the computational needs of the new domain called Artificial Intelligence. It is the first *functional programming* language, and as such mimics quite closely the language of mathematics. A purely functional programming language does not use variables or assignment statements, like an imperative language. Instead, programs written in a functional language are function definitions and function application specifications. Execution of such programs can be viewed as evaluation of function applications.

LISP, like a typical functional programming language, provides a set of primitive functions and a set of functional forms, which are used to construct more complex functions from the primitive ones. In addition, it possesses a function application operation, eval, and two specific structures for storing data, namely, atoms and lists. Its attractive functional programming character and its subsequent standardization in *Common Lisp*, have made LISP one of the most enduring languages, although its popularity is far from that of the dominant imperative languages. Its recent evolution to CLOS (Common List Object System) has integrated functional with object-oriented programming.

PROLOG - The Paradigm of Declarative Programming Languages

PROLOG (PROgramming LOGic) is a nonprocedural language. Each statement in a PROLOG program is a *declaration* of a logical relation, *proposition*, among variables, functions, or/and parameters. Predicate calculus is the notation used for the expression of logical statements, and provides the *resolution*, which is the inference technique during the execution of a PROLOG program. A product of joint work of the Artificial Intelligence groups at the Universities of Edinburgh and Aix-Marseille, PROLOG was conceived to support the logical inferencing required in AI applications (Roussel, 1975). It is the pre-eminent example of a purely *declarative programming* language, and allows a straightforward juxtaposition to imperative and functional languages.

Other Milestones in the Evolution of Programming Languages

Languages like, BASIC, PL/I, COBOL and others, have played an important role in the saga of programming languages and the evolution in the use of computers. BASIC (Mather and

Waite, 1971), for example, was designed in the early 1960s to be simple and easy to be used by non-computer science oriented students from remote terminals in a *time-shared* environment. Shunned over 20 years by programmers, it has recently been revived, as *Visual Basic* with tremendous popularity, thanks to the marketing genius of Microsoft. COBOL is a very interesting example of a deliberate design effort. Designed by a committee over a short period of time to address the computing needs of business applications (Department of Defense, 1960), COBOL became an ANSI standard in 1968, and today is still one of the most widely used languages. It introduced a number of innovations, which subsequently were adapted by other languages, e.g. hierarchical data structures.

PL/I was the product of an effort to produce a language which would serve the needs of a broad spectrum of applications. Known initially as NPL (New Programming Language), PL/I included the best features of FORTRAN IV, COBOL, and ALGOL60. It introduced facilities to handle a large number of run-time errors, create concurrently executing tasks, carry out recursion, etc. PL/I has been a convincing example of the assertion that complexity and redundancy in the formulation of a language are serious drawbacks. As a result, most of the present-day programming languages are based on a logically firm foundation, which serves efficiently a set of similar-type inferencing, symbolic manipulations, etc.

Ada (Goos and Hartmanis, 1983) has come as the product of the "history's largest design effort" (Sebesta, 1993). Developed by the Department of Defense, it involved hundreds of participants from several dozens of organizations, and went through a 3-phase process before its design was selected and frozen for development. Named after the mathematician Augusta Ada Byron (the poet's daughter), considered by historians as the first programmer, Ada is another example of exercise in complexity. It offers a tremendous set of facilities with imperative, functional and object-oriented programming features, but its future viability is in doubt.

From Programming Languages to Modeling Languages

Engineering is the task of creating new or revamping existing artifacts, using the fundamental analytical insights offered by science and relying on the synthetic skills of the human engineer. If computers are to play a significant role in engineering work, the man-machine interaction must change. Central to this change is the role of *modeling languages*.

Every task that an engineer is involved in includes *modeling*, i.e. the activity of creating representations of the artifacts under construction. Models are needed to represent (a) molecules, (b) mixtures of chemicals, (c) physical and chemical properties, (d) processing units, (e) operations, (f) control loops, etc. Such *declarative models* provide the list of parameters and variables that reveal the state of an artifact (data models), or the set of declarative relationships (mathematical models) which describe the behavior of the artifacts. The imperative, or procedural character of the dominant programming languages, e.g. FORTRAN, has had a profound effect on the style of models created by chemical engineers. They have all been procedural models, and as a result they can have a limited scope of applicability, covering only the intended domain of use. Thus, we can explain the explosive proliferation of different models for the same artifact.

The discussion in the previous section has hinted that *declarative programming* has the ability to generate generic data models and mathematical models with virtually unlimited scope of applicability. This observation has led to the creation of modeling languages, as natural descendants of the declarative programming languages.

A. N. Whitehead observed that: "*By relieving the brain of all unnecessary work, a good notation sets it free to concentrate on more advanced problems ....*" Indeed, a chemical engineering student or practitioner should not need to master all intricacies of FORTRAN 90 or ANSI C in order to create and run computer programs. The programming language "*... is not just a way of getting a computer to perform operations, but rather it is a novel formal medium for expressing ideas about methodology. Thus, programs must be written for people to read, and only incidentally for machines to execute*" (Abelson et al., 1985). Thus, by defining appropriate primitives, means of combination, and means of abstraction, we can create domain-specialized, problem-oriented languages which can provide high-level modeling facilities. Such languages already exist and form the intellectual backbone of all engineering domains, e.g. the languages of: (i) "electrical networks" for modeling devices in terms of discrete electrical elements; (ii) "civil engineering structures," using girders, rods, plates, beams, shells; (iii) "unit operations" in chemical engineering. Today's complexities of the processing systems have induced a refinement of the language of unit operations into that of "*foundation phenomena*." The computer-aided deployment of such domain-specific languages has led to a new generation of programming languages, called *modeling languages*.

MODEL.LA. (Stephanopoulos, et al., 1990a and b) is a computer-aided implementation of the phenomena-based language for chemical engineers. It is characterized by its capacity to support the articulation and accept the student's declarations of his or her knowledge of the model context, including assumptions, simplifications, and scope of task. The language is, for the user, fully declarative. In the next section we will discuss its design characteristics that support the articulation of procedural tasks. The user of MODEL.LA. "writes" programs in an, almost, English-like, natural language, employing known syntax and chemical engineering-oriented vocabulary. Interpretation of the "sentences," composed by the user, allows MODEL.LA. to identify the boundaries of a system, relevant streams and the chemicals flowing with them, the set of physical and chemical phenomena occurring in the system (e.g. reaction, separation, heat exchange), the mechanisms and the expressions that should be used to describe rate-based phenomena (e.g. chemical reaction, diffusion, heat transfer), and equilibrium conditions, the expressions that should be used to compute physical properties etc.

Modeling languages, like MODEL.LA., do relieve the user's brain from unnecessary work, so that the user can concentrate on the engineering problem and its requirements. During the last 5 years we have witnessed a significant increase in the number of modeling languages, all of which, with variable degree of success, provide the declarative paradigm that will dominate the future work. Typical examples of such languages are the following: ASCEND (Piela et al., 1991), Omola (Nilsson, 1993), Modass (Sorlie, 1990), HPT (Woods, 1993), gPROMS (Barton, 1992; Pantelides and Barton, 1993), DIVA (Kröner et al., 1990), LCR (Han et al., 1995a).

Generic modeling languages should conform with the *declarative and functional programming paradigms*. In the previous paradigms we discussed the need for the former, but the requirement for functional programming is almost as important. In assembling models from the elementary foundation phenomena, the primary task for a computer is *symbolic manipulations*. The functional programming paradigm maintains the consistency between the "computational artifact" called model and its mathematical (arithmetic and logical) counterpart. Thus, we can maintain the *referential transparency* between mathematics and computer programs, a feature which is lost with procedural programming. Deployment of modeling languages through a procedural programming language, e.g. FORTRAN, C, Pascal, is either limited in scope or/and prone to failures, as the complexity of the language increases to serve the ever expanding needs.

MODEL.LA. was developed in an object-oriented dialect of LISP, a language which through its conformity to LISP is functional, while its object-oriented character provides the declarative programming facility.

Task-Oriented Languages

If modeling languages elevate the forms, used to describe engineering artifacts, to the level of a natural language, *task-oriented languages* expand those facilities and allow the high-level description of engineering methodologies, all of which are procedural in character. For example, the conceptual design of a chemical process (Douglas, 1988), involves the transformation of functional specifications into realizable physical objects (i.e. processing units), properly interconnected and operated to yield the desired process. The methodology to carry out the above transformation is clearly a procedure, and involves a series of executable tasks. It is very natural to consider imperative programming as the natural setting for creating a program that would implement the design methodology.

Upon closer inspection of Douglas' hierarchical design methodology, we realize that the resulting software would be a typical spaghetti-like design, unmaintainable, hard to expand with new design knowledge, and virtually impossible to track its logic and thus guarantee its integrity. Here is where a significant lesson can be learned from our experience with programming languages, namely, "the essential material (of an introductory course in computer science) is not the syntax of particular programming-language constructs, nor clever algorithms for computing particular functions efficiently, nor even the mathematical analysis of algorithms and the foundations of computing, but rather the techniques used to control the intellectual complexity of large software systems" (Abelson et al., 1985).

Consequently, as the intellectual complexity of engineering methodologies increases, it is important that we develop programming media, i.e. languages, which allow us to control the complexity and maintain their integrity. Here is where the task-oriented languages have started playing a significant role. Consider, for example, the *design-agents* used by Han (Han, 1993; Han et al. 1995b and 1995c), to model the design activities during the conceptual design of a chemical process. Each design-agent is a computational object with a specific set of data and a list of procedures, which, upon execution, define what the design-agent can do. By stringing generic design-agents together, we can create very complex programs, which emulate very complex design methodologies, and still maintain control of complexity and easy verification of the programs' (i.e., of the methodologies) integrity. As new knowledge becomes available, requiring a change on how a design-agent carries out its tasks, one only needs to modify the set of attributes and the algorithms of the procedures, characterizing the "skills" or "behavior" of the design-agent. Similarly, one can refine a design-agent by another one carrying out similar tasks by different approaches, expand the design methodology with new design-agents, or "re-wire" the overall design methodology by reconnecting the agents in different ways.

The modularized construction of complex procedures that started with the concept of a "subroutine" and a "function," has been brought up to levels of advanced sophistication with the object-oriented character of *task-oriented languages*. These languages provide linguistic constructs for the definition of "tasks" as generic programming elements that possess specific behavior. Any engineering methodology that requires the solution of a set of linear equations, will simply send the appropriate message to the "task" called *LINEAR-EQUATION-SOLVER*. Similarly, engineering methodologies may invoke instances of the generic tasks *SEQUENCE-DISTILLATION-COLUMNS*, *SELECT-REACTOR*, *FORMULATE-BALANCES*, *COST-HEAT-EXCHANGER*, etc.

So, task-oriented languages provide the means for the description of engineering procedures, while modeling languages support the declarative description of engineering artifacts. When taken together as the two complements of one *engineering programming language*, they provide the paradigm of future programming. A paradigm which frees the engineer from the drudgery of lower-level programming and allows him/her to concentrate at the engineering task at hand.

The first examples of such languages are already here (although not fully declarative or functional): the graphic user interfaces that support the creation of programs around relational database management systems using SQL; the visual programming in Visual Basic or Visual C++; the script-based language of MATLAB; the algebraic language, GAMS; the symbolic language of MACSYMA; the object-oriented language of G2.

Paradigms of Software Design

Software productivity has progressed much more slowly than the hardware developments. Despite the development of high-level, and then higher-level languages, structured methods for developing software, and tools for configuration management and tracking requirements, bugs, and changes, the net gain is nowhere close to that realized in hardware developments. This is due to the fact that programming remains basically unchanged, and remains as it has been formed by the needs of numerical computations, i.e. a person-centered activity which results in a highly detailed formal object. Figure 1 illustrates the philosophy of the traditional paradigm for software development. It starts with an informal statement of the desired requirements, e.g. schedule batch operations in an open flow-shop with intermediate storage, using heuristic implicit enumeration according to branch-and-bounding conditions A and B. It proceeds to analyze these informal requirements and define formal specifications which will be converted into a software system. During this state we have the formal statement of the solution methodology, as this is expressed by the resulting logical flow diagram. As soon as this step has been completed, the software designer has locked himself/herself into a specific problem-formulation and problem-solution and the coding can start, leading to the final source program. This manual conversion from informal requirements to programs is error prone and labor intensive. Once the source program has been completed, validation, testing and tuning begins. These activities are carried out at the source-code level, and experience has amply demonstrated that they are quite labor-intensive and painful. Any change in the problem-formulation or problem-solving strategy requires major surgery of the source code, amplifying the inefficiency of the software development process.

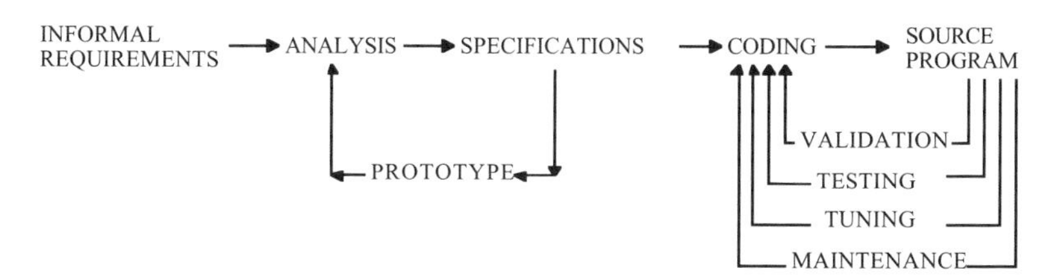

Figure 1. The traditional paradigm of software design.

Such philosophy for software design and development, although it may be acceptable for numerically intensive computer programs, is very poor and unacceptable for applications, which involve problem-specific contextual interaction between the human and the computer. For example, depending on the impact that various operating conditions have on the economics of process operations, the designer will determine the next state of activities during the design of control configurations for a chemical plant. Furthermore, the scheduling of batch operations is strongly influenced by plant-specific, operations-specific, and product-specific considerations. Consequently, we cannot have an all purpose problem-formulation and as a result the problem-solving methodology varies. Similar aspects one encounters in the formulation and solution of other problems such as; synthesis of operating procedures, diagnosis of plant or operating faults, identification of potential hazards in chemical plants, etc. Of course, one could try to capture all possible scenarios for all possible types of problem-formulations and direct each to a separate branch involving a problem-solving methodology with distinct features, thus creating a more complex structure of the logical flow diagram for a, say, FORTRAN program. But, this alternative is almost, by definition, impossible. Even more, such a system will lack facilities to "explain" the rationale of the problem-solving methodology (e.g. design, planning, diagnosis), and does not allow the human's own knowledge from past experience to be used effectively for the simple reason that they have not been anticipated by the program and have not been included in it.

By way of a summary, all the above discussion indicates that programs designed and developed using a model of fixed procedural knowledge are not satisfactory to easily capture varying problem-solving algorithms. A different paradigm is needed, along the lines of that shown in Figure 2. This flexible paradigm carries out the validation, testing, adaptation and maintenance of the software system within a symbolic, interpretive environment. Consequently, these tasks can be carried out at a higher level of abstraction, before the formal specifications have been cast in concrete. The human user's intervention to offer subjective preferences and decisions is allowed before the coding starts (see Figure 2), thus influencing decisively the set of formal specifications upon which the code will be generated. Thus, only the tuning of the code is done at the program's source level, in order to improve speed and efficiency. Such a programming paradigm allows a rather explicit description of what the resulting program does and why it does not require extensive work if changes must be made and provides a fairly precise framework for the simultaneous development of several of its parts by different programmers. Almost all software systems are presently designed and developed using the flexible paradigm of Figure 2. Finally, it should be noted that as the technology of automatic programming is advancing, allowing the automatic computer-driven conversion of formal specifications into a source code, the paradigm of Figure 2 will dominate over the traditional one of Figure 1 even for the development of self-standing numerically intensive software systems.

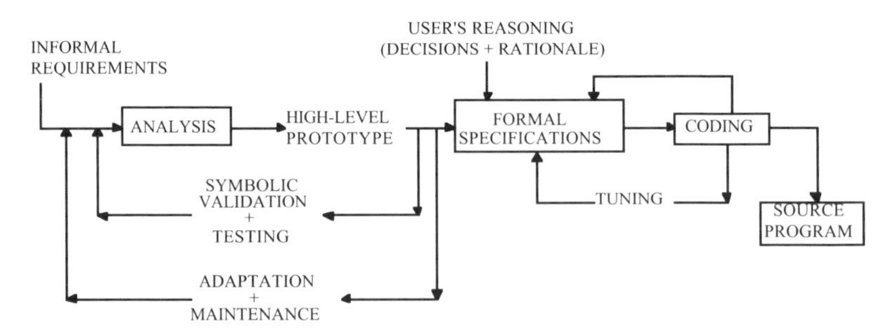

Figure 2. The evolving new paradigm for software design.

Programming and Computer Science for Chemical Engineers?

Chemical engineers are supposed to work on the creation of new products and processes and the optimal revamping of existing ones. Through the use of computers, they are expected to enhance their skills and abilities to carry out their main tasks, as described above. So, is there any role that programming and computer-science could play in their education and professional advancement? We believe that the answer is yes, but the rationale is more important than the answer itself.

The vast majority of chemical engineers are not and will not be writing programs, using the prevailing programming languages, as a regular part of their responsibilities. They will be using the computers, though, far more tomorrow than today, tackling a far broader number of problems than they do today. The implication is two-fold: (1) Since they will not be writing programs, they do not need to be taught computer-science and programming. (2) Some one must create the high-level languages that will support modeling and task-oriented description of engineering methodologies. Let us look at each one a little more closely.

The first implication is correct if programming in, say, C was the objective. But, "every computer program is a model, hatched in the mind, of a real or mental process" (Abelson et al., 1985) and through each program one must express in a disciplined and formal manner, poorly-understood and sloppily-formulated ideas. Such discipline is an essential ingredient of good engineering and therein lies the value of computer-science for a chemical engineer. In building such discipline one learns the concepts of; building abstractions with data or procedures, the interchangability of data and procedures, recursion, or iteration, modularity, reasoning in logic, representation of data, controlling complexity, and many others all of which play a central role in every engineering activity beyond programming itself. All concepts, taught in such a computer-science course, help strengthen the dual pivot of engineering work; problem formulation and problem-solving. Therefore, current curricula emphasizing the syntax of particular language or giving the students an uncritical exposure to the use of spreadsheets, databases, graphics, and prepackaged numerical algorithms, provide a short-term skill and not long-term sustenance in a computer-driven professional world.

The second implication has some interesting consequences. Chemical engineers and not computer-scientists must drive the process for the creation of chemical engineering-oriented modeling and task-oriented languages. They are the ones who know what these languages ought to be. But, chemical engineers do not, normally, possess the education and sophistication to create formal languages. Collaboration with computer scientists is not only necessary, but also desirable; being forced to formalize our ideas on modeling and engineering methodologies, we may learn a better process to teach and practice modeling and problem-solving.

References

Abelson, H. and G. Sussman (1985). *Structure and Interpretation of Computer Programs*, Cambridge, MA, MIT Press.

Barton, P.I. (1992). *The Modeling and simulation of combined discrete/continuous processes*. Ph.D. thesis, University of London.

Dahl, O.-J. and K. Nygaard (1967). *SIMULA 67 Common Base Proposal*, Norwegian Computing Center Document, Olso.

Department of Defense (1960). *COBOL, Initial Specifications for a Common Business Oriented Language*.

Douglas, J. M. (1988). *Conceptual Design of Chemical Processes*, McGraw-Hill.

Goos, G. and J. Hartmanis (eds.) (1983). *The Programming Language Ada Reference Manual*, American National Standards Institute, ANSI/MIL-STD-1815A-1983, Lecture Notes in Computer Science 155, Springer-Verlag, New York.

Han, C. (1993). *Human-Aided, Computer-Based Design Paradigm: The Automation of Conceptual Process Design*, Ph.D. Thesis, Department of Chemical Engineering, Massachusetts Institute of Technology, Cambridge, Massachusetts.

Han, C., C. Nagel, and G. Stephanopoulos (1995a). Modeling Languages: Declarative and Imperative Descriptions of Chemical Reactions and Processing Systems, *Paradigms of Intelligent Systems in Process Engineering*, Academic Press.

Han, C., J. Douglas, and G. Stephanopoulos (1995b). Automation in Design: The Conceptual Synthesis of Chemical Processing Schemes, in *Paradigms of Intelligent Systems in Process Engineering*. (G. Stephanopoulos and C. Han, Eds.), Academic Press.

Han, C., J. Douglas, and G. Stephanopoulos (1995c). Agent-based approach to a design support system for the synthesis of continuous chemical processes, ESCAPE-95.

Kay, A. (1969). *The Reactive Engine*, Ph.D. Thesis, University of Utah, September.

Kröner, A., P. Holl, W. Marquardt and E. D. Gilles (1990). DIVA - An open architecture for dynamic process simulation, *Comput. Chem. Engng*, **14**, 1289-1295.

Mather, D.G. and S. V. Waite (Eds.) (1971). *BASIC*, 6th ed. University Press of New England, Hanover, NH.

Meyer, B. (1990). *Introduction to the Theory of Programming Languages*, Prentice Hall Inc., Englewood Cliffs, NJ 07632.

Naur, P. (ed.) (1960). Report on the Algorithmic Language ALGOL 60, *Commun. ACM*, **3**(5), 299-314.

Nilsson, B., *Object-Oriented Modeling of Chemical Processes*, Doctoral thesis, Department of Automatic Control, Lund Institute of Technology (1993).

Pantelides, C.C. and P. Barton (1993). Equation-oriented dynamic simulation: current status and future perspectives, *Comp. Chem. Engng.*, **17S**, 263-285.

Piela, P. C., T. G. Epperly, K. M. Westerberg, and A. W. Westerberg, ASCEND: An Object-Oriented Computer Environment for Modeling and Analysis: The Modeling Language, *Comp. Chem. Engng*, **15**(1) p.53 (1991).

Roussel, P. (1975). *PROLOG: Manual de Reference et D'utilisation, Research Report*, Artificial Intelligence Group, Univ. of Aix-Marseille, Luming, France.

Sebesta, R. W. (1993). *Concepts of Programming Languages*, 2nd ed., The Benjamin/Cummings Publishing Co.

Sørlie, C. F., *A Computer Environment for Process Modeling*, Ph.D. thesis, Department of Chemical Engineering, The Norwegian Institute of Technology (1990).

Stephanopoulos, G., G. Henning and H. Leone (1990a). MODEL.LA.: A Modeling Language for Process Engineering. Part I: The Formal Framework, *Comput. Chem. Engng.*, **14**, 813-846.

Stephanopoulos, G., G. Henning and H. Leone (1990b). MODEL.LA.: A Modeling Language for Process Engineering. Part I: The Formal Framework, *Comput. Chem. Engng.*, **14**, 847-869.

Stroustrup, B. (1983). Adding Classes to C: An Exercise in Language Evolution, *Software-Practice and Experience*, **13**.

Wirth, N. (1971). The Programming Language Pascal, *Acta Informatica*, **1**(1), 35-63.

Wexelblat, R. L. (ed.) (1981). *History of Programming Languages*, Academic Press, New York.

Woods, E. A., *The Hybrid Phenomena Theory: A framework Integrating Structural Descriptions with State Space Modeling and Simulation*, Ph.D. thesis, Department of Engineering Cybernetics, The Norwegian Institute of Technology (1993).

Zuse, K. (1972). Der Plankalkül, Manuscript prepared in 1945, published in *Berichte der Gesellschaft fur Mathematik und Datenverarbeitung*, No. 63 (Bonn, 1972); Part 3, 285 pp. English translation of all but pp. 176-196 in No. 106 (Bonn, 1976), pp. 42-244.

VISUALIZATION

Andrew N. Hrymak and Patricia Monger
McMaster University
Hamilton, Ontario L8S 4L7 Canada

Abstract

Visualization is the use of computer graphics for scientific and engineering data analysis. This chapter highlights the current directions for scientific visualization software on personal computer and workstation platforms.

Introduction

Scientific visualization does not require a computer, as demonstrated by many examples in Edward Tufte's milestone book, *The Visual Display of Quantitative Information*. However, since the late 1980s, the term has come to be associated with the use of computer graphics to represent relationships in scientific data, especially for systems with very large quantities of information.

The use of computers as visualization tools began with techniques such as contour plotting, line graphics, and 2-D image display. Two factors combined to give this field much more prominence: an increase in the amount of data (due to computer simulation, image capture, etc.) which has overwhelmed the user's ability to process the information, and technological advances in computers that have made much more sophisticated data representation possible and affordable (Schmitz, 1994).

Early scientific graphics applications used *de facto* standard line graphics software libraries, almost all written in Fortran. The most widespread of these was the PLOT subroutine library written by Calcomp to drive their pen plotters. This same library was then adapted to support other hardcopy devices, e.g. electrostatic plotters, dot matrix printers, and graphics terminal devices. The first graphics terminal to come into widespread use was the Tektronix 4010. Almost all subsequent line graphics display systems supported Tektronix protocols, and even on modern systems, e.g. X window terminals, one finds client software that accepts Tektronix graphical escape sequences.

2-D data were visualized using contour or vector field representations, which relied on the line graphics subroutine libraries. Image display routines were tailored to the specific display hardware, so that use of this technique required writing interfaces to the graphics screen or hardcopy device drivers.

As recently as 15 years ago, the use of color hardcopy in data analysis was rare, and techniques such as animation virtually unknown. This began to change in the mid 1980s, when researchers gained access to higher performance computers and supercomputers, with concurrent improvements in data gathering instrumentation. This presented the researchers with the challenge of finding means to interpret these huge quantities of data. This challenge was brought to public attention with the 1987 report by the National Science Foundation, entitled *Visualization in Scientific Computing*. At about the same time, computer hardware supporting megapixel bitmapped displays in 8-bit color became widely available. Higher-end graphics workstations with specialized graphics hardware, while still largely out of reach of individual researchers, began to appear as departmental or group resources.

Software development accelerated with the advent of graphics standards and vendor-supplied libraries for simplifying the use of more sophisticated graphics techniques. The early standardization efforts developed on three fronts: attempts to define a standard plot file format that hardware vendors could then support for their display technologies, graphics systems standards to provide an interface layer between the hardware drivers and the application code at the level of graphics "primitives," and higher-level libraries that use the primitives and produce standard plot files. CGM, the computer graphics metafile, is the standard file format surviving from these efforts. GKS, Graphical Kernel System, remained as the 2-D graphics primitives system. No clear winner emerged from the attempts to standardize the higher level libraries, but examples of systems that developed from these efforts are NCAR, DISSPLA, and PLOT10.

As graphics technology developed, the need for standardization became more apparent. The development of laser printers and color hardcopy devices paralleled the spread of PostScript. The advent of bitmapped terminals brought the X Windowing System, and as 3-D graphics became more feasible, the standardization impetus that created GKS led to the effort to define PHIGS. At the same time, developers of the 2-D high level libraries extended these systems to provide support for 3-D data representation (e.g. hidden or ruled-surface plots, 3-D line plots), as well as animation.

The acceptance of standardized systems in 3-D display was hampered by the fact that the vendors provided their own libraries that were in general more efficient than, and at least as easy to use as, the standard libraries. For higher end visualization applications, the fact that only a very small number of vendors had hardware to support those applications meant that portability was not a driving consideration. The dramatic drop in the cost of the hardware also made software developers (especially those producing freeware) less hesitant to target the software to a particular hardware platform. This is less of a problem than appears at first glance, because of the trend for hardware vendors to license their software for other platforms, and because of the standardization of windowing systems on workstations. Most modern visualization systems ultimately rely on X window protocols, which are portable.

The next set of improvements in visualization built upon the development of windowing systems and graphical user interfaces to create data exploration systems that attempt to simplify the visualization process. These systems are founded on two ideas; that the focus of the visualization process should be on the flow of data through the stages of the process, and that the visualization program itself can be constructed using a visual programming model. The researcher selects the appropriate visualization modules, drags the module icons onto a workspace area, selects the data inputs, links the modules together in the appropriate order, accesses the data, and waits for the process to display. An example of a visual programming system, Explorer from Silicon Graphics Inc., is shown in Figure 1.

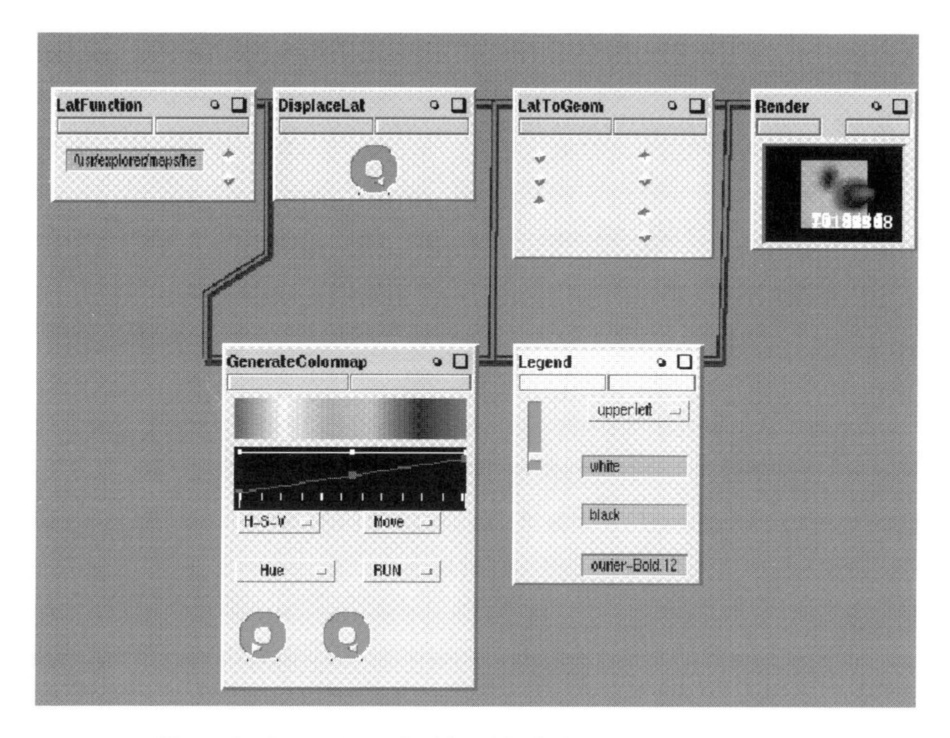

Figure 1. Screen image building blocks from the Explorer package (Silicon Graphics Inc.).

Of course, more traditional scripting languages for visualization also exist, e.g. PV-Wave (Precision Visuals, Inc.). Other visualization packages are customized for the needs of a particular class of problems, for example, Tecplot™ (Amtec Engineering, Inc.) or Sterling SSV™ (Sterling Federal Systems, Inc.) which have capabilities for visualizing data from finite difference and finite element calculations in computational fluid dynamics. A key feature within advanced visualization packages are cutting tools which allow the user to cut sections away from the object, model or graph to see inside the outer surface. A variant on arbitrary cutting planes, are isosurface plots of variables such as temperature, streamlines or stress. For an interesting example of visualization of a complex flow see Göde and Cuénod (1990). Some packages, such as AVS (Advanced Visual Systems), Explorer (Silicon Graphics) and the Visualization Data Explorer (IBM), are general purpose packages which provide visualization "environments" to allow animation and visualization of machinery, as well as derived data from computer simulations. The visualization software packages cited above are not an exhaustive list and are given as examples from classes of visualization tools. Interested readers should consult hardware and software vendors for more information.

Most recently, developers have been experimenting with ways to display data of more than 3 dimensions, using non-visual cues such as sound, and more flexible and intuitive ways of interacting with the data, e.g. with stereo viewers or virtual reality systems wherein the researcher uses a head-mounted display or other motion-sensitive detector to "move" the data. The goal of visualization software development is to integrate it with the data generating process, so that the researcher can explore the data interactively and use that information to modify the numerical computations.

Script language visualization systems increasingly incorporate numerical functions. Similarly, numerical and symbolic manipulation programs have increasingly sophisticated graphics capabilities.

Graphics systems on personal computers have not traditionally been tailored to scientific visualization, because these machines did not have the CPU power for complex calculations or complex graphics. Numerical and statistical systems with 3D graphics support, as well as image display and animation programs now run on personal microcomputer platforms. A number of the more specialized visualization packages for the workstation platform have versions, sometimes for limited problem classes, for personal computers.

The term scientific visualization has evolved to refer exclusively to more complex scientific graphics: animation and/or exploration of data in 3 or more dimensions. The discussion of visualization in this chapter is therefore also focussed on higher-end visualization requirements.

The Visualization Process

With all the impressive increase in hardware and software capabilities, it must be acknowledged that the techniques of data visualization have become more difficult for the user. In the past, most researchers wrote their own programs to call the subroutine libraries for line graphics. Putting together a program for interactive manipulation of 3-D fluid flows is quite another matter! Furthermore, software developers also find it difficult to create a generic program that will do this kind of visualization for arbitrary data types, and of course programs tailored to the users' specific data requirements are costly. The result is that the user may be faced a vast array of generic tools for gridding, color map manipulation, geometrical mapping, etc., but no clear idea of how the data fits into this toolset.

Two things aid the user in getting started with visualization. One is that the most common visualization tasks have been encapsulated into modules, for example: 3D plotting and animation in a projection calculation module, accumulation of single frames with playback into a movie module, or object rotation and zooming within a rendering module. The second helpful development is the explosion of freeware, user contributed modules, and user newsgroups for all visualization systems.

The visualization process is well represented by the dataflow programming model. The result at each stage depends strictly on the data input, not on the results of a previous execution, so that the process can be envisioned as data traveling in a pipeline through various processing steps.

The basic steps are:

- Normalization: Scale the data to the values expected by the display system (e.g. 0-255 for 8 bit display devices). This step also includes any scaling required to better analyze trends in the data (e.g. converting to logarithmic values, filtering).

- Mapping: Define the display image according to the normalized data and the computational grid - regrid from computational space to display space.
- Rendering: Add color, lighting, shading, projection information, display the image and store it in display memory.
- Interaction and/or animation using the rendered images.

The rendering step is often very computationally intensive, involving calculating projection information, calculating and removing occluded surfaces, and calculating the effect of light sources on the object's appearance. Despite the complexity of this step, it is of less concern for the user because it does not involve the scientific data, but rather the computer graphics image. Most visualization systems provide modules to handle rendering as a "black box" from the user's standpoint.

The mapping step is the point at which the computer graphics is added to portray the scientific data. The user will find this the most difficult step, because it requires an understanding of both the science and the graphics. First, one must decide what graphical representation best conveys the information about the data. For example, is a 3-D velocity field best rendered as a stream of particles in a 3-D coordinate space, as vectors on a plane, or as streamlines? What color scheme is best to display data values? Once these issues have been decided, the user must determine how to use the visualization system to convert the data into the desired display representation. The 3-D velocity field, for example, may not be defined over a regular 3-D matrix, but regridding will be required in order that the data be converted to graphics primitives for the renderer. Consider the set of mapping steps for the visualization of the energy conservation equation for a heat transfer problem. The heat flow is evaluated over a regular grid, and then output as a one-dimensional array, or lattice. This is then converted to a geometrical representation as expected by the renderer. A colormap is also attached when the geometrical representation is generated, as is a text legend for the color bar.

A crucial requirement in the visualization process is appropriate hardware. The rendering and mapping steps can both require significant CPU power and memory. Animation is very demanding of disk space. Higher end graphics workstations, supporting display and interaction of 3D solids, are distinguished from general purpose workstations by having options such as 24 or more bitplanes (memory specific to the storage of display frames), extra RAM for storing distance information (often called a z-buffer), and specialized ASICs for encoding graphical primitive instructions in hardware. Output of color images or animation will also require the appropriate hardware to interface to the computer (color printers, video output, CD-ROM, etc.).

Visualization Resources

In compiling a list of useful guides in visualization, one must cite in particular the US National Center for Supercomputer Applications. Modules for visualization software systems on PCs, Macintoshes, and workstations are all available here, as are programs developed in-house and made available to the research community. More information as well as the tools themselves are available by anonymous ftp from *ftp.ncsa.uiuc.edu*.

Usenet groups for computer graphics, visualization, and specific visualization software also exist. Users of Mosaic or Netscape and the World Wide Web can make use of the CUI World Wide Web Catalog. A keyword search of the catalog for "visualization" yields a number of entries. In particular the user should consult the NASA Ames annotation bibliography of scientific visualization URLs for visualization examples and references to freeware.

The usenet group *comp.graphics* publishes a biweekly resource listing for computer graphics systems that includes visualization software. This resource listing is also available at usenet archive sites, for example *rtfm.mit.edu*, and includes information on public domain and commercial visualization software for all hardware platforms.

Finally, as more of the technical tasks in visualization get absorbed into automated procedures, what the researcher will really need to know is how to make an effective visual presentation; e.g. what colors to use, how much information can be conveyed without cluttering the visualization, etc. For insight into that process, we would strongly recommend the books by Edward Tufte mentioned previously, *The Visual Display of Quantitative Information*, and the second book, *Envisioning Information*. These books make extensive use of examples of good and bad visualizations, which, though static, provide valuable lessons that can be extended to animations as well.

The collection of papers edited by Cunningham and Hubbold (1992) provides an interesting cross-section of experiences in education with various visualization tools. While there are many exciting possibilities with multimedia and even hypermedia, certain common problems recur in the use of visualization in education. The general availability of reasonable hardware with adequate graphics performance cannot be taken for granted, especially in developing countries. More importantly, the plethora of specialized graphics platforms means that the learning module must be portable for any kind of extended life in the fast changing world of computer hardware. The time needed to prepare an instructional module is considerable and there is a lack of authoring tools that are flexible and allow portability of the module. The instructional module must be accessible by students who are at different levels of ability with the scientific subject at hand so that there is help for the novice and depth for the student who wishes to try more difficult cases. The use of visualization in education is definitely limited in the traditional lecture mode, but is particularly well suited for self-study and enrichment exercises.

An Example of Visualization Techniques - Flow in a Driven Cavity

As an illustration of a simple method of getting useful graphical information for a problem of interest in chemical engineering, we take the program *cavpit*, part of the MINPACK-2 test problem collection from the Army High Performance Computing Research Center and Argonne National Laboratory (Brett et al., 1992). Program *cavpit* solves the steady-state Navier-Stokes equation for the 2-D driven cavity flow problem. The modified program uses MATLAB to display the streamline contours, velocity contours, and velocity vectors at several values of the Reynolds number (Kus, 1994).

The program makes use of the MATLAB sockets toolkit written by Barry Smith. The toolkit uses Internet protocols to establish socket-based communication between an application program and a MATLAB session. The toolkit provides an object library of routines for passing a matrix (the streamline values) to MATLAB, and a MEX function that waits for the matrix and then returns it to the MATLAB macro. The macro then calculates the vorticity contours and the velocity vectors from the streamlines, and uses the MATLAB graphics functions to plot these quantities. The macro for calculating and plotting the data is given in Figure 2. Plots of vorticity, streamlines and velocity vectors for Reynolds number 400 are shown in Figure 3 (a - c).

```
function cav(portnumber)
%
% Waits at socket portnumber 5005 until a matrix arrives
%
% This file is for the cavity program
%
%

if ( nargin == 0 ) portnumber = 5001; end;
for ii=0:10000
 m = receive(portnumber);
 if ( sum(size(m)) ~= 0 )
  nvar = size(m,1)
  n = nvar-1
  nx = sqrt(n)
  ny = n/nx
  for j = 1: nx
   for i = 1:ny
    psi(i+1,j+1) = m((i-1)*nx+j,1) ;
   end
  end
  for j = 1:nx+2
   psi(1,j) = 0.0 ;
   psi(ny+2,j) = 0.0 ;
  end
  for i = 1:ny+2
   psi(i,1) = 0.0 ;
   psi(i,nx+2) = 0.0 ;
  end

  xc = m(nvar);
  tc = num2str(xc)

%    Streamlines plots

  V = [ -.11 -.1 -.08 -.06 -.04 -.02 -.01 ...
    -1.0e-5 1.0e-6 1.0e-5 1.0e-4 1.0e-3 2.0e-3 ] ;

  figure(1);
```

```
    contour(psi,V);
    title([ 'Streamline Contours for Flow in a Driven Cavity for Reynolds Number R  =  ',tc]);

%   Equivorticity plots

    hx = 1/(nx+1); hy = 1/(ny+1);
    for j = 1:nx-1
     for i = 1:ny-1
      v(i+1,j+1) = -(psi(i+2,j+1)-2*psi(i+1,j+1)+psi(i,j+1))/(hx*hx) ...
                 -(psi(i+1,j+2)-2*psi(i+1,j+1)+psi(i+1,j))/(hy*hy) ;
     end
    end
    V = [ -5.0 -4.0 -3.0 -2.0 -1.0 0.0 1.0 2.0 3.0 4.0 5.0 ] ;

    figure(2);
    contour(v,V);
    title([ 'Vorticity Contours for Flow in a Driven Cavity for Reynolds Number R  =  ',tc]);
    for j = 1:nx
     x(j) = j*hx;
     y(j) = j*hy;
     for i = 1:ny
      vx(i,j) = (psi(i,j+1)-psi(i,j))/hy ;
      vy(i,j) = -(psi(i+1,j)-psi(i,j))/hx ;
     end
    end
%
%   transpose the x and y matrices to take account of matlab's coordinate
%   system conventions for the quiver function
    vx = vx';
    vy = vy';

    figure(3);
    quiver(x,y,vx,vy,4);
    title([ 'Velocity Vector for Flow in a Driven Cavity for Reynolds Number R  =  ',tc]);
  else
   break;
  end;
end;
```

Figure 2. Matlab code for generating graphics for driven cavity problem.

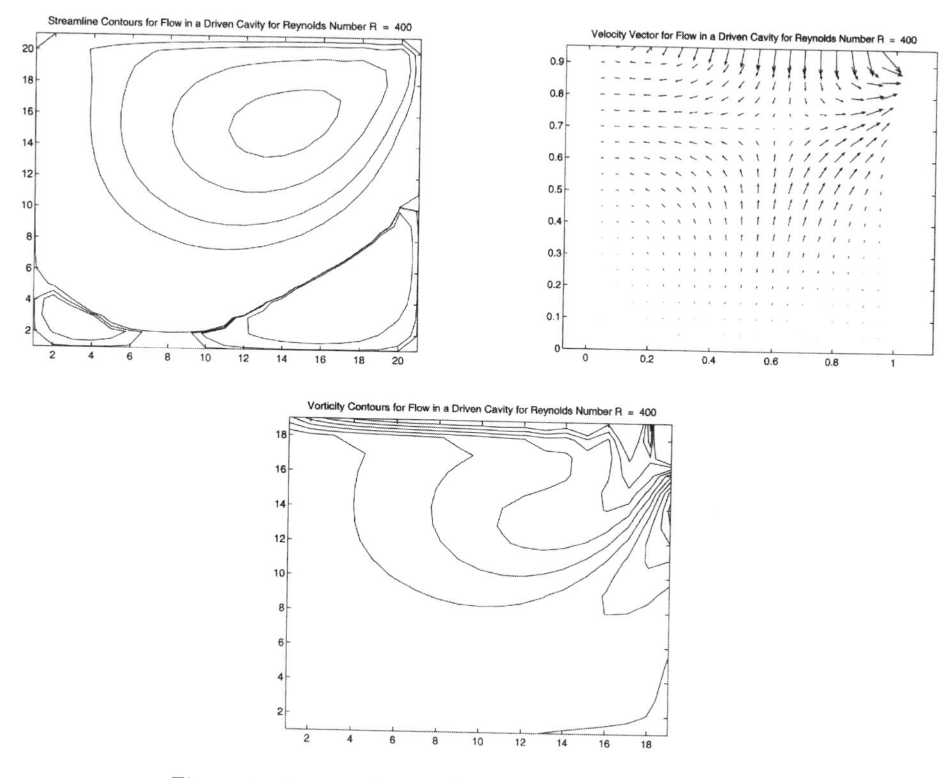

Figure 3. Images of streamlines, velocities and vorticity for driven cavity problem.

MATLAB is then an example of a numerical computation program that has incorporated graphical capabilities into its language. The actual plotting is done with a few simple commands; *contour* for the streamline and vorticity contours, and *quiver* to generate the velocity vector plots, plus commands for labeling the axes and putting titles on the graphs. Simple 3-D plots are also done by invoking only a few commands; for regridding the domain over which the function is defined if needed, and for display the data or function as a scatter plot, wire frame plot, or surface plot.

CACHE Activities

Elements of visualization are seen in a number of CACHE Corp. products for use in undergraduate education. An effective use of visualization is seen in applications where there is a large amount of data generated at one time or over a number of simulations which must be compared. The level of interaction between the user and the software is important to effectively use graphics to aid the student in understanding a concept.

Interactive computer modules for chemical engineering instruction by Fogler and coworkers (1993) is an example of interactive use of graphics to aid the student to explore the use of simulation models. Collections of modules exist for material and energy balances, fluids and transport, separations and reaction kinetics.

Visualization and graphics can also be used to provide a simulation of an industrial process experiment. Students can work with realistic models of existing industrial processes to develop strategies for operation and optimization through the use of dynamic simulation. The Purdue-Industry Simulation Modules (Squires et al.,1991) provide modules, developed through the support and collaboration of industrial sponsors, for hydrodesulfurization, catalytic reforming, emulsion polymerization, methyl acetate production and process heat transfer.

The Chemical Reactor Design Tool (see p. 121) and the Transport Module (see p. 129) both make extensive use of graphics. The data entry is done in X-windows, which provides a screen-oriented input. The power of these educational tools, however, is in the graphical output. Indeed, the programs are designed so that the graphs are the only output the user studies; a text file is created only for trouble shooting and special purposes. Line plots, 2D contour plots, and 3D perspective views can be chosen for the solution variables as well as derivatives, such as the difference between the catalyst temperature and fluid temperature, the diffusion term, or in the case of the Navier-Stokes equation, the vorticity.

Thermodynamic Visualization

West (1992) argues that a visual thinking approach to problem solving is particularly useful in fields where large amounts of information have underlying physical relationships. An example of such a discipline is thermodynamics. Professor Kenneth Jolls and co-workers at Iowa State University have developed visualization tools tailored to the needs of thermodynamics problems, especially those involving equations of state. Thermodynamic concepts are difficult for the average student to assimilate on the first pass and learning is made more difficult with the complex mathematical formulae which are necessary to describe vapor-liquid or reaction equilibria. Jolls et al. (1991) effectively show how thermodynamic information can be captured in graphical images in a package called *Equations of State*, which is available and used by a number of universities. Modern computing environments allow one to tackle more complex multicomponent equilibrium problems using color and shading techniques to capture added dimensionality which is lost in wireframe drawings. As the number of species goes beyond three, a decision must be made on what specific features need to be rendered since it becomes counter-productive to superimpose too much information on a diagram. Binary and ternary systems, common to problems in the undergraduate thermodynamics course, are especially amenable to visualization.

Thermodynamic visualization can be used as a tool for studying more advanced phenomena such as phase stability (Jolls and Butterbaugh, 1992) or process unit modeling (Jolls et al., 1994).

Conclusions

Visualization using computer graphics systems has undergone a remarkable change in the last thirty years. The student can use a number of packages which have many graphics modules already included to facilitate the analysis of data. Advanced visualization packages allow the use of more complex data representations on both personal computer and workstation platforms. A major need, for educational uses, remains authoring systems that provide flexible use of graphics while not burying the underlying scientific concepts that are being addressed. For research uses, visualization is a necessary component for large data set analysis and is addressed by a number of software vendors on various workstation platforms. The frontiers of visualization have pushed beyond what can be expressed as a simple two-dimensional monochrome line images to include interaction modes for color, sound, motion and touch. As data sets become larger, choices need to be made about the manner in which to represent (or interact with) the data which means that users cannot avoid learning about the opportunities, as well as the limitations, of the visualization process.

References

Brett M. Averick, Richard G. Carter, Jorge J. More, and Guo-Liang Xue, *The MINPACK-2 Test Problem Collection*, Argonne National Laboratory, June 1992 (preprint MCS-P153-0692)

Cunningham, S. and R.J. Hubbold (Eds.)., *Interactive Learning Through Visualization*, Springer-Verlag, New York, 1992.

Fogler, H.S., and S.M. Montgomery (1993). Interactive Computer Modules for Chemical Engineering Instruction. *CACHE News*. Fall, 1-5.

Göde, E. and R. Cuénod (1990). Numerical Simulation of Flow in a Hydraulic Turbine. *Chemical Engineering Progress*, **86**(8), 35-41.

Jolls, K.R., M.C. Schmitz, and D.C. Coy (1991). Seeing is believing: a new look at an old subject. *The Chemical Engineer*. 30 May, 42-16.

Jolls, K.R. and J.L. Butterbaugh (1992). Confirming Thermodynamic Stability. *Chemical Engineering Education*, **26**(3) 124-129.

Jolls, K.R., M. Nelson, and D. Lumba (1994) Teaching Staged-Process Design Through Interactive Computer Graphics, **28**(2) 110-115.

Kus, F., Personal Communication, McMaster University, 1994.

Schmitz, B. (1994). Engineering Visualization: Not Just Pretty Pictures. *Computer-Aided Engineering*, March, 44-50.

Squires, R.G. et al (1991). Purdue-Industry Computer Simulation Modules: The Amoco Resid Hydrotreater Process. *Chemical Engineering Education*, Spring, 98-101.

Tufte, Edward R., *The Visual Display of Quantitative Information*, Graphics Press, Cheshire, Connecticut, 1983

West, T. (1992). A Return to Visual Thinking. *Computer Graphics World*, November, 116-115.

2001

Brice Carnahan
University of Michigan
Ann Arbor, MI 49109

Abstract

1996 is a special year in the history of computing. It marks the 50th anniversary of the public announcement of the ENIAC, the first general-purpose electronic digital computer, the 25th anniversary of the first commercial microprocessor, the Intel 4004, and the 15th anniversary of the first IBM PC, whose basic structure and open architecture set the standard for subsequent personal computer development.

Most technological developments of consequence pass through the familiar S curve of slow initial growth, then a period of rapid acceleration, and finally another slow-growth phase of important, but marginal, improvement. For the digital computer, the slow growth period lasted about 15 years. The acceleration phase began with the introduction of the transistor in the late 1950s, received additional thrust with each new transforming technology (time-sharing, integrated circuits, real-time minicomputers, networking, the microprocessor, interactive graphical operating systems and programming environments, supercomputers and parallel machines, high speed communications, the Internet, the World Wide Web, etc.), and continues unabated to this day.

We describe past developments and current trends in the areas of computing hardware, communication, and software and make a few predictions about the state of computing in general and in chemical engineering education in the year 2001.

Introduction

Only "yesterday" those two most prominent dates of 20th century fiction, Orwell's *1984* and Clark's *2001*, seemed far into the future. Today, 1984 (do you remember your IBM PC AT or Macintosh 128?) is long past, and Big Brother is (almost) nowhere in sight.

2001 looms over the horizon, a mere half-decade away. And Clark's HAL, malevolent and almost infallible, also seems nowhere in sight. But, of course, he won't be "born" until 1997, so there is still time! One certainty is that HAL will look nothing like Kubric's 1969 vision of him. No astronaut in full space suit will float among HAL's innards. More likely, all of HAL's parallel processing capability will be encompassed in many microprocessors, housed in a "box" of quite modest size.

Neither Orwell nor Clark should be faulted for his vision. Predictions about the future of technology have not been particularly accurate. Verne was right about moon travel, and in his 1863 novel *Paris in the Twentieth Century* correctly imagined the automobile, electric lights, and the fax machine, but was wrong about many others. Edward Tenner, in his recent *Why Things Bite Back,* argues convincingly that most predictions about technological developments turn out to be either off the mark or just dead wrong.

Two of my favorite predictions, both from Vannevar Bush (inventor of the analog computer and World War II production czar), show how long-term right and short-term wrong the *same* person can be about the *same* subject:

> " ... The MEMEX will be for individual use, about the size of a desk with display and keyboard that will allow quick reference to private records, journal articles and newspapers, and perform calculations ..."
>
> in *As We May Think* (1945)
>
> " ... Will we soon have a personal machine for our own use? Unfortunately not!"
>
> in *MEMEX Revisited* (1967)

Incidentally, Bush, a consultant for IBM, is reputed to have told the IBM Board in 1955 that no more than 100 IBM 650s should be built, since they could do all the computing that the World needed done!

Even short-term prediction is fraught with peril. Few would have predicted in 1993 that a hyperlinked World Wide Web would transform computing from a machine-centered view to a network-centered one almost overnight. Despite such revolutionary surprises, technology instills a sense of unfolding promise, making prediction almost impossible to resist. Hence my title, *2001* (granted, I'm not looking all *that* far into the future!).

Hardware

Hennessy and Patterson, in their superb 1995 text *Computer Architecture* (a must-read for anyone interested in low-level computer structure and functionality), briefly outline the history of computer hardware and architectural development, noting that a 1995 desktop PC (costing say $4,000 in 1995 dollars) had more main and disk memory, and better performance, than a million-dollar (1965 dollars) 1965 main-frame.

Processors and Memory

During the twenty five years following ENIAC, actual computer "performance" (Hennessy and Patterson), a rough measure of ability to process a comprehensive mixed set of test programs, improved by about 25% per year for main-frames, year in and year out. That rate improved to between 25% and 30% during the 1970s, primarily because of the introduction of minicomputers into the hardware mix. Since 1980, following widespread production of microprocessors, the overall performance growth rate has climbed to roughly 35% per year. The Intel CISC (complex instruction set computer) microprocessor family, used in perhaps 85 percent of the approximately 75 million machines sold worldwide in 1996, is primarily responsible for this acceleration in average performance.

RISC (reduced instruction set computer) processors were first introduced in 1984. Since then, overall performance of computers incorporating this class of processors has improved at about a 50% annual rate. Although used in only 10 to 15 percent of computers in the current market, RISC processors play a significant role in engineering computing, since they are used primarily in high-performance engineering workstations and parallel machines (also in Power Macintoshes).

Figure 1 (data from URL *infopad.eecs.Berkeley.edu/CIC*) shows that raw peak processor capability (rather than observed average performance for assembled computers), estimated roughly as proportional to the product of processor transistor count and clock speed, has doubled about every eighteen months over a 20 year period for the Intel family of CISC microprocessors. The Power PC and DEC Alpha RISC microprocessor families show even shorter peak performance doubling times of 12 to 14 months.

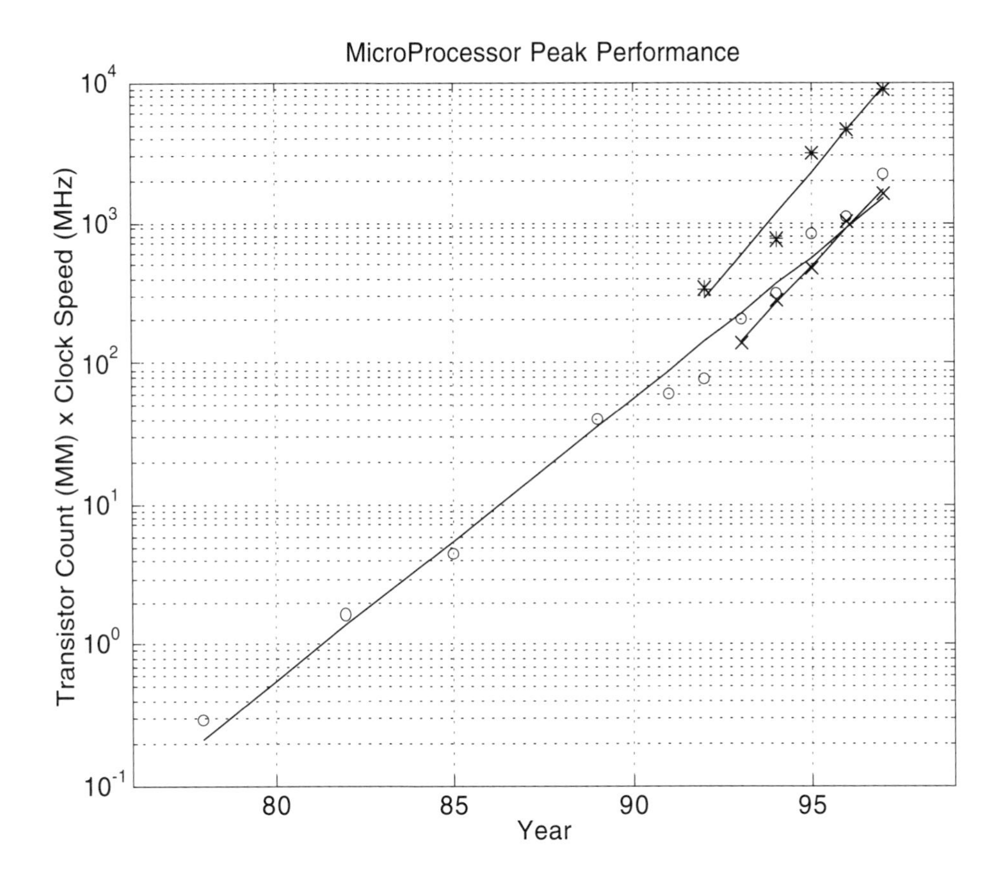

Figure 1. Peak performance estimated from the product of transistor count and clock speed for Intel (o), Power PC (x) and DEC Alpha (*) microprocessors.

Three current trends in chip and magnetic disk technologies are (Hennessy and Patterson):

1. Transistor density for integrated circuit logic (microprocessors) is increasing at about 50% per year; growth rates in transistor count per chip are in the range 60% - 80% per year.

2. Dynamic (main memory) RAM bit density is increasing at about 60% per year, resulting both from higher transistor densities and design improvements requiring fewer transistors per memory cell. The result is RAM of much larger capacity and much lower cost than was available even two or three years ago. 16 Mb chips are now standard, and 32 Mb and 64 Mb chips are widely available. [NEC has recently reported development of an experimental 4 Gbit RAM chip (256 times the capacity of the current standard chip), with production planned for around 2005.]

3. Prior to 1990, magnetic disk densities increased at about 25% per year. Since 1990, the rate has accelerated significantly to about 50% per year.

What about costs? Here, the impact of the technical developments alluded to in the preceding paragraphs, improved manufacturing performance (yield of good chips), and economies of scale resulting from volume production and commoditization of microcomputers, have led to incredible reductions in the cost/performance ratio.

Figure 2 shows a retail price curve for 16 Mb DRAM (dynamic RAM) chips between 1993 and 1996. The cost reduction by a factor of 8 to 10 over the 3 to 4 year life cycle of this DRAM chip is typical of that for prior standard chip sizes as well. With the compounding of cost reductions for standard chips, the cost per megabyte of DRAM has dropped incredibly from over $17,500 in 1977 to about $5 in 1996 (both in 1996 dollars). Thus real RAM costs have been reduced by a factor of 3500 in about 20 years!

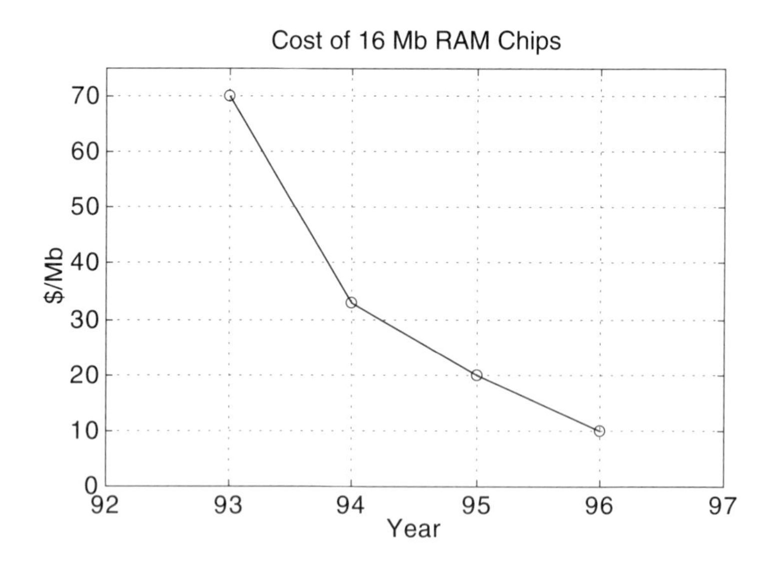

Figure 2. Cost of 16 Mb RAM chips - 8 required for 16 MB (data from Hennessy and Patterson).

What are the prospects for continuation of the performance trends for processor and memory chips described above? Very good, at least until 2001 and probably for several years beyond. Processor chips with a billion transistors and memory chips with many billions of transistors are certain to appear before the end of the first decade of the next millennium.

However, physical limitations of current photolithographic processes will force major changes in chip production technology before then. As shown in Fig.3, the minimum feature size (e.g., width of "wire" traces) of chip-making photolithographic processes on silicon has fallen from 12 microns in 1970 to 3.5 microns in 1980 to 0.8 microns in 1990 to 0.3 microns in 1996; at least two new 0.25 micron fabrication facilities will be in full production by mid-1997. The apparent limit of this technology, 0.365 microns, the wavelength of the ultraviolet light currently used for photoresist exposure, will in fact be reduced by more than half, to 0.1 microns, with lasers emitting 0.248 micron or 0.193 micron ultraviolet light and clever use of phase-shifted interference lithography. However, barring some even more clever (and unlikely) improvement within the next very few years, the minimum feature size of photolithographic technology is probably about 0.1 microns, likely to be in common use during the first half of the next decade (Stix).

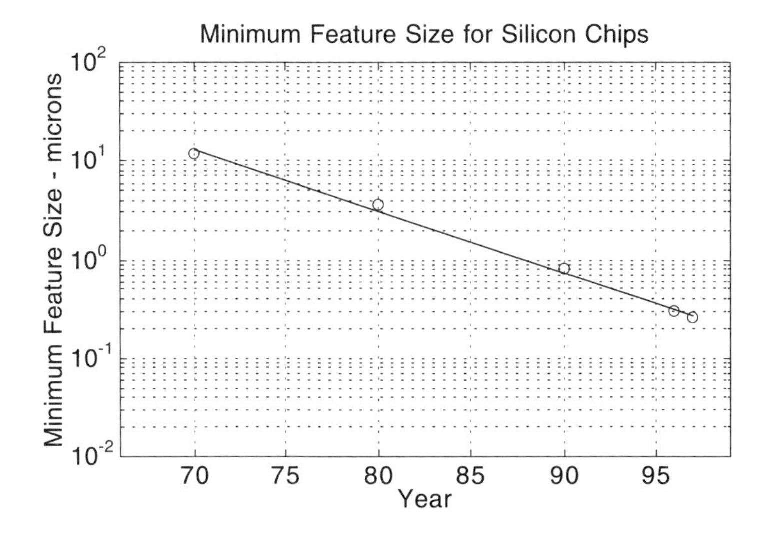

Figure 3. Minimum feature size for silicon chips.

Thus new technologies using photolithography with much shorter wavelengths or, more likely, completely new technologies not involving the interplay between light and photoreactive polymer resists, will be required. The new technologies may involve very-difficult-to-focus x-ray or rather slow electron beam processing (used by NEC in development of the experimental 4 Gbit DRAM chip mentioned earlier).

IBM researchers have reported creation of microstructures on silicon as small as 30 nanometers, and many university and corporate researchers worldwide are working on development of the ultimate smallest silicon component, the single-electron transistor, for which the minimum feature size is about 7 nanometers. Thus, for transistor-based micro-electronic silicon chips there *will* be an ultimate maximum transistor density and minimum feature size, whatever chip fabrication technology supplants photolithography. If the trend toward smaller feature size illustrated in Figure 3 were to continue unabated (improbable), that ultimate limit, two orders of magnitude smaller than 0.1 micron, would be reached sometime around 2025.

From the numbers cited in the previous paragraphs, it appears likely that the rapid increases in transistor density (hence computer performance) we have seen in past decades will continue for at least one more decade. Whether the pace continues for the decade after that will depend as much on the development of cost-effective manufacturing techniques as on theoretical possibilities. The sheer cost of a new "fab" (chip fabrication plant) may, in fact, determine when Moore's Law (transistor density doubles every eighteen months), first postulated in the mid-1960's by a founder of Intel, finally breaks down. The capital investment in a typical fab has climbed from about $1 million in 1970 to about $1 billion in 1994 to about $2.5 billion for a new Texas Instrument 0.25 micron fab scheduled to become operational in 1997.

Given that gross revenues for the largest chip maker, Intel, will be approximately $25 billion in 1996 (and growing at a compounded rate of about 35% per year) it seems doubtful that more than a very few companies worldwide will have the resources to create fabs based on entirely new technologies that may well cost $10 billion each by 2005 or so when the current photolithographic process will have run its course. $10 billion is not small change for any company; it is about 10% of the current *combined* market capitalization of the Big Three US automakers.

Parallel Processing

This is not to say that computational power of typical, affordable computers will at some point fail to grow at the exponential rate we are accustomed to. What cannot be accomplished with speedier individual processors will be made possible by processing in parallel on more than one.

The supercomputing and parallel processing paradigm is currently going through perilous times, as several once-prominent companies have either gone bankrupt or been taken over by larger ones whose major revenues come from more conventional computer and/or workstation sales. One current trend that seems certain to continue involves use of off-the-shelf mass-produced (typically RISC) processors, rather than extremely fast but very expensive custom-designed processors, such as have traditionally been used in the fastest supercomputers with pipelined vector architectures. At least one experimental parallel machine with over 9000 processors (from Intel, using Pentium II CISC processors) has already performed useful calculations at speeds exceeding 1 Teraflop (trillions of floating-point operations per second), almost three orders of magnitude faster than the fastest current custom-built supercomputer processor made by Fujitsu. By 2001 a 10-teraflop machine (to be built by IBM for the Energy Department) will be in operation at the Los Alamos or Livermore National Laboratories.

Currently, two general architectures predominate, one called SYM (symmetric multiprocessing), involving a shared memory accessible (either directly or indirectly) by all processors, and the other called MIMD (multiple instruction, multiple data) involving networked processors, each having its own "private" memory, inaccessible to the other processors.

Each of the architectures has its strengths and weaknesses. Programming of SYM machines is simpler than for MIMD machines (though explicit synchronization is required to avoid data access "races" by contending processors). For machines with a small number of processors (for example, just a few for desktop machines), the SYM architecture is likely to dominate. Unfortunately, as the number of processors grows, the bus structures for the shared memory SYM machines can become extremely complicated, so scaleup to massive parallelism (beyond say a few hundred processing nodes) is difficult technically and very expensive.

On the other hand, the networked individual processor MIMD model typically requires message-passing software such as PVM (Parallel Virtual Machine) or MPI (Message Passing Interface) to explicitly request transfer of information from one processor's memory to the memory of another processor that needs it. This complicates software development. The individual processors on these parallel machines behave like individual workstations with each node, normally running its own copy of the (Unix) operating system, attached to a local area network (albeit a very, very fast one, called a *switch*). Because of the architecture, MIMD machines are inherently scalable to quite large numbers (many thousands) of processors.

Parallel programs for tightly-switched MIMD machines can often be run as easily (though not as quickly) on networks of workstations. My guess is that this model will be the predominant one for engineering computing by 2001. Software to ease the programming problem is coming to market, and there will be a vast resource of unused compute cycles on very high performance networked workstations already in place in most engineering-oriented businesses (and universities), just waiting to be exploited for solution of computationally complex problems.

Communications

Without doubt, communications is the "computing" area experiencing the greatest acceleration in growth and greatest decrease in price/performance ratio. Daniel Atkins, Dean of the School of Information and Library Sciences and Professor of Computer Engineering at Michigan, estimates (personal communication) that the price/performance ratio for communication bandwidth is currently decreasing at a rate ten times that for computing hardware.

Tables 1 and 2 show the astounding impact of the capacity of storage media and network bandwidth on the storage and transmission of typical documents (here for uncompressed text only). Note also that the recently released single-side/single-layer DVD (Digital Versatile Disk) has about eight times (4.7 GBytes) the capacity of the conventional CD ROM listed in the table. If two layers per side (read by adjusting the focus of the laser) and both sides are used, then a single DVD, physically the size of a CD, can store a total of 18.8 GBytes, equivalent in capacity to 32 current CD ROMs. So all the text (but not the graphics!) in all the documents in the Library of Congress could be stored in fewer than 1000 such DVDs. (Don't expect to see this anytime soon! The Library of Congress is currently creating a digital library by scanning at high-resolution about 1 million new and recent documents per year; the library will be available to K12 and university libraries as an on-line resource. Digitizing all 20 million documents in the Library would be a massive and expensive undertaking, and is unlikely to be funded.)

The fastest local area network data-transmission rates, about 5 Mbps (millions of bits per second) in 1980, will certainly be greater than 1 Gbps by 2001 (1 Gbps Ethernet equipment will be available from manufacturers sometime in 1997). Milti-gigabit rates for high-performance switched networks, such as those used in massively parallel MIMD machines, will also be common by 2001.

Table 1: Sizes of Typical Stored Documents (Lucky)

Document	Bytes	Floppy Disks	CD ROMs
Page	2,400	0	1
Report	72,000	0	1
Book	720,000	1	1
Dictionary	60,000,000	43	1
Encyclopedia	130,000,000	93	1
Local Library	70,000,000,000	50,000	108
College Library	700,000,000,000	500,000	700
Library of Congress	18,000,000,000,000	12,900,000	26,000

Table 2: Transmission Times for Typical Documents (Lucky)

Document	Modem 28,000 bps	T1 (1.5Mbps)	Fiber Optic (1.7 Gbps)
Page	0.68 sec	0.013 sec	0.0000113 sec
Report	20.5 sec	0.38 sec	0.000339 sec
Book	3.42 min	3.84 sec	0.0034 sec
Dictionary	4.76 hr	5.3 min	0.28 sec
Encyclopedia	10.32 hr	11.6 hr	0.61 sec
Local Library	232 days	4.32 days	5.49 min
College Library	6.34 yr	43.2 days	0.92 hr
Library of Congress	163 yr	3.04 yr	23.5 hr

In late 1996, fast backbone Internet lines operate at about 45 Mbps. By the end of 1997, the facilities of a yet faster Internet, called vBNS (very high speed backbone network service) or Internet II, will connect about 100 research Universities and the national supercomputing sites with optical backbone connections operating at 155 Mbps. No doubt Internet II will, in turn, be replaced by Internet III, having a substantially greater bandwidth. Gigabit Internet backbone service will almost certainly be available for at least some users by 2001.

Most current modems attached to commercial cable television networks support Internet transmissions at T1 rates (1.5 Mbps), some 50 times typical telephone modem rates (28.8 kbps). Potentially, however, broadband cable modems with optical cable lines are capable of very much higher data rates, in the 10 to 40 Mbps range, at least when downloading (transfers from the Net to local PC); uploading (PC to Net) is normally slower, but still several times faster than for current telephone modem communication. Unfortunately, only about 10% of current commercial cable systems support two-way communication at all (so that PC to Net transmissions require a separate telephone modem), a situation that will probably be remedied by 2001 as companies upgrade their equipment and replace copper lines with optical ones.

Wireless services are now pervasive, and relatively inexpensive. Wireless local area networks are proliferating, and represent a cost-effective alternative to copper or optical cabling as a way of "wiring" a building (though the data rates cannot compete with those for high-bandwidth optical fiber). Several different sets of globe-girdling LEO (low-earth-orbit) satellite systems are about to be put into place by consortiums of computing and communication companies. They will provide high speed wireless services from virtually any place on the earth's surface. For example, 28 Mbps wireless Internet connections are planned for the $9 billion Microsoft/McCaw/Boeing Teldesic system of 288 satellites, scheduled to be fully operational in 2001.

The Web Changes Everything

Local (LAN) and wide area (WAN) networks have been with us for about twenty five years, starting with the Ethernet LAN, developed at XEROX PARC in the early 1970s, and the wide-area ARPANet, for which the first machine to machine transmission (from the University of Southern California to the University of California at Berkeley) took place in September 1969. Prior to the development of these technologies, most remote computing was performed at dumb terminals connected by telephone modem to mainframes operating in time-sharing mode.

Local area networks (LANs) made possible the high-speed sharing of resources such as mass storage units and printers. With the development of Unix in the late 1960s and early 1970s, remote computational resources on the network could also be tapped by multiple users directly from their own networked machines.

ARPANet allowed for remote messaging (EMail) and file transfer, and the sharing of ideas, data, research results and computational resources among users of machines that were widely separated geographically. A key breakthrough came with the adoption of the Internet transmission protocol (TCP/IP) and Internet domain addressing scheme in 1974. TCP/IP created a universally recognized standard for transmission of information "packets" over the network, and the addressing standard supported creation of world-wide directories, allowing unique identification of every networked machine and the speedy delivery of information packets to their intended target machines.

With support from the US Defense Department, ARPANet grew at a steady but unspectacular pace during the 1970s from 24 sites in 1971 to about 200 in 1981. In 1986 ARPANet was replaced by a less "military," more academic, network called NSFNet. One of the principal functions of NSFNet was to provide high bandwidth access between University campuses and five National Supercomputing sites (reduced to two in 1996). NSFNet was in turn replaced by the loosely (some believe chaotically) managed, rather amorphous Internet (a network of networks) in the late 1980s. As of late 1996, the world-wide Internet ties together more than 100,000 individual networks with more than 50 million attached (either directly or through Internet service providers) individual machines.

The incredible recent growth in Internet infrastructure and use can be attributed in large part to the development of the World Wide Web (*www* or simply *the Web*), which allows for easy access to *hyperlinked* documents stored on any accessible machine located anywhere on the Internet. Hypertext, which permits the nonlinear linking (essentially without restriction) of one piece of information with another (presumably related) one, was first suggested by Vannevar Bush (see earlier) in 1945, and was promoted by researchers at XEROX PARC in the early 1980's as a way of referencing and retrieving related information.

Tim Berners-Lee, an English physicist working at CERN in Zurich, is credited with developing three key Web concepts in the late 1980s and early 1990s:

A standard language for encoding hypertext documents.

A standard Internet protocol for linking and transmitting hypertext documents.

An addressing system for locating documents.

The language HTML (hypertext markup language), protocol HTTP (hypertext transfer protocol), and addressing system URL (universal resource locator) form the underpinnings of the current Web. One of the key decisions made by Berners-Lee was that a central directory (which would allow for deleting and updating links as hyperlinked documents were removed or modified) was infeasible for a "World-Wide" information web, i.e., that "dangling" hyperlinks to a removed document would simply have to be tolerated for the greater good of allowing independent actions by individual Web document owners.

In 1990, Berners-Lee developed a browser/editor at CERN for creating, accessing and displaying hypertext documents (principally research papers and data at CERN) using his HTML, HTTP, and URL concepts. However, it was not until 1993 when the Mosaic browser was created at NSCA (University of Illinois), that the potential of the Web became apparent (at least to some). With the formation of Netscape, Inc. and the release of the Netscape browser in 1994, the Web "took off".

The principal resources that can be hyperlinked on the web go far beyond mere text, indeed can be almost anything (text, graphics, sound, animation, video, virtual reality images, signals from an instrument on a remote experiment, software, ...); essentially any information that can be digitized can, or soon will be, Web-linkable.

It is, I think, fair to say that the Web changes everything. If, as McLuhan suggested in 1967, the medium is the message, then here the message is the Web. Some (myself included) think the Web may be as revolutionary an invention affecting the way we communicate and access and disseminate information as was the printing press.

The Web is the first truly new medium since television; it is, in fact, a multi-medium because the range of resources it can access potentially includes essentially all of the other mass media (newspapers, movies, radio, television). But in addition to providing a vehicle for "push" technology or "webcasting," in which the Net delivers information to the user (much as broadcasters do), the Web (1) allows an individual user to *interact* with resources and other users on the Web, and (2) allows an individual user to *create* resources, in essence to become a "publisher," without formal approval by an oversight authority. The Web can be an extraordinarily liberating medium for individual creativity (much that is worthless gets created in the process as well). The Web (or whatever supplants it) is clearly a multi-medium destined to have enormous impact on information dissemination, education, and commerce.

The Web's growth during the past three years is simply phenomenal. Virtually every corporation, academic institution, government agency, and uncounted individuals now have a presence on the Web, at least in the form of a hyperlinked home page.

Commercial Web activity is growing apace. It is hard to believe that the first commercial advertisement on the Web (for a lawyer's services) (1) raised a furor among Web users as a violation of Web etiquette, and (2) happened as recently as April 19, 1994. Contrast this with an estimate by Forrester Research that commercial Web transactions will exceed $10 billion in 1999, with computer hardware and software sales comprising a large piece of the commercial pie.

The recent flurry of multi-billion dollar buyouts and takeovers involving television, telephone, wireless, cable, computing and on-line Net-access companies attests to the fluidity of the situation and to the uncertainty about which technologies or combinations of them will eventually win in the "interactive" educational, entertainment, and commercial marketplace. My guess is that by 2001 commercial and educational interactivity will be centered on the Web and computer (with commercial cable systems providing Net connectivity for a substantial fraction of home users), rather than on interactive cable and the television set. News and entertainment currently delivered by mass media will stay in the domains of newspapers, radio, movies, and television, but the Web will play a role in these areas too (on-line magazines, webcast stock market reports and breaking news, etc.).

Perhaps the Web's greatest impact has been on the nature of "computing" itself. The center of gravity for computing activity is shifting perceptibly from individual computers and workstations to the network. Major corporations such as IBM and Microsoft have reoriented their missions to focus on delivery of "Net-centric" services and "Net-aware" application software. The distinction between what's done locally and what's done remotely will blur substantially in the coming years. Many corporations have already taken advantage of the net-centric Web model by creating internal internets called *intranets*, mini-Internets that use standard Web software such as browsers for accessing and updating corporate data and information; firewalls (software to isolate the internal intranet from the Internet) protect corporate information from the outside world.

Because the Web is so new and has grown so rapidly, it has many rough edges and problems. Service can be slow; breakdowns do occur (and will surely continue to occur). However, the Net's distributed nature and redundancy of communication paths virtually insures that only parts of it will be affected by a particular failure; a catastrophic Net breakdown comparable to a nation-wide power outage seems unlikely. On balance, the commercial carriers have done a remarkable job in installing new lines and equipment in response to rising demand, and will undoubtedly continue to do so, as a wholly new optical fiber and wireless communication infrastructure takes shape over the next decade.

Two issues that need to be addressed before the Web can reach its full potential in the commercial arena are: (1) security of Web-based financial transactions, and (2) copyright protection for Web-accessible intellectual property. The problems in these areas are challenging both technically and legally, and too involved to discuss here (and I don't know much about them!).

Substantial work has been done on financial security issues, mostly involving some form of encryption, and the Web is already being used fairly heavily for on-line sales and financial transactions, with apparently little fraud or error (but the occasional horror story shows that problems exist). US copyright laws are under active discussion in the Congress, and an International Copyright Convention of 160 nations, meeting during 1996 and 1997, will attempt to reach global agreement on the handling of intellectual property, particularly computer-based intellectual property. Real copyright protection is more likely to come from the bottom up (in the form of technical safeguards on the Web) than from the top down (international treaties or US copyright law). Because these issues are so important to the future of the Web and to the use of Web-based information for commerce and education, I believe that some reasonable and effective solutions for current security and copyright problems are likely to be in place by 2001.

Software

The most important software for engineering students, faculty, and professionals can be broadly classified by general function: (1) system software, (2) network services, (3) programming language translators/compilers, (4) general productivity tools, (5) problem, task, or discipline-oriented applications, and (6) multi-media instructional aids.

System software includes the base operating system for the computer and a myriad of programs for providing communication and access to system resources. The first operating system, for controlling the processing of batches of programs on punched cards on the IBM 704, was released in 1957. Subsequent systems, for time-shared access to mainframes, appeared in the mid-1960s. With each new class of hardware (minicomputers, personal computers, workstations, etc.), new and usually more sophisticated operating systems appeared and then slowly evolved over the years.

Currently, the most important operating systems are Unix, used on engineering workstations with RISC processors, Windows95/DOS and Windows NT for personal computers with Intel and compatible processors, and the Macintosh OS for Macintosh and clones with Power PC processors. All now have graphical user interfaces that are descendants of windowed interfaces developed at XEROX PARC in the early 1970s and popularized with release of the Macintosh in 1984; the interfaces appear remarkably alike to the user, making switching from one machine to another much less onerous than in earlier times (but still not without pain!).

New features such as support for multiple users (Unix), preemptive multi-tasking (timesharing of the local processor among several running programs), threads (parallel minitasks within a single application), and network services (e.g., remote file transfer, electronic mail,

Internet and Web access) have over the years been incorporated into or made available to operating systems.

Future versions of the predominant operating systems will almost certainly be much more "network-aware" than current ones, i.e., it is likely that the user interface (probably with a three-dimensional appearance) will eventually make little distinction between what is local to the user's workstation and what is on the network or Web.

The desktop metaphor with its emphasis on files and folders has been dominant since the introduction of the Macintosh in 1984, but is giving way to new metaphors, in particular that of the compound document as a container for objects - textual, graphical, sound, video, computational. Group interfaces for conferencing and for working collaboratively over the network are under development, and likely to be very important in both the academic and industrial workplace by 2001. Virtual reality will almost certainly play a role as a 3-dimensional imaging tool for navigating in the net-centered operating systems of the future. New multi-sensory tools involving gesture, eye and body motion, voice, smell (maybe not!) will eventually supplement the keyboard and mouse for interacting with those future operating systems.

Stephanopolous and Han (see p. 239) have described programming language developments in lucid detail, so I won't elaborate on their observations here, except to note that there are strong trends toward development of modular, structured, and more easily maintained and reusable object-oriented programs (the latter allow for hierarchical decomposition, and encapsulate both data structure and algorithm). My guess is that by 2001 most chemical engineering students will be trained to use high-level and visually-oriented computational and "programming" tools rather than traditional procedure-oriented languages such as FORTRAN and C.

In general, software improvements come much more slowly than those for hardware described earlier. This results from the fact that much code is still "handcrafted," that many popular applications were originally written years ago with unstructured code, that subsequent "improvements" have been made in ad hoc fashion, and that programs are often quite large (many are enormous), hence difficult to validate.

Nevertheless, there is hope from a new computing discipline, software engineering, from object-oriented programming technology, from the trend toward open systems, cross-platform (different kinds of computers) compatibility, emerging international standards for communications and hardware interfacing, and, in some cases, cooperation among software developers and hardware manufacturers.

One of the most exciting recent software developments is the emergence of Sun Microsystems' Java, an object-oriented language for creating programs that reside on network servers. When a networked user requests a service or a calculation or a display provided by a Java program (small ones are called applets), the Java code is downloaded to the user's workstation from the server, and either immediately interpreted locally by a *Java virtual machine* (JVM) translator or compiled into machine code for the workstation and executed. JVMs are available for essentially all personal computers and workstations and are being incorporated into Web browsers such as Netscape and Microsoft Explorer and other application programs.

This means that a Java program automatically has cross-platform compatibility (Nirvana for a programmer), since it can be run without modification on any computer, such as a Unix workstation, Macintosh, or "Wintel" (Windows operating system and Intel processor) machine with a JVM or Java-compliant browser. The user need not store the program locally, and the Java programmer can change the program without concern for the machine it will eventually be run on. Java, or some successor to it, is central to the notion of Net-centric computing.

Sun and a few other hardware/software vendors view Java as the tool for creating network-centered operating systems to replace conventional workstation operating systems such as Windows and the Macintosh OS, treating individual machines as network computers. My guess is that this will not happen, and that conventional operating systems will instead become much more Java-compliant by incorporating JVMs and Java compilers as core components.

Many other new software developments are also Web oriented, especially for creating and improving search engines for locating and cataloging URLs of interest on the Web, and for creating intelligent Web agents (small programs that move about the Web searching data bases and filtering information to carry out specific tasks of interest to the user). Current Web search engines are quite primitive, and identify URLs of interest based on text matching only. Engines available in 2001 will likely allow for searching of text in context and searching of non-textual information, in particular, images (e.g., find pictures containing sunsets) and sounds (e.g., find soundbites with references to global warming).

Implications for Chemical Engineering Education

How do (and will) all these intense (and accelerating) computing activities affect academia in general and engineering education in particular? In many respects, the impact of computing on education and academic life has paralleled that in the world outside the academy. Substantial basic research that feeds the computer revolution is performed by academics and their students. The University of the present looks quite different from the University of even a decade ago. Virtually every desk supports a networked desktop computer, the University library is "on-line," and every dorm room is (or soon will be) connected to the rest of the electronic world. The computer has brought with it systemic changes in the ways the University conducts its business and research and interacts with its students, graduates, faculty, and staff.

The impact of the computer *in* the classroom has, to date, been much less dramatic than in other areas of the academy. An 1896 still photo of an engineering classroom, professor lecturing with chalk in hand, would look remarkably similar to most engineering classrooms in 1996. Will that paradigm last for yet another century? Not likely.

Despite its apparent lack of impact to date in most classrooms, the computer has certainly affected what we teach and how students learn, as is evident from the many earlier papers in this monograph. Most chemical process design [Biegler, Seader and Seider (p. 153), Grossmann (p. 171), Grossmann and Morari (p. 185)] and many control (Arkun and Garcia, p. 193), and separations processes (Taylor, p. 139) courses have already been transformed into computationally centered ones. Most undergraduate laboratories have also undergone major upgrading, and the computer is now an integral part of experimental systems, both for data acquisition and data processing, as described by Mellichamp and Joseph (p. 203).

Interactive multi-media instructional tools, such as those described by Fogler and Montgomery (p. 57) and Fogler (p. 103), allow faculty to create remedial and supplemental tools for self-paced learning of course material that are particularly effective for students whose learning styles differ from the linear, analytical learning and teaching styles of most faculty. Numerical mathematical tools, described by Shacham, Cutlip, and Brauner (p. 73), allow students to solve nonlinear model equations and to carry out simulations that could not easily have been included in core engineering science coursework in thermodynamics (p. 85), transport processes (Finlayson and Hrymak, p. 125), and reaction engineering (Fogler, 103), as recently as a few years ago.

Nevertheless, the overall impact of the computer in our undergraduate curricula, particularly in our core chemical engineering science courses, has been less than might have been expected, given the computing resources now available on most engineering campuses. As pointed out by Kantor and Edgar (p. 9), most of our standard textbooks do not yet have a significant computer-orientation. I foresee the standard textbooks of the next century as much different from those of the past. The future text will not be structured as the linear sequence of static information found in our current texts, but will contain hyperlinked navigational aids, animation and sound, tools for manipulating data, and programs for carrying out simulations with user-supplied inputs. By definition, they will require the computer for their use, and may even be completely Web based, with small charges assessed for each access. (No doubt students will still want to print many screen images as they appear, so even these "electronic" texts are likely be accompanied by hard copy materials).

Most classrooms are not yet equipped with ready-to-use networked computing and projection facilities needed for doing good in-class demonstrations. Such equipment represents a significant investment by the university, but it will be essential for engineering classroom instruction in the next decade. Classroom visualization (Hrymak and Monger, p. 253) of the unfolding dynamic solution of a fluid mechanics problem is worth much more than a thousand faculty words.

Creation of good computer-oriented instructional materials and stimulating in-class demonstrations (even with the best computing and display equipment in place) is challenging, time-consuming, and expensive. Unfortunately, such efforts by faculty often receive little recognition by department chairs and deans, particularly at research universities, so we may continue to see fairly slow progress in fully integrating computing work into our core engineering science courses.

On the other hand, the Web could revolutionize the way we manage course-related activities. By 2001, virtually every course will have a home page on a Web site, with links to class schedules, problem assignments and solutions, interactive messaging, and student records. Driven by faculty interest and commitment, some courses will be strongly Web-centered, with facilities for student collaboration and group interaction, in-class note taking (using laptops with wireless communication?), interactive multi-media tools that integrate materials from local machines, CD ROM and the network, on-line examinations, and access to a wide array of data base resources and numerical and discipline-specific software; classical lecture presentations in these courses will diminish in importance. Some departments may have full-time Webmasters whose principal duties involve assisting faculty in bringing the curriculum to the Web.

The Virtual University?

Viewed more broadly, the new computing technologies, and particularly the Web, promise to transform the way Universities keep in touch with their alumni and deliver continuing education to professionals. As the bandwidth of Internet trunk lines rises dramatically in the next decade, videoconferencing, on-line interaction and collaboration, and Web access to digital data bases and libraries will make cost-effective delivery of distance learning a reality.

Some Internet-based experimental distance learning experiments are already underway. One, a joint venture of the University of Michigan, the State of Michigan and the auto industry called the "Virtual Automotive College," is being directed by James Duderstadt, a former Dean of Engineering and now President Emeritus of the University. This virtual college will deliver its first all-electronic courses to engineers in the automobile industry over the Internet during the Winter term of 1997.

In theory, a "virtual university" could be created using the new computing and communication tools and instructional technology with the Internet as its infrastructure. The driving force behind such a university would be delivery of instruction to large numbers of students and professionals in their homes or workplaces at much lower cost than with traditional campus-based courses. However, unless students have access to substantial interaction, group collaboration, and the opportunity to form learning communities, I'm skeptical of the near-term success of such a university, at least for young (18 to 24 year old) students. A college education involves much more than passive learning in near isolation. As suggested by Brown and Duguid, social experience, groups joined, scholars worked with, friendships made, prestige and marketability of degrees earned, and a sense of place play powerful roles in the College experience. It seems unlikely that a virtual university will ever be viewed with nostalgia as *alma mater*.

Summary

I have attempted to give an overview of past developments and current trends in digital computing and communications (in particular of networks, the Internet, and the Web), and to make some predictions about what the computing world will look like half a decade hence, in 2001. Many of the facts and figures are unattributed; for the most part, they come from extensive notes taken and articles (from newspapers, trade magazines, technical journals) clipped over many, many years.

The conjectures about the future are mostly my own; fortunately I can't be proved wrong until 2001, at which point the whole exercise can be done again!

References

Brown, J.S., and P. Duguid (1995). Universities in the Digital Age. Position paper, XEROX PARC, Palo Alto.

Hennesey, J.L., and D.A. Patterson (1995). *Computer Architecture: A Quantitative Approach*. Morgan Kaufmann, San Francisco.

Stix, G. (1995). Toward "Point One." *Scientific American*, (2), 90-95.

Lucky, J. (1989). *Silicon Dreams*. St. Martin's Press, New York.

AUTHOR INDEX

SUBJECT INDEX

H

I

J

K

L

M

N

O

P

Q

R